Tropical Lichens:
Their Systematics, Conservation,
and Ecology

The Systematics Association
Special Volume No. 43

# Tropical Lichens: Their Systematics, Conservation, and Ecology

*Edited by*

D.J. GALLOWAY

*Department of Botany,
The Natural History Museum, London*

Published for the SYSTEMATICS ASSOCIATION by
CLARENDON PRESS · OXFORD
1991

Oxford University Press, Walton Street, Oxford OX2 6DP
Oxford New York Toronto
Delhi Bombay Calcutta Madras Karachi
Petaling Jaya Singapore Hong Kong Tokyo
Nairobi Dar es Salaam Cape Town
Melbourne Auckland
and associated companies in
Berlin Ibadan

Oxford is a trade mark of Oxford University Press

Published in the United States
by Oxford University Press, New York

© The Systematics Association, 1991

All rights reserved. No part of this publication may be reproduced,
stored in a retrieval system, or transmitted, in any form or by any means,
electronic, mechanical, photocopying, recording, or otherwise, without
the prior permission of Oxford University Press.

British Library Cataloguing in Publication Data
Tropical lichens.
1. Tropical regions. Lichens
I. Galloway, D. J.   II. Systematics Association   III. Series
589.10913
ISBN 0-19-857720-6

Library of Congress Cataloging in Publication Data
Tropical lichens : their systematics, conservation, and ecology /
edited by D.J. Galloway.
(Systematics Association special volumes ; no. 43)
Includes index.
1. Lichens—Tropics—Congresses. I. Galloway, D. J., 1942-
II. Series.
QK597.7.T76 1991    589.1'0913—dc20    90-44333
ISBN 0-19-857720-6

Typeset by Downdell Limited, Oxford
Printed in Great Britain by
St Edmundsbury Press Limited
Bury St Edmunds, Suffolk

*Dedicated to*
*Mason E. Hale Jr.*
*1928–1990*

*A valued member of the international lichenological community, and a tireless worker for systematic and tropical lichenology.*

# Preface

This conference on tropical lichens originated from a lunchtime conversation in 1987 with Hildur Krog and Dougal Swinscow, who were then in London working on the final chapters of their book, *Macrolichens of East Africa*. Enthused as they were with the lichens of tropical East Africa, they felt, quite rightly, that the time was ripe for a wider assessment of tropical lichenology and that a conference bringing together a wide spectrum of lichenologists interested in tropical problems would allow a synthesis of what was currently known, but more importantly, would highlight areas where work was urgently required. The aim of the conference, which was held at the Natural History Museum in London (4–8 September 1989), was to discuss the state of knowledge on conservation, systematics, and ecology of tropical lichens and in this aim I was readily assisted by the following convenors: Professor P.M. Jørgensen for the Asian tropics, Dr L. Arvidsson for the American tropics, and Professor H. Krog for the African tropics, while I assumed responsibility for the Pacific tropics. We were fortunate to have stimulating introductory lectures from Professor G.T. Prance of the Royal Botanical Gardens, Kew on 'Diversity in tropical plants and vegetation: problems facing systematists in recording and conservation of tropical vegetation'; and from Dr S.R. Gradstein of the University of Utrecht on 'Biogeography of tropical rainforest bryophytes', both of which set the ensuing lichenological papers and discussions in the right regional and global perspectives.

The present volume includes most of the presentations from the London meeting and deals with systematic, ecological, biogeographical, and conservation problems of lichen floras in all the major tropical areas of the world and I must thank the contributors for meeting editorial demands so promptly. A number of contributions presented at the conference were not able to be offered for publication, but nevertheless provoked much stimulating discussion. These included: Dr A. Aptroot on 'Destruction of lichen habitats in Asia'; Dr E. Serusiaux on 'The foliicolous flora in central East Africa'; Mr P.W. James on 'Macrolichen floras of mid-Atlantic islands'; Dr R. Moberg on 'Problems in tropical Physciaceae with special reference to Central and South

America'; Dr L. Tibell on '*Tylophoron moderatum* and its anamorph'; Professor G. Follmann on 'The lichen family Roccellaceae in the neotropics' and Mr A.S. George on 'Lichens in the *Flora of Australia* Project'.

The conference was held under the auspices of the Systematics Association and of the International Association for Lichenology and it is a pleasure to acknowledge additional financial support from the Royal Society and the Overseas Development Administration. I am also indebted to the Director of the Natural History Museum for the use of facilities at the Museum, to the Linnean Society of London for the use of their library, and to the British Lichen Society for hosting a reception to delegates. Dr L. Arvidsson, Dr D.H. Brown, Dr B.J. Coppins, Dr D.H. Dalby, Professor M. Galun, Professor D.L. Hawksworth, Professor P.M. Jørgensen, Professor H. Krog, Mr J.R. Laundon, and Dr T.D.V. Swinscow are thanked for chairing the sessions. In the final session, Mr N. deN. Winser, Deputy Director of the Royal Geographical Society introduced us to his enthusiastic concept of expeditionary science, 'The global taskforce', and Mr R.H. Kemp of the Overseas Development Administration spoke on 'Conservation and development aid'. Dr M.R.D. Seaward led the final discussion/forum on conservation of tropical lichens. I am especially grateful to Dr O.W. Purvis for his sterling help with many aspects of the organization of the conference.

The index was prepared by Dr B. Aguirre.

*London*  D.J.G.
August 1990

# Contents

| | | |
|---|---|---|
| List of contributors and participants | | xi |
| 1. | Biogeographical relationships of Pacific tropical lichen floras<br>*D.J. Galloway* | 1 |
| 2. | The lichen genus *Relicina* in Australasia<br>*J.A. Elix* | 17 |
| 3. | Lichen conservation in Hawaii<br>*C.W. Smith* | 35 |
| 4. | The tropical Pacific species of *Usnea* and *Ramalina* and their relationship to species in other parts of the world<br>*G.N. Stevens* | 47 |
| 5. | Lichens of Papua New Guinea<br>*P.W. Lambley* | 69 |
| 6. | Lichenological observations in low montane rainforests of eastern Tanzania<br>*H. Krog* | 85 |
| 7. | New and interesting records of Tanzanian foliicolous lichens<br>*E. Farkas* | 95 |
| 8. | Evolutionary rates in the Teloschistaceae<br>*I. Kärnefelt* | 105 |
| 9. | Lichenological studies in Ecuador<br>*L. Arvidsson* | 123 |
| 10. | Notes on the lichen flora of the Guianas, a neotropical lowland area<br>*H.J.M. Sipman* | 135 |

| | | |
|---|---|---|
| 11. | Aspects of the foliose lichen flora of the southern-central coast of São Paulo State, Brazil<br>*M.P. Marcelli* | 151 |
| 12. | Ultrastructure of subtropical crustose lichens<br>*S.C. Tucker, S.W. Matthews, and R.L. Chapman* | 171 |
| 13. | On the importance of botanical gardens for lichens in the Asian tropics<br>*L. Arvidsson* | 193 |
| 14. | Some foliicolous lichens in Xishuangbanna, China<br>*J.C. Wei and Y.M. Jiang* | 201 |
| 15. | Observations on the composition and distribution of the '*Lobarion*' in forests of South East Asia<br>*P.A. Wolseley* | 217 |
| 16. | Spore ontogeny of *Sphaerophorus diplotypus* and *S. fragilis*<br>*M. Wedin* | 245 |
| 17. | Tropical pyrenocarpous lichens, a phylogenetic approach<br>*A. Aptroot* | 253 |
| 18. | Epilogue<br>*T.D.V. Swinscow* | 275 |
| Index | | 279 |
| List of Systematics Association Publications | | 299 |

# Contributors and participants

\* = Contributors; ‡ = Chairmen

B. AGUIRRE
*36 Bearfield Road, Kingston, Surrey KT2 5ET, UK*

\*A. APTROOT
*Institute of Systematic Botany, Heidelberglaan 2, 3508 TC Utrecht, The Netherlands*

\*‡L. ARVIDSSON
*Naturhistoriska Museet, PO Box 7283, S-402 35, Göteborg, Sweden*

G. BARON
*73 Guibal Road, London SE12 9LY, UK*

U. BECKER
*Uerdinger Strasse 15, D-5000 Köln 60, West Germany*

B. BENFIELD
*Penspool Cottage, Plymtree, Cullampton, Devon EX15 2JY, UK*

F.H. BRIGHTMAN
*59 Rosendale Road, West Dulwich, London SE21 8DY, UK*

R. BRINKLOW
*Museums and Art Galleries, Albert Square, Dundee DD1 1DA, UK*

‡D.H. BROWN
*Department of Botany, University of Bristol, Bristol BS8 1UG, UK*

J.A. CAYTON
*New Cottage, Herringfleet Road, St Olave near Great Yarmouth NR31 9HW, UK*

\*R.L. CHAPMAN
*Department of Botany, Louisiana State University, Baton Rouge, Louisiana, 70803-1705, USA*

‡B.J. COPPINS
*Royal Botanic Gardens, Inverleith Row, Edinburgh EH3 5LR, UK*

‡D.H. DALBY
*Department of Botany, Imperial College, Prince Consort Road, London SW7 5BB, UK*

\*J.A. ELIX
Chemistry Department, The Faculties, Australian National University, GPO Box 4, Canberra ACT 2601, Australia

\*E. FARKAS
Institute of Ecology and Botany, Hungarian Academy of Sciences, Vácrátót, H-2163, Hungary

A. FLETCHER
Leicestershire Museums Service, 96 New Walk, Leicester LE1 6TD, UK

\*G. FOLLMANN
Botanical Institute, University of Cologne, Gyrhof Strasse 15, D-5000 Köln 41, West Germany

\* ‡D.J. GALLOWAY
Department of Botany, The Natural History Museum, Cromwell Road, London SW7 5BD, UK

‡M. GALUN
Department of Botany, Tel Aviv University, 69 978 Tel Aviv, Israel

\*A.S. GEORGE
Bureau of Flora and Fauna, GPO Box 1383, Canberra ACT 2601, Australia

\*S.R. GRADSTEIN
Institute of Systematic Botany, Heidelberglaan 2, 3508 TC Utrecht, Netherlands

‡D.L. HAWKSWORTH
CAB Mycological Institute, Ferry Lane, Kew, Surrey TW9 3AF, UK

D.J. HILL
Department for Continuing Education, University of Bristol, Bristol BS8 1HR, UK

\*P.W. JAMES
Department of Botany, The Natural History Museum, Cromwell Road, London SW7 5BD, UK

\*Y.M. JIANG
Systematic Mycology and Lichenology Laboratory, Institute of Microbiology, Academia Sinica, Beijing, China

‡P.M. JØRGENSEN
Botanical Institute, University of Bergen, Allégt. 41, N-5007 Bergen, Norway

K. KALB
Adalbert-Stifter Strasse 5b, D-8430 Neumarkt/Opf., West Germany

\*I. KÄRNEFELT
Department of Systematic Botany, University of Lund, Östra Vallgatan 18-20, S-223 61 Lund, Sweden

*R.H. KEMP
*Foreign and Commonwealth Office, Overseas Development Administration, Eland House, Stag Place, London SW1E 5DH, UK*

*‡H. KROG
*Botanical Museum, University of Oslo, Trondheimsveien 23B, N-0562 Oslo 5, Norway*

*P.W. LAMBLEY
*The Cottage, Elsing Road, Lyng, Norwich NR9 5RR, UK*

‡J.R. LAUNDON
*Department of Botany, The Natural History Museum, Cromwell Road, London SW7 5BD, UK*

*M.P. MARCELLI
*Instituto de Botânica, C.P. 4005, CEP 01051, São Paulo-SP, Brazil*

*S.W. MATTHEWS
*Department of Botany, Louisiana State University, Baton Rouge, Louisiana, 70803-1705, USA*

*R. MOBERG
*The Herbarium, University of Uppsala, PO Box 541, S-751 21 Uppsala, Sweden*

A.M. O'DARE
*Springfield, 13 Barrows Road, Cheddar, Somerset BS27 3AY, UK*

S. OHM
*Krefelder Strasse 48, 5 Köln 1, West Germany*

A. ORANGE
*Department of Botany, National Museum of Wales, Cathays Park, Cardiff, Dyfed CF1 3NP, UK*

J. PEINE
*Diepenbeekallee 14, D-5000 Köln 40, West Germany*

J. PICKERING
*Department of Botany, The Natural History Museum, Cromwell Road, London SW7 5BD, UK*

*G.T. PRANCE
*Royal Botanic Gardens, Kew, Surrey TW9 3AE, UK*

C. PRINTZEN
*Uerdinger Strasse 15, D-5000 Köln 60, West Germany*

O.W. PURVIS
*Department of Botany, The Natural History Museum, Cromwell Road, London SW7 5BD, UK*

‡M.R.D. SEAWARD
*University of Bradford, Bradford BD7 1DP, UK*

*E. SÉRUSIAUX
*Institute of Botany, Sart Tilman, B-4000 Liège, Belgium*

*H.J.M. SIPMAN
*Botanical Garden and Botanical Museum, Königin Luise Strasse 6–8, D-1000 Berlin 33, West Germany*

*C.W. SMITH
*Department of Botany, University of Hawaii at Manoa, 3190 Maile Way, St John 409, Honolulu, Hawaii 96822, USA*

*G.N. STEVENS
*Botany Department, The University of Queensland, St. Lucia, Brisbane, Queensland 4072, Australia*

*‡T.D.V. SWINSCOW
*24 Monmouth Street, Topsham, Exeter EX3 0AJ, UK*

*L. TIBELL
*Department of Systematic Botany, University of Uppsala, PO Box 541, S-751 21 Uppsala, Sweden*

*S.C. TUCKER
*Department of Botany, Louisiana State University, Baton Rouge, Louisiana 70803-1705, USA*

M.F. WATSON
*Plant Science Laboratories, University of Reading, Whiteknights, Reading RG6 2AS, UK*

*M. WEDIN
*Department of Systematic Botany, University of Uppsala, PO Box 541, S-751 21 Uppsala, Sweden*

*J.C. WEI
*Systematic Mycology and Lichenology Laboratory, Institute of Microbiology, Academia Sinica, Beijing, China*

*N.DEN. WINSER
*Royal Geographical Society, 1 Kensington Gore, London SW7 2AR, UK*

*P.A. WOLSELEY
*Nettlecombe Studios, Williton, Taunton, Somerset TA4 4HS, UK*

# 1. Biogeographical relationships of Pacific tropical lichen floras

D.J. GALLOWAY

*Department of Botany, The Natural History Museum, Cromwell Road, London, UK*

## Abstract

The Pacific Ocean comprises a vast area of open sea with an array of mainly small, scattered islands indicating the presence of inter-basin ridges and rises. The only continental areas in the region are Australia–New Guinea, on the western boundary Asia, and on the eastern boundary the Americas. Although modern lichenological studies in the tropical Pacific are still in their infancy, it is possible to detect several biogeographical elements in the region's lichen floras as presently known. These include: (1) pantropical, (2) palaeotropical, (3) endemic, (4) cosmopolitan, (5) antitropical, (6) Indo-Malayan, (7) Pacific–American, and (8) austral elements. The new combination *Pseudocyphellaria diplomorpha* (Müll. Arg.) D. Galloway, is proposed.

## Introduction

The Pacific Basin is a vast area of ocean, bordered to the west by the continent of Asia, to the north and east by North, Central, and South America, and to the south by Australia, New Zealand, the subantarctic islands, and Antarctica. Scattered throughout this oceanic expanse are hundreds of small islands, all resulting from a complex series of geological process which led to the formation of this enormous ocean basin.

### 1. Geology

The present day Pacific Ocean bed is formed of ocean crust which is considerably younger than the continental crust which surrounds it in present day land masses. Pacific Ocean crust was formed from well-defined spreading ridges at the margins of the mobile Pacific Plate, while older oceanic crust has been subducted back into the asthenosphere at destructive plate margins or trenches which are a prominent feature of the Pacific Ocean bed.

The Pacific Ocean opened during the drifting apart of components of the great southern continent Gondwanaland, a process still continuing today. The geology of the Pacific and circum-Pacific region is exceedingly complex, with Pacific Plate margins and the leading edge of the Australian–Indian Plate composed of complex mixtures of unrelated geological units known as terranes (Howell *et al.* 1985), considered by many geologists to be continental fragments rifted from both Gondwanaland and a now totally vanished continent, Pacifica, which was thought to front the warm temperate to subtropical Tethyan margin of Gondwanaland. The Tethys Sea being the warm shallow sea separating northern and southern parts of the Palaeozoic supercontinent Pangaea and spreading across tropical and equatorial latitudes. During the latter part of the Palaeozoic era, the great Gondwana continent with its southern cold interior stretched northwards into latitudes of subtropical seas and from late Palaeozoic until the mid-Cainozoic the Gondwana continental margin habitats were carried across a wide range of latitudes with consequent changes in sea levels and environments.

What we have in the Pacific today then, is a complex regional geology such as that found in New Guinea, for example, which consists of at least 32 terranes (Pigram and Davies 1987); island arcs such as New Caledonia; hot spot traces such as the Hawaiian islands and Juan Fernandez; continental areas such as northern Australia (Audley-Charles 1988); and the region of south-east Asia (Metcalfe 1988), a collage of continental fragments rifted from Gondwana to collide with the developing continent of Asia (Holloway 1986).

## 2. *Climate*

In western parts of the region the seasonal Asian monsoon is a major influence, elsewhere in the Pacific trade winds dominate the climate, blowing from the south-east in the Southern Hemisphere and from the north-east in the Northern Hemisphere, giving most of the Pacific islands ample precipitation (Whitmore 1988; Lauer 1989). Mist zones are found from 500 m to 2000 m leading to development of a characteristic cloud forest vegetation, while above 2000 m, humidity falls dramatically and the highest mountain tops in the region can become very dry and almost desert-like. In the Galápagos, however, the climate is semi-arid.

## 3. Vegetation

A variety of vegetation types are found in the Pacific tropics from coastal mangroves to montane forests and alpine grasslands and shrublands at high altitudes, with a range of forest types in between dependent on climate, soils, and geology (Whitmore 1973, 1988, 1989; Schmid 1989). Takhtajan (1986), in his arrangement of the floristic kingdoms of the world, has the vegetation of the Pacific disposed among two kingdoms namely: *Palaeotropic Kingdom* (Indomalesian subkingdom and Polynesian subkingdom); and the *Holantarctic Kingdom* (Fernandezian Region and Neozeylandic region). In all of these various vegetation formations lichens are present and show a number of biogeographical groupings or affinities, several of which are complex. A short, general account of tropical rainforest lichens is given in Sipman and Harris (1989).

## 4. History of lichen exploration

The earliest Pacific lichens are those collected from Tahiti during Captain Cook's circumnavigation in the 1770s. The Scottish botanist Archibald Menzies (1754–1842) made the most comprehensive eighteenth century collections from the Pacific tropics during several visits to Hawaii, Tahiti, and Java between 1786 and 1795. He visited Hawaii 5 times collecting 9 taxa in 7 genera including the endemic *Ramalinopsis* (Galloway and Groves 1987, unpublished observations). An increasing number of voyages to the Pacific were made in the nineteenth century by Austrian, British, Dutch, French, German, Italian, Swedish, and United States expeditions with the following areas being visited and botanized: Hawaii, Samoa, Java, New Caledonia, Fiji, Norfolk Island, New Guinea, the Galápagos Islands, and Juan Fernandez. Lichens were collected from all of these areas.

At the close of the nineteenth century a substantial literature existed on Pacific lichens, based partly on material brought back to Europe during the great period of Pacific exploration and partly on collections made by colonial botanists who sent material to the major European botanical centres for identification (Galloway 1985*b*).

In the latter part of the twentieth century interest in Pacific lichens is reviving, although the area still lacks keys to major genera and regional floras which would facilitate study of Pacific lichens on a wider base. Apart from New Zealand (Galloway 1985*a*), which anyway has only a small tropical element in its flora, few modern systematic treatments (e.g. Stenroos 1988) are available for most of the Pacific region. Several general works deal with very broad patterns of lichen distribution in the Pacific region (Galloway 1979, 1985*b*, 1987, 1988*a*, *b*, 1990;

Jørgensen 1983) and a rather uneven Pacific lichen bibliography is given in Hawksworth and Ahti (1990).

Studies on Pacific lichens at a regional level include: Amboina (Krempelhuber 1871), Australia (Filson 1986, 1988), Borneo (Krempelhuber 1875), Fiji (Crombie 1873; Krempelhuber 1873), Galápagos (Weber 1986), Hawaii (Krempelhuber 1877; Magnusson 1955; Klement 1966), Java (Montagne and van den Bosch 1857; Zahlbruckner 1943; Zahlbruckner and Mattick 1956), Juan Fernandez (Crombie 1877; Zahlbruckner 1924; Redon and Quilhot 1977), Philippines (Vainio 1909, 1913, 1921, 1923; Gruèzo 1979), New Caledonia (Nylander 1861, 1868; Müller Argoviensis 1893; Harmand 1911, 1912; Smith 1922), New Guinea (Szatala 1956; Streimann 1986), New Zealand (Galloway 1985a), Noumea (Müller Argoviensis 1887), Samoa (Krempelhuber 1873; Müller Argoviensis 1897; Zahlbruckner 1908), Tahiti (Guillemin 1836; Montagne 1848; Krempelhuber 1870; Crombie 1877; Vainio 1924).

A palaeotropical lichen flora, based on current concepts of the identity of lichen genera is urgently needed and a macrolichen flora as a first priority would at least introduce lichens to ecologists, foresters, reserve planners, and all those responsible for land management in the Pacific tropics. In these days of rapid mass air travel, the long haul jet has effectively tamed the Pacific, many island groups are now well-known tourist resorts and the press of tourism is only one of several substantial threats to tropical vegetation, including lichens, in the Pacific. Other environmental hazards leading to long-term change which must ultimately affect lichen floras are: atmospheric nuclear testing, logging of forests, agricultural monocultures of various kinds, and flooding of low-lying areas consequent on global warming.

## Biogeography of tropical Pacific lichens

Biogeography, simply defined as the study of the geographical distribution of organisms, both present and past, involves considerations both of patterns of distribution shown by organisms and also of the processes whereby such distributions have come about. For recent general reviews of various competing theories of distribution of taxa see Humphries and Parenti (1986), Brundin (1988), and Craw (1988). Although detailed knowledge of lichen distribution in the Pacific region is still scanty (Jørgensen 1983; Galloway 1990), sufficient is known to be able to discern a number of biogeographical affinities or elements. These are briefly outlined below.

## 1. Pantropical element

Pantropical taxa, in contrast to southern cool temperate pan-austral taxa, have warm temperate affinities and are found on all major land masses in tropical regions. They are probably derived from equatorial Tethyan environments and habitats and in the Pacific region are best developed in lowland habitats. *Coccocarpia palmicola* (Arvidsson, 1983, Fig. 53) and *Pseudocyphellaria clathrata* (Galloway and Arvidsson 1990, Fig. 2) illustrate such a pantropical distribution. Other common pantropical taxa are: *Bulbothrix goebelii*, *B. isidiza*, *B. tabacina* (Hale 1976*b*); *Calicium hyperelloides* (Tibell 1987); *Cladia aggregata* (Filson 1981; Stenroos 1988); *Coccocarpia erythroxyli*, *C. pellita* (Arvidsson 1983); *Collema kauaiense*, *C. leptaleum* (Degelius 1974); *Cryptothecia candida* (Santesson 1952); *Dimerella epiphylla* (Santesson 1952); *Dirinaria applanata*, *D. confluens*, *D. picta* (Awasthi 1975); *Glyphis cicatricosa*, *Heterocyphelium leucampyx* (Tibell 1987); *Hypotrachyna formosana*, *H. revoluta*, *H sinuosa* (Hale 1975*a*); *Letrouitia domingensis*, *L. subvulpina*, *L. transgressa*, *L. vulpina* (Hafellner 1981); *Mazosia melanophthalma*, *M. phyllosema* (Santesson 1952); *Megalospora sulphurata*, *M. tuberculosa* (Sipman 1983); *Myriotrema compunctum* (Hale 1981); *Ocellularia aurata* (Hale 1981); *Parmotrema austrosinense*, *P. chinense*, *P. crinitum*, *P. cristiferum*, *P. dilatatum*, *P. permutatum*, *P. praesorediosum*, *P. reticulatum*, *P. tinctorum*, *P. ultralucens*, *P. zollingeri* (Hale 1965; Krog and Swinscow 1981; Swinscow and Krog 1988); *Phyllopsora corallina* var. *ochroxantha* (Brako 1989); *Physma byrsaeum*, *Porina epiphylla* (Santesson 1952); *Pseudoparmelia texana* (Hale 1976*a*); *Pyrgillus javanicus* (Tibell 1987); *Pyxine berteriana*, *P. cocoes*, *P. minuta* (Rogers 1986; Sammy 1988); *Ramalina celastri*, *R. peruviana* (Stevens 1987); *Sphaerophorus formosanus* (Tibell 1987); *Sphinctrina tubaeformis* (Tibell 1987); *Stirtonia macrocephala* (Santesson 1952); *Thelotrema coccineum*, *T. lacteum*, *T. monosporum* (Hale 1981); *Tylophoron moderatum* (Tibell 1987).

## 2. Palaeotropical element

These are also warm temperate taxa which have a distribution including Africa, India, south-east Asia, and the Pacific region. *Pseudocyphellaria argyracea* (Galloway and James 1986; Galloway 1988*b*) has such a distribution, although many other taxa have a narrower distribution in the Pacific. *Heterodea muelleri* (Filson 1978) and *Leioderma duplicatum* (Galloway and Jørgensen 1987), for example, being restricted to northern New Zealand, eastern Australia, and Norfolk Island or New Caledonia. *Erioderma sorediatum* (Galloway and Hayward 1987) originally described from far northern New Zealand, is now known from the Philippines to Oregon, Chile, and even Brazil, so is a good example of a wide-ranging palaeotropical taxon. Others include: *Anzia formosana*, *A. japonica*, *A. madagascarensis*, *A. semiteres* (Galloway 1978, 1985*b*; Yoshimura 1987); *Coccocarpia adnata*, *C. dissecta*, *C. pruinosa*, *C. rottleri*, *C. smaragdina* (Arvidsson 1983); *Collema coilocarpum* (Degelius 1974); *Dirinaria caesiopicta*, *D. subconfluens* (Awasthi 1975); *Leioderma erythrocarpum*, *L. sorediatum* (Galloway and Jørgensen 1987); *Letrouitia muralis* (Hafellner 1981); *Lobaria retigera* (Yoshimura 1971); *Megalospora atrorubicans* (Sipman 1983); *Myriotrema cinereoglaucescens*, *M. microporum*, *M. subconforme* (Hale 1981); *Nadvornikiana hawaiensis* (Tibell 1987); *Neophyllis melacarpa* (Galloway 1979, 1985*a*—as *Gymndoderma melacarpum*); *Pannaria brisbanensis*, *P. fulvescens*, *P. mariana*, *Parmotrema andina* (Hale 1965; Krog and Swinscow 1981); *Peltigera dolichorhiza* (Galloway 1985*b*); *Peltula decorticans*, *P. subglebosa* (Büdel 1987); *Pseudocyphellaria poculifera* (Galloway 1988*b*); *Ramalina leiodea*, *R. nervulosa* var. *dumeticola*, *R. tenella* (Stevens 1987); *Relicina samoensis* (Hale 1975*b*); *Sphaerophorus diplotypus*, *S. kinabalensis* (Tibell 1987); *Thelotrema pilulifera*, *T. platysporum*, *T. weberi* (Hale 1981).

## 3. Endemic element

Endemic taxa are of considerable biogeographical interest and may represent either the evolution of new genera or species from ancestors over a period of isolation, or else they reflect an old or relict distribution surviving after widespread extinctions as a consequence of geographic or climatic change.

The tropical Pacific region has a number of interesting endemic lichen taxa, some of limited geographical extent, others ranging more widely in the Pacific but not known elsewhere. Preliminary work on insect biogeography in the Pacific recognized a number of areas of endemism (Duffels 1986; Schuh and Stonedahl 1986, Fig. 1), and lichen taxa also fall into similar areas of endemism. At the generic level may be mentioned *Ramalinopsis* from Hawaii; *Myelorrhiza* (Verdon and Elix 1986) from northern Queensland; *Sarrameana* (Vězda and James 1973) from New Caledonia [a second species now also from Tasmania (Vězda and Kantvilas 1988)]; *Calathaspis devexa* (Lamb *et al.* 1972; Stenroos 1988), and *Compsocladium archiboldianum* (Lamb 1956) from New Guinea.

At the species level taxa may range more widely in the Pacific; *Bryoria indonesica* (Jørgensen and Galloway 1983) and *Pseudocyphellaria sulphurea* (Galloway 1985*c*) range from Sri Lanka to north-eastern Australia while *Sticta samoana* is found in Samoa, Fiji and Hawaii, while others have narrower geographical ranges such *Coccocarpia fulva* (Arvidsson 1983) in New Caledonia or *Megalospora granulans* and *M. weberi* in New Guinea (Sipman 1983). Taxa endemic to the Pacific area include: *Arthonia pellicula* (Müller Argoviensis 1883); *Coccocarpia endoferruginea* (Arvidsson 1983); *Cladia retipora* (Filson 1981; Stenroos 1988); *Graphina insulana* (Müller Argoviensis 1883); *Letrouitia pseudomuralis* (Hafellner 1981); *Megalospora albescens, M. halei, M. hillii, M. sulphureorufa* (Sipman 1983); *Parmotrema austrocetratum* (Elix and Johnston 1988); *Pseudocyphellaria godeffroyi, P. poculifera, P. reineckeana, P. semilanata, P. stenophylla* (Krempelhuber 1873; Galloway 1985c, 1988b); *Relicina fluorescens* (Hale 1975b; Elix, this volume); *Sticta boschiana, S. heppiana, S. pedunculata* (Krempelhuber 1873).

## 4. Cosmopolitan element

The cosmopolitan element contains taxa which are generally widespread being present on all land masses and on oceanic islands. The extremely wide distribution of the *Xanthorion*, especially *Xanthoria parietina, Physcia adscendens*, and *P. stellaris*, is already known from Australasia and South America. Other widely distributed taxa which are also found in the Pacific area include: *Calicium abietinum, C. glaucellum, C. salicinum* (Tibell 1987); *Candelaria concolor, Catapyrenium lachneum, Chaenotheca brunneola* (Tibell 1987); *Chaenothecopsis pusilla* (Tibell 1987); *Chrysothrix candelaris* (Laundon 1981); *Cladonia pleurota* (Stenroos 1989); *Diploschistes scruposus* (Lumbsch 1989); *Mycocalicium subtile* (Tibell 1987); *Peltula euploca* (Büdel 1987); *Porpidia crustulata* (Rambold 1989); *Pseudocyphellaria aurata, P. crocata, P. intricata* (Galloway 1988b); *Sphaerophorus melanocarpus* (Tibell 1987); *Teloschistes flavicans* (Almborn 1989, Kärnefelt 1989); *Tephromela atra* (Rambold 1989).

## 5. Antitropical element

Antitropical, amphitropical or bipolar taxa have intrigued botanists and biogeographers for many years (Humphries and Parenti 1986; Briggs 1987). Briggs (1987) discusses three current vicariance biogeography propositions to account for antitropical distributions.

(1) Allopatric speciation resulting from some kind of geographic barrier separating a formerly continuous population. A connection between vicariance and plate tectonics is established by assuming some kind of primitive cosmopolitanism of taxa which contrasts with Darwin's centre of origin concept where taxa are thought to have arisen in a limited area and to have subsequently dispersed outwards from it.

(2) An island integration hypothesis (Rotondo *et al.* 1981) which assumes that '. . . on the Pacific Plate, an original uniform biota was broken up by plate movement and by the emergence of new islands and the subsidence of old ones. As old islands and island chains were moved or replaced by newer ones, their biotas were

supposedly carried along or passed to succeeding islands . . .' (Briggs 1987 p. 239). The island integration hypothesis attempts to account for present-day biogeographic patterns in the Pacific Basin without having to resort to ocean currents, or other dispersal mechanisms.

(3) The Pacifica model (Nelson 1985), which assumes the existence of Pacifica, an early Mesozoic continental eastwards extension of Gondwanaland bordering the Tethys Sea. The Pacific Ocean was formed as the result of the fragmentation and dispersal of Pacifica to the far reaches of the Pacific, in south-east Asia, Asia proper, and North America. As the fragments of Pacifica dispersed they took cargoes of living taxa with them.

However, Briggs (1987) feels that

> Antitropical distributions of continental shelf, Indo–West Pacific species are probably not due to transgression of the tropics during the glacial periods, isothermic submergence, island integration, rising Neogene temperatures, or the Mesozoic dispersal of fragments from a Pacific continental mass. Characteristics of common antitropical patterns, plus information from systematic works on a variety of animal and plant groups, indicate that Théel's long discarded 'relict theory' appears to best fit the evidence, for it provides a mechanism whereby antitropical distributions may be brought about. The relict theory is compatible as the Indies part of the Indo–West Pacific has been functioning as a centre of evolutionary origin. It suggests that antitropical and associated disjunct patterns are produced as an older species, that has spread out to occupy a broad range, loses ground and gradually becomes supplanted by a younger species that had subsequently evolved in the East Indies. As this process goes on, the older species becomes restricted to a few isolated localities on the fringe of its original range. These isolates are often found to the north and south of the equatorial region but may include relict populations at the western edge of the Indian Ocean (Briggs 1987 p. 237).

Lichens also show distinctive antitropical disjunctions and several well-known boreal lichens such as *Aspicilia alpina*, *A. subsorediza*, *Carbonea vorticosa* (Rambold 1989); *Pannaria hookeri*, *Pertusaria dactylina*, *Solorina crocea*, *S. spongiosa*, and *Xanthoria elegans*, for example (Galloway and Bartlett 1986), are also known from subalpine to high alpine habitats in the Southern Hemisphere, most notably in cool temperate regions. However, two antitropical taxa are known with high altitude Pacific tropic stations, these are *Arthrorhaphis citrinella* (Galloway and Bartlett 1986) and *Leproloma vouauxii* (Laundon 1989).

*6. Indo-Malayan element*

Taxa in this element are most strongly represented in south-east Asian lichen floras but some extend northwards to Japan, westwards to India or even Africa and, occasionally, south and east to north-east Australia and even to New Zealand. In many ways palaeotropical taxa have this sort of distribution, but generally range more widely eastwards into the Pacific, often as far east as the Galápagos or Juan Fernandez. Among the Indo-Malayan element in Pacific tropic lichens may be mentioned: *Cetrelia braunsiana* (Culberson and Culberson 1968); *Coccocarpia aeruginosa*, *C. glaucina*, *C. pruinosa*, *C. rottleri*, *C. smaragdina* (Arvidsson 1983); *Collema actinoptychum*, *C. japonicum*, *C. rugosum* (Degelius 1974); *Heterodermia japonica*, *Letrouitia bifera*, *L. corallina*, *L. flavocrocea*, *L. leprolyta* (Hafellner 1981); *Megalospora coccodes* (Sipman 1983); *Nephromopsis*, *Parmelia erumpens* (Hale 1987); *Peltigera nana*, *Pseudoparmelia intertexta* (Hale 1976*a*); *Pseudocyphellaria diplomorpha* (Müll. Arg.) D. Galloway **comb. nov.** [Basionym: *Stictina diplomorpha* Müll. Arg., *Flora* **65**, 301 (1882). Type: Ceylon. *Thwaites*, 1876 (G 001975 ! - holotype)]; *P. junghuhniana*, *P. sulphurea* (Galloway 1985*c*); *Ramalina pacifica* (Stevens 1987); *Relicina connivens*, *R. malesiana*, *R. relicinula* (Hale 1975*b*); *Thysanothecium scutellatum* (Galloway and Bartlett 1983).

*7. Pacific–American affinities*

A number of lichen taxa have distribution patterns that include both Pacific and American stations, a fact earlier noted by Magnusson and Zahlbruckner (1945 p. 75) and Weber (1986). The following examples may be cited here: *Cladina confusa*, *C. halei* (Weber 1986; Stenroos 1988); *Coccocarpia domingensis*, *C. erythrocardia* (Arvidsson 1983); *Dirina catalinariae* f. *catalinariae* and f. *sorediata* (Tehler 1983); *Erioderma sorediatum* (Galloway and Hayward 1987); *Hypotrachyna bogotensis* (Hale 1975*a*); *Leioderma sorediatum* (Galloway and Jørgensen 1987); *Letrouitia parabola* (Hafellner 1981); *Myriotrema album*, *M. glaucophaenum*, *M. terebratulum* (Hale 1981); *Roccellina badia*, *R. limitata*, *R. nigrocincta*, *R. suffruticosa*, *R. terrestris* (Tehler 1983); *Thelotrema platycarpoides*, *T. porinoides* (Hale 1981). Taxa in the genera *Coccotrema*, *Placopsis*, and *Turgidosculum*, have a circum-Pacific distribution (Galloway 1985*b*, 1990).

*8. Austral element*

There are also affinities in climate, vegetation, and soils between upper timberlines of the austral cool temperate zone and the cool tropical highlands, with the prevailing life forms in both subantarctic and tropical high altitudes being cushion plants, dwarfed shrubs, and tussock grasses (Troll 1973). Lichen genera most richly speciating in cool temperate austral habitats and reflecting a Panthalassic Gondwanan origin, also extend into tropical areas at high altitudes—such genera including *Psoroma*, *Pseudocyphellaria*, *Sphaerophorus*, and *Siphula* amongst others (Galloway 1987, 1988a, b). Cloud forest vegetation may have lichens from this element represented in their floras. There is a distinctive austral element in the lichen flora of Juan Fernandez (Zahlbruckner 1924; Redon and Quilhot 1977) with *Parmeliella nigrocincta*, *Pseudocyphellaria physciospora*, and *Psoroma sphinctrinum* being examples of austral taxa widespread in the South Pacific Ocean.

## References

Almborn, O. (1989). Revision of the lichen genus *Teloschistes* in central and southern Africa. *Nordic Journal of Botany*, **8**, 521–37.

Arvidsson, L. (1983). A monograph of the lichen genus *Coccocarpia*. *Opera Botanica*, **67**, 1–96.

Audley-Charles, M.G. (1988). Evolution of the southern margin of Tethys (north Australian region) from early Permian to late Cretaceous. In *Gondwana and Tethys*, Geological Society Special Publication, No. 37 (ed. M.G. Audley-Charles and A. Hallam), pp. 79–100. Oxford University Press, Oxford.

Awasthi, D.D. (1975). A monograph of the lichen genus *Dirinaria*. *Bibliotheca Lichenologica*, **2**, 1–108.

Brako, L. (1989). Re-evaluation of the genus *Phyllopsora* with taxonomic notes and introduction of *Squamicidia*, gen. nov. *Mycotaxon*, **35**, 1–19.

Briggs, J.C. (1987). Antitropical distribution and evolution in the Indo-West Pacific Ocean. *Systematic Zoology*, **36**, 237–47.

Brundin, L. (1988). Phylogenetic biogeography. In *Analytical biogeography* (ed. A.A. Myers and P.S. Giller), pp. 343–69. Chapman and Hall, London.

Büdel, B. (1987). Zur Biologie und Systematik der Flechtengattungen *Heppia* und *Peltula* im südlichen Afrika. *Bibliotheca Lichenologica*, **23**, 1–105.

Craw, R.C. (1988). Panbiogeography: method and synthesis in biogeography. In *Analytical biogeography* (ed. A.A. Myers and P.S. Giller), pp. 405–35. Chapman and Hall, London.

Crombie, J.M. (1873). Lichenes. In *Flora Vitiensis* . . . (ed. B. Seemann), pp. 419–21. Reeve and Co., London.

Crombie, J.M. (1877). The lichens of the 'Challenger' expedition (with a revision of those enumerated by Dr Stirton in the Linnean Journal of Botany **XIV**, pp. 366–75). *Botanical Journal of the Linnean Society*, **16**, 211–31.

Culberson, W.L. and Culberson, C.F. (1968). The lichen genera *Cetrelia* and *Platismatia* (Parmeliaceae). *Contributions from the United States National Herbarium*, **34**, 449–558.

Degelius, G. (1974). The lichen genus *Collema* with special reference to the extra-european species. *Symbolae Botanicae Upsalienses*, **20**, 1–215.

Duffels, J.P. (1986). Biogeography of Indo-Pacific Cicadoidea: a tentative recognition of areas of endemism. *Cladistics*, **2**, 318–36.

Elix, J.A. and Johnston, J. (1988). New species in the lichen family Parmeliaceae (Ascomycotina) from the Southern Hemisphere. *Mycotaxon*, **31**, 491–510.

Filson, R.B. (1978). A revision of the genus *Heterodea* Nyl. *Lichenologist*, **10**, 13–25.

Filson, R.B. (1981). A revision of the lichen genus *Cladia* Nyl. *Journal of the Hattori Botanical Laboratory*, **49**, 1–75.

Filson, R.B. (1986). *Index to type specimens of Australian lichens: 1800–1984.* Bureau of Flora and Fauna, Canberra.

Filson, R.B. (1988). *Checklist of Australian lichens.* 3rd edn. National Herbarium of Victoria, Melbourne.

Galloway, D.J. (1978). *Anzia* and *Pannoparmelia* (Lichenes) in New Zealand. *New Zealand Journal of Botany*, **16**, 261–70.

Galloway, D.J. (1979). Biogeographical elements in the New Zealand lichen flora. In *Plants and islands* (ed. D. Bramwell), pp. 201–24. Academic Press, London.

Galloway, D.J. (1985a). *Flora of New Zealand Lichens.* P.D. Hasselberg, Wellington.

Galloway, D.J. (1985b). Lichenology in the South Pacific, 1790–1840. In *From Linnaeus to Darwin: commentaries on the history of biology and geology*, SHNH Special Pubs No. 2 (ed. A. Wheeler and J.H. Price), pp. 205–14. London.

Galloway, D.J. (1985c). Nomenclatural notes on *Pseudocyphellaria*. II. Some Southern Hemisphere taxa. *Lichenologist*, **17**, 303–7.

Galloway, D.J. (1987). Austral lichen genera: some biogeographical problems. *Bibliotheca Lichenologica*, **25**, 385–99.

Galloway, D.J. (1988a). Plate tectonics and the distribution of cool temperate Southern Hemisphere macrolichens. *Botanical Journal of the Linnean Society*, **96**, 45–55.

Galloway, D.J. (1988b). Studies in *Pseudocyphellaria* (lichens). I. The New Zealand species. *Bulletin of the British Museum (Natural History) Botany*, **17**, 1–267.

Galloway, D.J. (1990). Phytogeography of Southern Hemisphere lichens. In *Advances in quantitative phytogeography* (ed. T. Crovello and P.L. Nimis), pp. 233–62. Kluwer Academic Publishers, Dordrecht.

Galloway, D.J. and Arvidsson, L. (1990). Studies in *Pseudocyphellaria* (lichens). II. Ecuadorean species. *Lichenologist*, **22**, 103–35.

Galloway, D.J. and Bartlett, J.K. (1983). The lichen genus *Thysanothecium* Mont. and Berk., in New Zealand. *Nova Hedwigia*, **36**, 381–98 ['1982'].

Galloway, D.J. and Bartlett, J.K. (1986). *Arthrorhaphis* Th.Fr. (lichenized Ascomycotina) in New Zealand. *New Zealand Journal of Botany*, **24**, 393–402.

Galloway, D.J. and Groves, E.W. (1987). Archibald Menzies MD, FLS (1754–1842), aspects of his life, travels and collections. *Archives of Natural History*, **14**, 3–43.

Galloway, D.J. and Hayward, B.W. (1987). Lichens from the Three Kings Islands, northern New Zealand. *Records of the Auckland Institute and Museum*, **24**, 197–213.

Galloway, D.J. and James, P.W. (1986). Species of *Pseudocyphellaria* Vainio (lichenes) recorded in Delise's 'Histoire des Lichens: Genre *Sticta*'. *Nova Hedwigia*, **42**, 423–90.

Galloway, D.J. and Jørgensen, P.M. (1987). Studies in the lichen family Pannariaceae. II. The genus *Leioderma* Nyl. *Lichenologist*, **19**, 345–400.

Gruèzo, W.S. (1979). Compendium of Philippine lichens. *Kalikasan, Philippine Journal of Biology*, **8**, 267–300.

Guillemin, J.B.A. (1836). Énumeration des plantes découvertes par les voyageurs dans les îles de la Société, principalement dans celle de Taîti. *Annales des Sciences Naturelles, Paris. Botanique séries 2*, **6**, 297–320.

Hafellner, J. (1981). Monographie der Flechtengattung *Letrouitia* (Lecanorales, Teloschistineae). *Nova Hedwigia*, **35**, 645–729.

Hale, M.E. (1965). A monograph of *Parmelia* subgenus *Amphigymnia*. *Contributions from the United States National Herbarium*, **36**, 193–358.

Hale, M.E. (1975a). A revision of the lichen genus *Hypotrachyna* (Parmeliaceae) in tropical America. *Smithsonian Contributions to Botany*, **25**, 1–73.

Hale, M.E. (1975b). A monograph of the lichen genus *Relicina* (Parmeliaceae). *Smithsonian Contributions to Botany*, **26**, 1–32.

Hale, M.E. (1976a). A monograph of the lichen genus *Pseudoparmelia* Lynge (Parmeliaceae). *Smithsonian Contributions to Botany*, **31**, 1–62.

Hale, M.E. (1976b). A monograph of the lichen genus *Bulbothrix* Hale (Parmeliaceae). *Smithsonian Contributions to Botany*, **32**, 1–29.

Hale, M.E. (1981). A revision of the lichen family Thelotremataceae in Sri Lanka. *Bulletin of the British Museum (Natural History)* Botany, **8**, 277–332.

Hale, M.E. (1987). A monograph of the lichen genus *Parmelia* Acharius *sensu stricto* (Ascomycotina: Parmeliaceae). *Smithsonian Contributions to Botany*, **66**, 1–55.

Harmand, J. (1911). Lichens recueillis dans la Nouvelle-Calédonie ou en Australie par le R.P. Pionnier, missionaire. *Bulletin des séances de la Société des Sciences de Nancy séries 3*, **12**, 124–44.

Harmand, J. (1912). Lichens recueillis dans la Nouvelle-Calédonie ou en Australie par le R.P. Pionnier, missionnaire. *Bulletin des séances de la Société des Sciences de Nancy séries 3*, **13**, 37–64.

Hawksworth, D.L. and Ahti, T. (1990). A bibliographic guide to the lichen floras of the world (second edition). *Lichenologist*, **22**, 1–78.

Holloway, J.D. (1986). Origins of lepidopteran faunas in high mountains of the Indo-Australian tropics. In *High altitude tropical biogeography* (ed. F. Vuilleumier and M. Monasterio), pp. 533–56. Oxford University Press, Oxford.

Howell, D.G., Jones, D.L., and Schermer, E.F. (1985). Tectonostratigraphic terranes of the circum-Pacific region. In *Tectonostratigraphic terranes of the circum-Pacific region* (ed. D.G. Howell), pp. 3–30. Circum-Pacific Council for Energy and Mineral Resources, Houston.

Humphries, C.J. and Parenti, L.R. (1986). *Cladistic biogeography*. Clarendon Press, Oxford.

Jørgensen, P.M. (1983). Distribution patterns of lichens in the Pacific region. *Australian Journal of Botany*, Supplement, **10**, 43–66.

Jørgensen, P.M. and Galloway, D.J. (1983). *Bryoria* (lichenized Ascomycotina) in New Zealand. *New Zealand Journal of Botany*, **21**, 335–40.
Kärnefelt, I. (1989). Morphology and phylogeny in the Teloschistales. *Cryptogamic Botany*, **1**, 147–203.
Klement, O. (1966). Zur Kenntniss der Flechtenflora und—vegetation des Hawaii Archipels. I. Lanai. *Nova Hedwigia*, **11**, 245–83.
Krempelhuber, A. von (1870). Lichenes. In *Reise der österreichischen Fregatte Novara um die Erde in den Jahren 1857, 1858, 1859*. Botanischer Theil. Band 1 (ed. E. Fenzl), pp. 107–29. Wien.
Krempelhuber, A. von (1871). Flechten aus Amboina. *Verhandlungen der Zoologisch–Botanischen Gesellschaft in Wien*, **21**, 861–72.
Krempelhuber, A. von (1873). Beitrag zur Kenntniss der Lichenen-Flora der Südsee-Inseln. *Journal des Museum Godeffroy*, **4**, 93–110.
Krempelhuber, A. von (1875). Lichenes quos legit O. Beccari in insulis Borneo et Singapore annis 1866 et 1867. *Nuovo Giornale Botanico Italiano*, **7**, 5–67.
Krempelhuber, A. von (1877). Aufzählung und Beschreibung der Flechtenarten, welche Dr Heinrich Wawra Ritter von Fernsee von zwei Reisen um die Erde mitbrachte. *Verhandlungen der Zoologisch–Botanischen Gesellschaft in Wien*, **26**, 433–46.
Krog, H. and Swinscow, T.D.V. (1981). *Parmelia* subgenus *Amphigymnia* (lichens) in East Africa. *Bulletin of the British Museum (Natural History) Botany*, **9**, 143–231.
Lamb, I.M. (1956). *Compsocladium*, a new genus of lichenized Ascomycetes. *Lloydia*, **19**, 157–62.
Lamb, I.M., Weber, W.A., Jahns, H.M., and Huneck, S. (1972). *Calathaspis*, a new genus of the lichen family Cladoniaceae. *Occasional Papers of the Farlow Herbarium of Cryptogamic Botany*, **4**, 1–12.
Lauer, W. (1989). Climate and weather. In *Tropical rain forest ecosystems*, Ecosystems of the world, No. 14B (ed. H. Lieth and M.J.A. Werger), pp. 7–53. Elsevier, Amsterdam.
Laundon, J.R. (1981). The species of *Chrysothrix*. *Lichenologist*, **13**, 101–21.
Laundon, J.R. (1989). The species of *Leproloma*—the name for the *Lepraria membranacea* group. *Lichenologist*, **21**, 1–22.
Lumbsch, H.T. (1989). Die Holarktischen Vertreter der Flechtengattung *Diploschistes* (Thelotremataceae). *Journal of the Hattori Botanical Laboratory*, **66**, 133–96.
Magnusson, A.H. (1955). A catalogue of Hawaiian lichens. *Archiv för Botanik Series 2*, **3**, 223–402.
Magnusson, A.H. and Zahlbruckner, A. (1945). Hawaiian lichens. III. The families Usneaceae to Physciaceae. *Arkiv för Botanik*, **32A**, 1–89.
Metcalfe, I. (1988). Origin and assembly of south-east Asian continental terranes. In *Gondwana and Tethys*, Geological Society Special Publication, No. 37 (ed. M.G. Audley-Charles and A. Hallam), pp. 101–18. Oxford University Press, Oxford.
Montagne, J.F.C. (1848). Cryptogamae taitenses. *Annales des Sciences Naturelles. Paris. Botanique séries 3*, **10**, 124–34.

Montagne, J.F.C. and van den Bosch, R.B. (1857). Lichenes. In *Plantae Junghuhnianae* (ed. F.A.W. Miquel), pp. 427–94. Leyden.
Müller Argoviensis, J. (1883). Die auf der Expedition der *Gazelle* von Dr Naumann gesammelten Flechten. *Botanische Jahrbücher für Systematik, Pflanzengeschichte und Pflanzengeographie*, **4**, 53–8.
Müller Argoviensis, J. (1887). Énumération de quelques lichens de Nouméa. *Revue de Mycologie*, **34**, 77–82.
Müller Argoviensis, J. (1893). Lichenes Neo-Caledonici. *Journal de Botanique*, **7**, 51–5; 92–4; 106–11.
Müller Argoviensis, J. (1897). Lichenes. In *Die Flora der Samoa-Inseln*, (ed. F. Reinecke). *Botanische Jahrbücher für Systematik, Pflanzengeschichte und Pflanzengeographie*, **23**, 291–9.
Nelson, G. (1985). A decade of challenge: the future of biogeography. *Journal of the History of Earth Science*, **4**, 187–96.
Nylander, W. (1861). Expositio lichenum Novae Caledoniae. *Annales des Sciences Naturelles. Botanique, séries 4*, **15**, 37–54.
Nylander, W. (1868). Synopsis lichenum Novae Caledoniae. *Bulletin de la Société Linnéene de Normandie séries 2*, **2**, 39–140.
Pigram, C.J. and Davies, H.L. (1987). Terranes and the accretion history of the New Guinea orogen. *BMR Journal of Australian Geology and Geophysics*, **10**, 193–211.
Rambold, G. (1989). A monograph of the saxicolous lecideoid lichens of Australia (excl. Tasmania). *Bibliotheca Lichenologica*, **34**, 1–345.
Redon, J. and Quilhot, W. (1977). Los liquenes de las islas de Juan Fernandez 1: estudio preliminar. *Anales del Museo de Historia Natural de Valparaiso*, **10**, 15–26.
Rogers, R.W. (1986). The genus *Pyxine* (Physciaceae, lichenized ascomycetes) in Australia. *Australian Journal of Botany*, **34**, 131–54.
Rotondo, G.M., Springer, V.G., Scott, G.A.J., and Schlanger, S.O. (1981). Plate movement and island integration—possible mechanism in the formation of endemic biotas, with special reference to the Hawaiian Islands. *Systematic Zoology*, **30**, 12–21.
Sammy, N. (1988). The genus *Pyxine* (Physciaceae, Lichenes) in Western Australia. *Nuytsia*, **6**, 279–84.
Santesson, R. (1952). Foliicolous lichens. I. A revision of the taxonomy of the obligately foliicolous, lichenized fungi. *Symbolae Botanicae Upsalienses*, **12**, 1–590.
Schmid, M. (1989). The forests in the tropical Pacific archipelagos. In *Tropical rain forest ecosystems*, Ecosystems of the world, No. 14B (ed. H. Lieth and M.J.A. Werger), pp. 283–301. Elsevier, Amsterdam.
Schuh, R.T. and Stonedahl, G.M. (1986). Historical biogeography in the Indo-Pacific: a cladistic approach. *Cladistics*, **2**, 337–55.
Sipman, H.J.M. (1983). A monograph of the lichen family Megalosporaceae. *Bibliotheca Lichenologica*, **18**, 1–241.
Sipman, H.J.M. and Harris, R.C. (1989). Lichens. In *Tropical rain forest ecosystems*, Ecosystems of the world, No. 14B (ed. H. Lieth and M.J.A. Werger), pp. 303–9. Elsevier, Amsterdam.

Smith, A.L. (1922). Lichens. In A systematic account of the plants collected in New Caledonia and the Isle of Pines by Mr R.H. Compton, M.A., in 1914. Part III. Cryptogams (Hepaticae—Fungi). *Journal of the Linnean Society Botany*, **46**, 71–87.

Stenroos, S. (1988). The family Cladoniaceae in Melanesia. 4. The genera *Cladia*, *Cladina*, *Calathaspis*, and *Thysanothecium*. *Annales Botanici Fennici*, **25**, 207–17.

Stenroos, S. (1989). Taxonomy of the *Cladonia coccifera* group. 1. *Annales Botanici Fennici*, **26**, 157–68.

Stevens, G.N. (1987). The lichen genus *Ramalina* in Australia. *Bulletin of the British Museum (Natural History) Botany*, **16**, 107–223.

Streimann, H. (1986). Catalogue of the lichens of Papua New Guinea and Irian Jaya. *Bibliotheca Lichenologica*, **22**, 1–145.

Swinscow, T.D.V. and Krog, H. (1988). *Macrolichens of East Africa*. British Museum (Natural History), London.

Szatala, O. (1956). Prodrome de la flore lichénologique de la Nouvelle Guinée. *Annales Historico—Naturales Musei Nationalis Hungarici*, **7**, 15–50.

Takhtajan, A. (1986). *Floristic regions of the world*. University of California Press, Berkeley.

Tehler, A. (1983). The genera *Dirina* and *Roccellina* (Roccellaceae). *Opera Botanica*, **70**, 1–86.

Tibell, L. (1987). Australasian Caliciales. *Symbolae Botanicae Upsalienses*, **27**, 1–279.

Troll, C. (1973). The upper timberlines in different climatic zones. *Arctic and Alpine Research*, **5**, A3–18.

Verdon, D. and Elix, J.A. (1986). *Myelorrhiza*, a new Australian lichen genus from north Queensland. *Brunonia*, **9**, 193–214.

Vězda, A. and James, P.W. (1973). *Sarrameana paradoxa* Vězda et P. James gen. nov. et sp. nova, eine bemerkenswerte Flechte aus Neu Kaledonien. *Preslia*, **45**, 305–10.

Vězda, A. and Kantvilas, G. (1988). *Sarrameana tasmanica*, a new Tasmanian lichen. *Lichenologist*, **20**, 179–82.

Weber, W.A. (1986). The lichen flora of the Galápagos Islands, Ecuador. *Mycotaxon*, **27**, 451–97.

Whitmore, T.C. (1973). Plate tectonics and some aspects of Pacific plant geography. *New Phytologist*, **72**, 1185–90.

Whitmore, T.C. (1988). Phytogeography of the eastern end of Tethys. In *Gondwana and Tethys*, Geological Society Special Publication, No. 37 (ed. M.G. Audley-Charles and A. Hallam), pp. 307–11. Oxford University Press, Oxford.

Whitmore, T.C. (1989). South-east Asian tropical forests. In *Tropical rain forest ecosystems*, Ecosystems of the world No. 14B (ed. H. Lieth and M.J.A. Werger), pp. 195–218. Elsevier, Amsterdam.

Vainio, E.A. (1909). Lichenes insularum Philippinarum. I. *Philippine Journal of Science C*, **4**, 651–62.

Vainio, E.A. (1913). Lichenes insularum Philippinarum. II. *Philippine Journal of Science C*, **8**, 99–137.

Vainio, E.A. (1921). Lichenes insularum Philippinarum. III. *Annales Academiae scientiarum fennicae, A*, **15**(6), 1–368.

Vainio, E.A. (1923). Lichenes insularum Philippinarum. IV. *Annales Academiae scientiarum fennicae, A*, **19**(5), 1–84.

Vainio, E.A. (1924). Lichenes a A.W. Stechell et H.E. Parks in Insula Tahiti 1922 collecti. *University of California Publications in Botany*, **12**, 3–15.

Yoshimura, I. (1971). The genus *Lobaria* of eastern Asia. *Journal of the Hattori Botanical Laboratory*, **34**, 231–364.

Yoshimura, I. (1987). Taxonomy and speciation of *Anzia* and *Pannoparmelia*. *Bibliotheca Lichenologica*, **25**, 185–95.

Zahlbruckner, A. (1908). Die Flechten der Samoa-Inseln. *Denkschriften der Kaiserlichen Akademie der Wissenschaften Wien. Mathematisch—Naturwissenschaften Klasse*, **81**, 222–87.

Zahlbruckner, A. (1924). Die Flechten der Juan Fernandez-Inseln. In *The natural history of Juan Fernandez and Easter Island* Vol. 2 (ed. C. Skottsberg). pp. 315–408. Almqvist and Wiksells, Uppsala.

Zahlbruckner, A. (1943). Flechtenflora von Java. *Beihefte Feddes Repertorium*, **127**, 1–80.

Zahlbruckner, A. and Mattick, F. (1956). Flechtenflora von Java, 2. Teil. *Willdenowia*, **1**, 433–528.

# 2. The lichen genus *Relicina* in Australasia

J.A. ELIX
*Chemistry Department, The Faculties, Australian National University, GPO Box 4, Canberra, ACT 2601, Australia*

## Abstract

The lichen genus *Relicina* Hale, is characterized by narrow, adnate lobes, marginal bulbate cilia, and usnic acid in the upper cortex. Morphological and chemical evolutionary trends and the geographic distribution of the genus are discussed. Taxonomic problems arise in several species complexes. The occurrence of overlapping chemosyndromes of medullary metabolites makes accurate chemical determinations difficult, but of critical importance since there is a high degree of correlation between chemistry and morphological characters. Chemistry is important both as a species character and in the delimitation of evolutionary direction. The discontinuity of *Relicina* floras of Australasia and Indonesia is discussed, as well as a novel interpretation of the disjunct distribution of *R. abstrusa* (Australasia and the Caribbean). The majority of species of *Relicina* are restricted to trunks and canopy branches of tropical mangroves and primary coastal and lower montane rainforests; their major habitat is threatened by logging and forest clearance.

## Introduction

The lichen genus *Relicina* Hale, a segregate of *Parmelia* Acharius *sensu lato*, was erected by Hale (1974) to accommodate those species with narrow adnate lobes, marginal bulbate cilia, and the constant occurrence of usnic acid in the upper cortex.

In a subsequent monograph, Hale (1975) considered *Relicina* to comprise 24 species with the major centre of distribution being southeast Asia and Indonesia. At that time a total of 7 species were known from the Australasian region and a further 23 species have been recognized since (Elix and Johnston 1986, 1988a, b; Elix and Stevens 1979; Kashiwadani 1975; Kurokawa 1979, 1986; Stevens 1981). Hence, it is now apparent that the centre of the world distribution is south-east Asia and Australasia with 34 of the 38 known species of *Relicina*—the remaining four species are endemic to the neotropics.

These species possess a simple and relatively uniform morphology. The most significant morphological features include: the presence or absence of isidia and lobules, the colour of the lower surface and the type of rhizines, the structure of the upper cortex, the ornamentation of the apothecia, and the medullary chemistry.

In Australia, the majority of these species are restricted to trunks and canopy branches of tropical mangroves, coastal and hinterland rainforests of Queensland, and exhibit a significant overlap with the flora of New Guinea (where this genus is more common in montane rainforests). Three species however (*Relicina limbata*, *R. subnigra*, and *R. sydneyensis*), are essentially temperate and occur on both trees and rocks in moist areas.

Three main aspects of this lichen genus will be discussed here: (i) morphological evolutionary trends; (ii) chemical evolutionary trends; (iii) geographical distribution.

## Thallus morphology

The main lines of evolution in thallus morphology have been: (i) the production of vegetative propagules; (ii) the ornamentation of the apothecium; (iii) the structure of the upper cortex; (iv) the morphology of the lower surface.

### 1. *Vegetative propagules*

Soredia are unknown but isidia and lobules are common in the genus *Relicina* and are important characteristics for distinguishing taxa at the species level.

Regular cylindrical isidia (constricted at the base) are quite common in the genus with 10 isidiate species recognized from the region. These include *R. amphithrix*, *R. circumnodata*, *R. conglutinata*, *R. demethylbarbatica*, *R.* cf. *abstrusa*, *R. planiuscula*, *R. ramboldii*, *R. schizospatha*, *R. subnigra*, and *R. sydneyensis*. In three species, *R. amphithrix*, *R. schizospatha*, and *R. planiuscula*, isidia vary from regular and cylindrical to procumbent and spathulate or, in extreme cases, lobulate. In some specimens lobulae may actually be more numerous than isidia, but here lobulae appear to originate from or intergrade with isidia.

Two species, *R. gemmulosa* and *R. luteoviridis*, possess lobulae but lack isidia. These lobulae are actually dichotomously branched, diminutive lacinae which are dorsiventral, procumbent, and often highly imbricate — but distinctly larger (0.2–0.8 mm cf. <0.1 mm) than those referred to above and are never associated with isidia.

The concept of pairs of species that are chemically identical and morphologically similar except for the production of vegetative propagules by one, has received considerable attention (Poelt 1970, 1972; Tehler 1983).

Among the *Relicinae* examples of this phenomenon are quite common. Chemically identical pairs of species differing primarily by the presence or absence of isidia and lobules are listed in Table 2.1. However, it should be emphasized that these are not all morphologically identical but show various degrees of differentiation in addition to the presence or absence of vegetative propagules.

## 2. Apothecia

(*a*) *External characteristics* Apothecia are produced frequently in *Relicina* and have been observed in all Australasian species. However, apothecia occur erratically in some species, and their production is correlated inversely with the production of various asexual propagules, a relationship that has been observed in many other groups of lichens. The discs are generally 1–3 mm in diameter, imperforate, and may be ornamented in several ways. The rim may be smooth or crenate, and either coronate, subcoronate or ecoronate. Coronate apothecia possess black bulbae around the inner margin of the apothecial rim—these bulbae are erect pycnidia with an apical pore rather than cilia and the presence or absence of such constitutes a key species character.

Confusion sometimes arises from the fact that such bulbae may be absent in immature or aged apothecia, even though these species are typically coronate (e.g. *R. samoensis*); while in several other species although consistently present (e.g. *R. retrospinosa*, *R. schizospatha*), the bulbae may be very rare (one or two per apothecium) and be accompanied by occasional bulbate cilia along the inner margin. Such apothecia are here termed subcoronate and the rhizines introrose.

Retrorose rhizines (bulbate cilia) often develop around the outer base of both coronate and ecoronate apothecia.

(*b*) *Internal characteristics* The internal apothecial characteristics and spore characters are typical of the Parmeliaceae. The asci normally contain 8 elliptical to ovoid spores between 2–6 μm wide and 3–10 μm

**Table 2.1.** Chemical and morphological data for comparable species-pairs of *Relicina* in Australia

| Chemistry | Primary species (Fertile–non-isidiate) | Secondary species | |
|---|---|---|---|
| | | Isidiate | Lobulate |
| Echinocarpic acid | R. cf. *fluorescens* | R. *planiuscula* | R. *gemmulosa* |
| | R. *samoensis* | R. *amphithrix* | |
| | [R. *echinocarpa*][a] | R. cf. *planiuscula* | R. *schizospatha* |
| Gyrophoric acid | R. *retrospinosa* | | R. *luteoviridis* |
| Norstictic acid | R. *subabstrusa* | R. cf. *abstrusa* | |
| Protocetraric acid | [R. *precircumnodata*][a] | R. *circumnodata* | |
| Fumarprotocetraric acid | R. *ramosissima* | R. *conglutinata* | |
| Stictic acid | R. *limbata* | R. *sydneyensis* | |

[a] Not present in Australasia.

long, although the South American, *R. relicinella* produces multispored asci with 16–32 spores. The other exceptions are *R. circumnodata* and *R. precircumnodata* which produce bicornute spores, $2-3 \times 10-13\,\mu m$ in size.

The microconidia are remarkably uniform throughout *Relicina*, usually being bifusiform and $6-9\,\mu m$ long and $c.1\,\mu m$ wide.

## 3. Epicortex and cortex

SEM studies by Hale (Hale 1973) confirmed that in *Relicina* the upper cortex has a basic palisade plectenchyma covered by a thin, pored epicortex. Later, more detailed work by the same author (Hale 1975) confirmed this structure and established that most species have an irregularly arranged palisade layer $10-15\,\mu m$ thick, rarely extending downward to $20\,\mu m$. The epicortex is approximately $0.6\,\mu m$ thick, which is similar to other epicorticate Parmelioid genera (Hale 1973). An upper cortex with a particularly well-developed palisade structure was termed a columnar cortex, consisting of packed columnar cells $20-25\,\mu m$ thick (overlain by the epicortex). Such columnar cortices are characteristic of eight Australasian species: *R.* cf. *fluorescens*, *R. fluorescens*, *R. gemmulosa*, *R. luteoviridis*, *R.* cf. *abstrusa*, *R. planiuscula*, *R. retrospinosa*, and *R. subabstrusa*.

## 4. Lower surface

(*a*) *Colour* The lower surface of the thallus in *Relicina* varies in colour from off-white (pale ivory) to jet black and such colour differences are a very important taxonomic character. The colour of the lower surface is usually very uniform. Even so, the pale brown lower surface of some species is occasionally obscured by a very dense layer of black rhizines, for example, *R. limbata*, so particular care must be taken to distinguish the colour of the rhizines and that of the lower surface.

Of the 30 species of *Relicina* known from the region, 21 have a black lower surface and 9 have a pale tan or brown lower surface. Only one pair of species is chemically and morphologically similar and differs primarily in the colour of the lower surface; *R. sydneyensis* (brown below) and *R. subnigra* (black below). The fact that most species lack a counterpart with a lower surface of the alternative colour seems to support the validity of this trait as a species-level character.

(*b*) *Rhizines* The rhizines of *Relicina* vary from simple, sparsely branched or tufted, to highly branched and agglutinated; they are an important species character. The type of rhizine is broadly correlated with the colour of the lower surface. Simple or sparsely branched rhizines often occur in species with a black lower surface and these are

common and widely distributed. In Australasia this group includes
*R. amphithrix, R.* cf. *planiuscula, R.* cf. *fluorescens, R. connivens, R. demethylbarbatica, R. fluorescens, R. fijiensis, R. gemmulosa, R. hirtifructa, R. luteoviridis, R.* cf. *abstrusa, R. nuiginiensis, R. planiuscula, R. relicinula, R. samoensis, R. schizospatha, R. subabstrusa,* and *R. terricrocodila.*

Another group of species has sparsely to densely branched, almost squarrose rhizines, but these rhizines are always brown-black to black, shiny, and not agglutinated. Four such species have a pale lower surface, *R. filsonii, R. limbata, R. ramboldii,* and *R. sydneyensis*; while the remainder have a black lower surface and include *R. malesiana, R. retrospinosa,* and *R. subnigra.*

A unique group of species with a pale lower surface has pale, highly branched and agglutinated rhizines. These rhizines form a woolly mat about 1 mm thick over the lower surface, have blunt tips and tend to adhere to one another and the adjacent lower surface.

The overall evolutionary trends (as we now perceive them) are summarized in Table 2.2.

**Table 2.2.** Summary of morphological evolutionary trends

| Primitive ---------------------- > Advanced |
| --- |
| Apothecia: Ecoronate -------- > Subcoronate -------- > Coronate |
| Spores: Ellipsoid -------- > Bicornute |
| Lower surface: Pale -------- > Black |
| Rhizines: Agglutinated -------- > Branched -------- > Simple |
| Propagules: Non-isidiate -------- > Lobulate -------- > Isidiate |
| Upper cortex: Columnar -------- > Palisade |

## Phytochemistry

### 1. Chemotaxonomy

The chemistry of genus *Relicina* is reasonably diverse and has proved to be of value here as in many other groups of Parmelioid lichens (Hale and Kurokawa 1964; Hale 1975; Kurokawa 1979, 1986; Elix and Johnston 1986, 1988a, b).

Indeed specimens previously thought to represent chemical races have, in some cases, been shown to have morphological or distributional characteristics which justify the rank of species. A list of the Australasian species of *Relicina* according to their diagnostic (invariant) secondary metabolites follows.

Alectoronic acid
  *R. fluorescens*
Barbatic acid (major), 4-*O*-demethylbarbatic acid (minor)
  *R. agglutinata*
4-*O*-Demethylbarbatic acid (major), barbatic acid (minor)
  *R. demethylbarbatica*
Echinocarpic acid, conechinocarpic acid
  *R. amphithrix*, *R. columnaria*, *R. gemmulosa*, *R. planiuscula*, *R. samoensis*, *R. schizospatha*
Echinocarpic acid, conechinocarpic acid, hirtifructic acid
  *R. hirtifructa*, *R. terricrocodila*
Fumarprotocetraric acid
  *R. malesiana*
Fumarprotocetraric acid, succinprotocetraric acid
  *R. conglutinata*, *R. ramosissima*
Gyrophoric acid
  *R. luteoviridis*, *R. retrospinosa*
Hirtifructic acid
  *R. fijiensis*, *R. niuginiensis*
Norstictic acid, connorstictic acid, hypostictic acid
  *R. filsonii*, *R. neoabstrusa*, *R. subabstrusa*
Protocetraric acid
  *R. circumnodata*, *R. sublanea*
Protolichesterinic acid
  *R. connivens*
Stictic acid, constictic acid
  *R. limbata*, *R. subnigra*, *R. sydneyensis*
Stictic acid, hypostictic acid, menegazziaic acid
  *R. ramboldii*
Unidentified aliphatic compounds
  *R. connivens*, *R. relicinula*

The unidentified echinocarpic acid complex (echinocarpic acid, conechinocarpic acid, and hirtifructic acid) and the $\beta$-orcinol depsidones form by far the most important group of secondary metabolites found in *Relicina*, with stictic acid, norstictic acid, fumarprotocetraric acid, protocetraric acid, and succinprotocetraric acid all being particularly common. The latter depsidones are widely distributed compounds occurring commonly in numerous lichen genera. Echinocarpic acid seems to be especially characteristic of the *Relicinae*, although it does occur rarely in other Parmelioid genera including *Hypotrachyna* (*H. dentella*, *H. subaffinis*, *H. thysanota*), *Parmotrema* (*P. dilatatum*) and *Parmelia* (*P. crambidiocarpa*). Species containing echinocarpic acid form the core of the genus in southeast Asia, while those with stictic acid are mainly Australian endemics.

## 2. Chemosyndromic variation

A chemosyndrome refers to a set of biogenetically related metabolites and in this pattern of chemical variation the major secondary metabolite in any one taxon is *invariably* accompanied by minor quantities of several biosequentially related substances which, in turn, are observed as major metabolites in related taxa. Species belonging to the *Relicina samoensis* complex exhibit both replacement compounds and chemosyndromic variations (Table 2.3).

## •3. Biosequential relationships

There have been relatively few biosynthetic studies on lichens, but evidence from those that have been made and pathways to similar products in non lichen-forming fungi and higher plants allow us to construct preliminary biogenetic hypotheses. From such simple hypotheses it is possible, in some cases, to establish biosynthetic sequences.

Some biosequential relationships appear self-evident, thus:

a. *O*-methylated derivatives can be considered to be more highly derived than the corresponding hydroxy compounds, e.g. methylation of norstictic acid will give stictic acid.

b. *C*-hydroxylated derivatives can be considered to be more highly derived than the corresponding parent compounds, e.g. hydroxylation of lecanoric acid will give diploschistesic acid.

c. Esterified products can be considered more highly derived than the parent hydroxy compounds, e.g. fumarprotocetraric acid can be considered to be more highly derived than protocetraric acid.

d. Sequential oxidations are also common in lichens, with the more highly oxidized substrate representing the more highly derived secondary metabolite, e.g. oxidation of norstictic acid would give the more highly derived salazinic acid.

e. *C*-methylated derivatives can be considered to be more highly derived than the corresponding parent compounds, e.g. 4-*O*-demethylbarbatic acid could be considered more highly derived than lecanoric acid.

Such biosequential relationships can then be used to identify primitive and derived chemical characters.

## Divergence of some species complexes

### 1. Relicina agglutinata *complex*

This group of species is restricted to lowland rainforests of south-east

Table 2.3. Chemosyndromic variation in the *Relicina samoensis* complex

| Species | Echinocarpic | Conechinocarpic | Hirtifructic | Gyrophoric | Fatty acids |
| --- | --- | --- | --- | --- | --- |
| *R. samoensis* (Pan-pac) | major | minor | — | — | — |
| *R. amphithrix* (Aust/Indo) | major | minor | — | — | — |
| *R. terricrocodila* (Aust) | major | minor | trace | — | — |
| *R. fijiensis* (Fiji) | — | — | major | — | — |
| *R. niuginiensis* (PNG) | — | — | major | trace | minor |
| *R. relicinula* (Indonesia) | — | — | — | — | major |

Asia and Australasia. As pointed out by Hale (1975) there appear to be two series of species. One series including *R. agglutinata*, *R. conglutinata*, *R. ramosissima*, and *R. sublanea*, has ecoronate apothecia and small ovoid spores while the other series including *R. precircumnodata* and its isidiate morph *R. circumnodata* is coronate and has bicornute spores. The main lines of evolution here seem to have involved differentiation in apothecial and spore characters and as Hale (1975) pointed out, it seems apparent that ecoronate species are more primitive than coronate species and that in the *R. sublanea–R. circumnodata* group, the coronate condition is paralleled by spore evolution (from simple to bicornute) but chemistry is not significantly changed. The divergence of this species complex is illustrated diagrammatically in Fig. 2.1.

## 2. Relicina limbata *complex*

Four of the five species of this complex are endemic to Australia; only the isidiate species, *R. sydneyensis* is more widely distributed extending to Papua New Guinea, India, Indonesia, Taiwan, and Japan. This species complex is further characterized by the presence of the stictic-norstictic acid chemosyndrome in the medulla (the only group of species with this chemistry), ecoronate apothecia and, generally, a pale

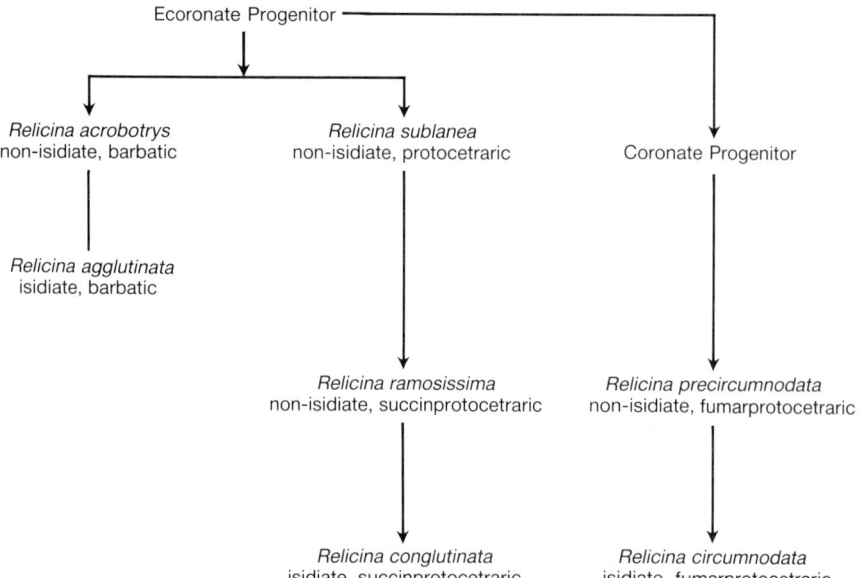

**Fig. 2.1.** Divergence of the *R. agglutinata* complex.

lower surface with simple to squarrose branched rhizines. Three representatives are unusual ecologically (for *Relicina* s.l.) in that they are very common on rock surfaces in temperate areas (*R. limbata*, *R. subnigra*, and *R. sydneyensis*). The divergence of this species complex is illustrated diagrammatically in Fig. 2.2.

*3. Relicina fluorescens complex*

This group of species is centred in Papua New Guinea and Indonesia. Morphological characteristics include very broad lobes (to 4.0 mm wide), ecoronate or subcoronate apothecia, a black lower surface with simple rhizines, and the characteristic thick, columnar upper cortex. Chemically the group is quite variable, exhibiting the presence of the echinocarpic acid complex, alectoronic acid and gyrophoric acid. The divergence of this species complex is illustrated diagrammatically in Fig. 2.3.

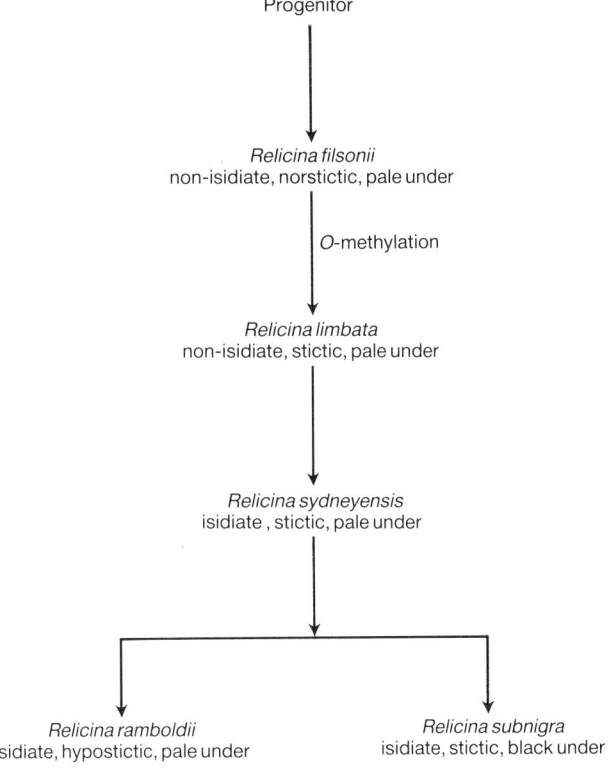

**Fig. 2.2.** Divergence of the *R. limbata* complex.

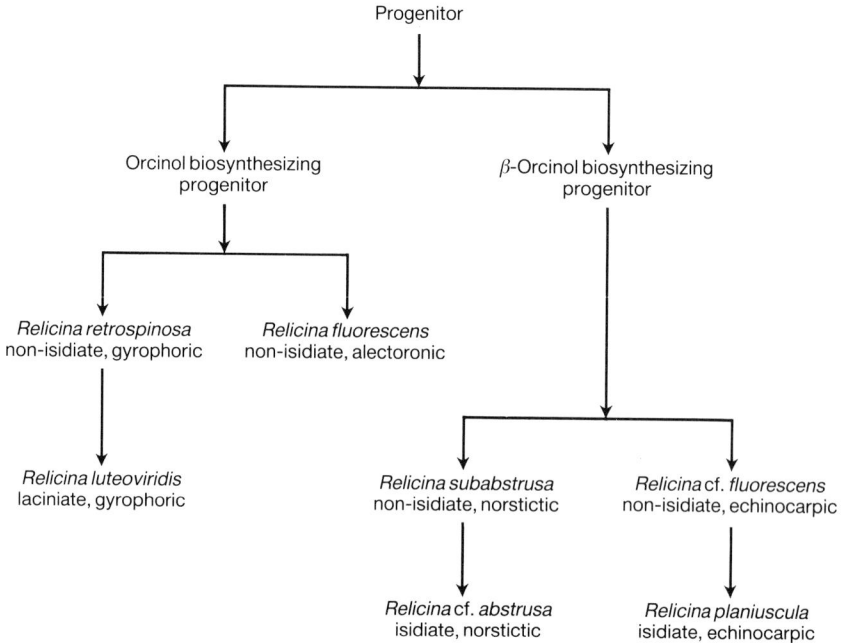

**Fig. 2.3.** Divergence of the *R. fluorescens* complex.

### 4. Relicina abstrusa *in Australasia*

Hale (1975) first recognized that neotropical and Asian–Australasian populations of the widely distributed *R. abstrusa* exhibited different ecological requirements and minor chemical differences. These differences are summarized in Table 2.4. We believe this combination of characters may warrant the recognition of the Asian–Australasian populations as a separate taxon.

### 5. Relicina samoensis *complex*

This group of species is distributed in northern Australia, Papua New Guinea, Indonesia and, particularly, Oceania. Morphological characteristics include very narrow lobes (0.2–1.0 mm wide), coronate apothecia, a black lower surface with simple rhizines, and the presence of aliphatic acids and the echinocarpic acid complex in the medulla. In this group, the production of apothecial bulbae is extremely variable— on the one thallus mature apothecia may appear ecoronate, subcoronate or highly coronate, so this character must be treated with caution.

**Table 2.4.** Ecological requirements and chemical differences of neotropical and Asian-Australasian populations of *R. abstrusa*

| Characteristic | *Relicina* cf. *abstrusa* | *Relicina abstrusa* |
| --- | --- | --- |
| Upper cortex | Columnar | Palisade |
| Elevational range | 100–1400 m | 700–2500 m |
| Apothecia | Rare (7%) | Common (53%) |
| Chemistry | Norstictic, connorstictic, Pcr-3 | Norstictic, connorstictic |
| Geographic distribution | Indonesia; Australasia | Southern USA; central and south America |

Species belonging to the *Relicina samoensis* complex exhibit both morphological (isidiate/non-isidiate) and chemosyndromic variations (see below). The divergence of this species complex is illustrated diagrammatically in Fig. 2.4.

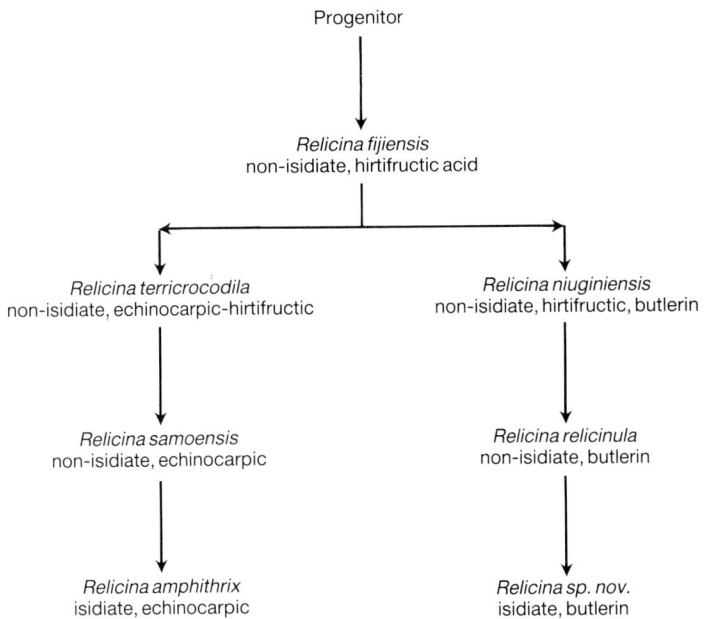

**Fig. 2.4.** Divergence of the *R. samoensis* complex.

## Geographic distribution

*1. World distribution*

A very distinctive feature of *Relicina* is the high degree of endemism in south-east Asia and Australasia. Apart from *R. subabstrusa* which occurs in the neotropics as well as Asia, and four neotropical endemics (*R. abstrusa, R. eximbricata, R. incongrua,* and *R. relicinella*) all the remaining 34 species occur only in Asia and Oceania. As yet, the genus is unknown in Europe or central Asia and is very rare in east Africa.

It would appear that the genus probably evolved in lowland dipterocarp forests of south-east Asia. The endemic neotropical species (all 4 of which are coronate) probably have evolved separately (from *R. subabstrusa* ?) but the conditions here have obviously not been as favourable for the rapid evolution and expansion of the genus.

*2. Regional distribution*

This is best summarized by reference to Table 2.5 which compares the species overlap between Indonesia, Papua New Guinea (PNG), and Australia. It can be seen from this Table that Indonesia and PNG have 14 species in common (and PNG 4 endemics); PNG and Australia 12 species in common (and Australia 7 endemics); while 11 species are common to all three regions. Thus, there is an overall continuum of species throughout the region, providing further evidence for the antiquity of the genus. Given their proximity and climatic similarities the greater number of shared species between Indonesia and PNG is not surprising.

Although there is a strong continuity in the *Relicina* floras of these areas, the long isolation of the Australian continent, together with the wider range of climatic variation, is reflected in the larger number of endemic species occurring here. In particular, in the *R. limbata* group 4 of the 5 species are endemic to Australia and occur more commonly in temperate areas. This is possibly the result of adaptation to an increasing aridity of climate and lighter, more open habitats.

## Systematic position of *Relicina*

The genera *Relicina* and *Bulbothrix* occupy an isolated position in the Parmeliaceae, because they are the only groups to exhibit the following characters.

*1. Production of bulbate cilia*

Bulbae are most commonly recognized in the cilia. On new lobes they

**Table 2.5.** *Relicina* species in Indonesia, PNG, and Australia

| Indonesia | Papua New Guinea (PNG) | Australia |
| --- | --- | --- |
| R. cf. *abstrusa* | R. cf. *abstrusa* | R. cf. *abstrusa* |
| R. *amphithrix* | R. *amphithrix* | R. *amphithrix* |
| R. *circumnodata* | R. *circumnodata* | R. *circumnodata* |
| R. *planiuscula* | R. *planiuscula* | R. cf. *planiuscula* |
| R. *ramosissima* | R. *ramosissima* | R. *ramosissima* |
| R. *relicinula* | R. *relicinula* | R. *relicinula* |
| R. *samoensis* | R. *samoensis* | R. *samoensis* |
| R. *schizospatha* | R. *schizospatha* | R. *schizospatha* |
| R. *subabstrusa* | R. *subabstrusa* | R. *subabstrusa* |
| R. *sublanea* | R. *sublanea* | R. *sublanea* |
| R. *sydneyensis* | R. *sydneyensis* | R. *sydeyensis* |
| **R. acrobotrys** | | |
| **R. precircumnodata** | | |
| **R. subconnivens** | | |
| R. *fluorescens* | R. *fluorescens* | |
| R. *gemmulosa* | R. *gemmulosa* | |
| R. *malesiana* | R. *malesiana* | |
| | **R. agglutinata** | |
| | **R. hirtifructa** | |
| | **R. niuginiensis** | |
| | **R. retrospinosa** | |
| | R. *connivens* | R. *connivens* |
| | | **R. conglutinata** |
| | | **R. demethylbarbatica** |
| | | **R. filsonii** |
| | | **R. limbata** |
| | | **R. ramboldii** |
| | | **R. subnigra** |
| | | **R. terricrocodila** |

begin as ordinary cilia but soon become basally inflated. The ciliate tip may be eroded or even become densely branched with agglutinated hyphae. Hale has shown that the mature globose bulbae are hollow and filled with a colourless fluid. The heavily carbonized walls are uniformly thickened and composed of densely conglutinated paraplectenchyma. A parallel-oriented prosoplectenchymatous structure, the normal characteristic of cilia, cannot be recognized at the base but occurs towards the apices where the cilia are tapered. Carbonized bulbae also occur at the base of occasional rhizines, as well as on the retrorose rhizines at the base of many apothecia.

It appears that the bulbae can act as propagules, a function first recognized by Hale (1975). Occasionally, one can see lobules or even lacinae originating from detached bulbae. Since they do not contain algae, presumably they function as giant fungal diaspores and encounter suitable photobionts before continuing growth.

## 2. Production of coronate apothecia

The presence of coronate apothecia, that is of erect, bulbate pycnidia around the inner margin of the apothecium (as well as having the normal immersed pycnidia, typical of Parmeliaceae) is another characteristic restricted to *Relicina* and *Bulbothrix*.

The only other segregate bearing a similarity to *Relicina* is *Relicinopsis*. These two genera have lobes that are typically linear-elongate and dichotomously branched, with analogous crowded or intricate lobe configuration, and which form thalli with similar overall gross morphology. However, the species of *Relicinopsis* lack the characteristic bulbae and have cylindrical to weakly fusiform conidia (conidia are bifusiform in *Relicina*).

A comparison of the morphology of these three segregates is given in Table 2.6. So, *Bulbothrix* differs from *Relicina* in: (i) the size of the

**Table 2.6.** A comparison of *Relicina*, *Bulbothrix*, and *Relicinopsis*

| Character | *Relicina* | *Bulbothrix* | *Relicinopsis* |
|---|---|---|---|
| Bulbae | Present | Present | Absent |
| Coronate apothecia | Present | Present | Absent |
| Spores ($\mu$m) | 3–10 × 2–6 | 12–19 × 5–10 | 5–8 × 3–5 |
| Conidia ($\mu$m) | Bifusiform 6–9 | Cylindrical 6–10 | Weakly fusiform 5–7 |
| Distribution (centres) | South-east Asia Neotropics | Africa South-east USA | South-east Asia |
| Habitat | Tropical rainforest | Subtropical savannah | Tropical rainforest |
| Colour | Yellow-green | Grey | Yellow-green |
| Epicortex | Pored | Pored | Pored |
| Cortex | Usnic acid | Atranorin | Usnic acid |
| Medullary chemistry | $\beta$-orcinol depsides, and depsidones | Orcinol depsides, and depsidones | $\beta$-orcinol depsides, and depsidones |

spores and the shape of the conidia; (ii) habitat preference (drier subtropical regions); (iii) centres of distribution (only 7 of the 35 known species of *Bulbothrix* occur in the study region); (iv) cortical chemistry—atranorin rather than usnic acid; (v) medullary chemistry. On the basis of such evidence Hale concluded that there was little chemical evidence of any significant chromosomal exchange between *Relicina* and *Bulbothrix*—and that as they presently stand they are not closely related biologically. However, this does not rule out the possibility that they had a distant common origin and that subsequent divergence was reflected in the atranorin–usnic acid split.

## Conservation

The majority of species of *Relicina* have the misfortune of occupying an endangered habitat, consequently they also are endangered. The most common habitats in Australia and Papua New Guinea are canopy branches of montane 'oak' or *Castanopsis* forests in PNG and the lowland dipterocarp forests in PNG and tropical rainforests in Australia. Both forest types are prime areas for logging operations. Peter H. Raven, the Director of the Missouri Botanical Garden made a recent statement in respect to Amazonia which is equally valid here:

> It appears likely that no fewer than 1.2 million species, at least a quarter of the biological diversity existing in the mid-1980s, will vanish during the next quarter-century or soon thereafter, and that a much higher proportion of the total will follow by the second half of next century, as the remaining forest refuges are decimated.

Forestry operations in PNG are a particular worry. Clear-felling of forests in mountainous areas by Malaysian and Japanese wood-chipping interests are particularly destructive because of subsequent extensive soil erosion. However, in many other areas such problems have been exacerbated by population growth. This has meant that the traditional slash and burn agriculture has been occurring at shorter intervals, resulting in poorer regrowth and, in the longer term, forest degradation with far fewer mature trees with their extensive lichen and bryophyte floras. Similar forest degradation has also occurred in Fiji, but here this has occurred over a longer time period, with virtually no commercial logging operations.

North Queensland, on the other hand, is much better served by National Parks in the afforested areas and, even in State Forests where forestry operations are permitted, logging is very selective, with no more than 10–15 per cent of the trees being removed and these being the mature or over-mature specimens. With such management practices, although there may be a temporary diminution in lichen habitats and ecological niches, the longer term survival of most species is not threatened.

## References

Elix, J.A. and Johnston, J. (1986). New species of *Relicina* (Lichenized Ascomycotina) from Australasia. *Mycotaxon*, **27**, 611–16.

Elix, J.A. and Johnston, J. (1988a). New species of Parmeliaceae (Lichenized Ascomycotina) from the Southern Hemisphere. *Mycotaxon*, **31**, 491–510.

Elix, J.A. and Johnston, J. (1988b). New species of *Relicina* and *Xanthoparmelia* (Lichenized Ascomycotina) from the Southern Hemisphere. *Mycotaxon*, **33**, 353–64.

Elix, J.A. and Stevens, G.N. (1979). New species of *Parmelia* (lichens) from Australia, *Australian Journal of Botany*, **27**, 873–83.

Hale, M.E. Jr. (1973). Fine structure in the cortex in the lichen family Parmeliaceae viewed under the scanning electron microscope. *Smithsonian Contributions to Botany*, **10**, 1–92.

Hale, M.E. Jr. (1974). *Bulbothrix*, *Parmelina*, *Relicina*, and *Xanthoparmelia*, four new genera in the Parmeliaceae (Lichenes). *Phytologia*, **28**, 334–9.

Hale, M.E. Jr. (1975). A monograph of the lichen genus *Relicina* (Parmeliaceae). *Smithsonian Contributions to Botany*, **26**, 1–32.

Hale, M.E. Jr. and Kurokawa, S. (1964). Studies on *Parmelia* subgenus *Parmelia*. *Contributions from the United States National Herbarium*, **36**, 121–91.

Kashiwadani, H. (1975). Enumeration of *Anaptychiae* and *Parmeliae* of Papua New Guinea. In *Reports on the cryptogams in Papua New Guinea* (ed. Y. Otani), pp. 75–85. National Science Museum, Tokyo.

Kurokawa, S. (1979). Enumeration of species of *Parmelia* in Papua New Guinea. In *Studies on cryptogams of Papua New Guinea* (ed. S. Kurokawa), pp. 125–50. Academia Scientific Book Inc., Tokyo.

Kurokawa, S. (1986). Further notes on *Parmelia* (Parmeliaceae) of Papua New Guinea. *Annals of the Tsukuba Botanical Garden*, **5**, 1–15.

Poelt, J. (1970). Das Konzept der Artenpaare bei den Flechten. *Vorträge aus dem Gesamtgebeit der Botanik*, **4**, 187–98.

Poelt, J. (1972) Die taxonomische Behandlung von Artenpaare bei den Flechten. *Botaniska Notiser*, **125**, 77–81.

Stevens, G.N. (1981). The macrolichen flora on mangroves of Hinchinbrook Island. *Proceedings of the Royal Society of Queensland*, **92**, 75–84.

Tehler, A. (1983). The genera *Dirina* and *Roccellina* (Roccellaceae). *Opera Botanica*, **70**, 1–86.

# 3. Lichen conservation in Hawaii

C.W. SMITH

*Department of Botany, University of Hawaii at Manoa, 3190 Maile Way, Honolulu, HI 96822, USA*

## Abstract

Magusson's *Catalogue of Hawaiian Lichens* listed 678 species, 38 per cent of which he considered endemic. Subsequent work suggests that there may be over 800 species present but that the level of endemism is much lower. Only one genus, *Ramalinopsis* is endemic to the islands and that is monotypic. Though most rocks are volcanic in origin, there are small pockets of coralline limestone on which six lichen species are found very sparsely. Concrete pavements are colonized by a few of these species. In contrast, the vast lava flows are quickly colonized in wet areas by an assemblage of lichens, principally species of *Stereocaulon*. In drier areas rates of colonization are orders of magnitude slower but species diversity is higher. Above the inversion layer at 2250 m, the lichen cover drops off dramatically and above 3250 m most species are confined to cracks or the lower surfaces of rocks. Sea cliffs have their own unique flora. The soil lichen flora is poorly developed and rarely a significant element. The phorophyte lichen flora is richest in the cloud forests above 1000 m. It is here that the bulk of the endemic species are found. The lichen flora is poorest in the rainforest where the bark is soon colonized and dominated by hepatics, for example, *Bazzania* and mosses, for example *Macromitrium*. Foliicolous species are common in suitable habitats. Below 1000 m, disturbance from agriculture and housing developments has had a significant negative impact on lichen habitats probably due to desiccation and loss of substrate. The lichen flora in this area is characterized by species common throughout the Pacific. Yet, there is no evidence that alien lichens have invaded the area and displaced native species, a feature of all other plant groups. The Hawaiian lichen flora is threatened mainly by the destruction of forest understories by feral pigs, replacement of mixed native forests by monotypic stands of weedy alien plants, fires carried by alien grasses, and the continuously increasing exploitation of land. Air pollution is not a major factor.

## Introduction

Hawaii, the most isolated island group in the world, is over 3000 km from the nearest substantial land mass. The endemism of its terrestrial fauna and flora provide the best example of Darwinian evolution. For example, 90 per cent of the native angiosperm flora and 32 genera are endemic (Wagner *et al.* 1990). Unfortunately, Hawaii is a ravaged landscape where there are approximately as many naturalized species of flowering plants as native species. A place where 27 per cent of the flora is considered endangered and 10 per cent of the species are thought to be extinct (D.R. Herbst, personal communication). The situation for the native fauna and flora continues to deteriorate as the consequences of human activity spread and intensify in all areas of the Islands. There are no known alien lichen species in the Islands and none are known to have been extirpated. Indeed, none are known to be endangered. However reassuring that may sound, it is matter of ignorance of the lichen flora not its resilience. There are 723 recorded species to date, 240 of which are supposedly endemic. There is only one endemic genus, the monotypic *Ramalinopsis*.

## Lichen communities

### 1. Windward communities

(*a*) *Coastal cliffs* Coastal cliffs are on the windward side of the Islands where rain and ocean swells erode the soft lava rock. Trees and large shrubs do not attain much stature, breaking away and exposing new rock surfaces for lichen colonization. The dominant lichens are *Dirinaria aegialita*, *D. applanata*, and *Xanthoparmelia subramigera*. Areas close to the ocean are characterized by *Caloplaca poliotera*, *C. inconstans*, *Dirina catalinariae*, and *Ramalina microspora*. Further inland *R. umbilicata* becomes common. Lichens are the dominant plants in this environment with more than 35 species present, over half of which are endemic. There are few threats to the habitat. Feral goats are a problem in some isolated areas where they cause landslides and erosion.

(*b*) *Lowland coastal forest* Little native forest remains; much of it was destroyed during the Hawaiian colonization. *Pandanus* forest occurs in some areas and there are one or two other relict forests, for example, *Pritchardia* palm forest on an inaccessible islet at Kalaupapa National

Park, Molokai. The lichen flora is characterized by cosmopolitan tropical species, including, but not limited to, *Anthracothecium subochraceum*, *Buellia* spp., *Candelaria concolor*, *Coccocarpia erythroxyli*, *Dirinaria aegialita*, *D. applanata*, *Hyperphyscia adglutinta*, *Lecanora leprosa*, *Parmotrema tinctorum*, and *P. cristiferum*. Lichen biomass and diversity are low due to the continuous disturbance. Less than 5 per cent of the species are endemic.

These areas have been severely disturbed since the initial Hawaiian occupation of the Islands. Resident populations have always favoured these areas but the current popularity of beach-front and ocean view property has transformed the once-forested areas into 'city-desert' labyrinths. Even coastal estuarine communities have been drained and developed or replaced by alien mangroves which do not support many lichen species.

(c) *Lower elevation forest*  These areas have been very badly disturbed by sugar and pineapple plantations. The decline of these industries has led to their abandonment, conversion to macadamia orchards or housing developments. Even areas unsuitable for agriculture rarely sustain native vegetation but have been occupied by alien trees and shrubs.

Drier areas are characterized by *Caloplaca inconstans* and *Xanthoparmelia subramigera* on rocks, and *Catillaria ochraceonigra*, *Dirinaria applanata*, *Lecanora leprosa*, and *Pyxine cocoes* on bark. *Buellia* spp., *Coccocarpia erythroxyli*, *Collema rugosa*, *Dirinaria aegialita*, *D. applanata*, *D. picta*, *Glyphis cicatricosa*, *Hyperphyscia adglutinata*, *Megalospora sulphurata*, *Parmotrema cristiferum*, *P. reticulatum*, *P. tinctorum*, *Pyxine retirugella*, *Ramalina exiguella*, *Sticta wiegelii* as well as a number of species of Graphidaceae and Pyrenulaceae are frequent in mesic areas. *Coccocarpia erythroxyli*, *Leptogium* spp., *Pannaria mariana*, and *Physma pseudoisidiatum* dominate in lowland rainforest (more than 100 mm of rain each month) habitats on bark and *Parmotrema cristiferum* and *P. tinctorum* on rock. Over 150 species have been collected in this general area to date but less than 10 per cent are endemic. Lichen cover is high on favourable hosts but the overall biomass is low because of disturbance and the high frequency of myrtaceous species. There is a dismaying uniformity to the lichen communities in alien forests. The lack of old-age stands and the monotypic character of many of these forests precludes much diversity in the lichen flora. Myrtaceous species, with their sloughing bark, are widespread and provide a poor habitat for lichens. Yet, *Dimerella zonata* is always best developed on *Eugenia cuminii* and the endemic *Heterodermia obesa*, is common on *Psidium guajava* in some areas. The rare endemic, *Ramalinopsis mannii*, is now apparently confined to the Polynesian introduction *Aleurites moluccana*. Yet, macadamia trees are good hosts for foliicolous and cosmopolitan tropical corticolous lichens and woody legumes host a good variety of crustose species.

Fire is a continuous threat in drier communities because of the establishment of alien bunchgrasses which substantially increase the fuel load (Smith and Tunison 1990). Each succeeding fire intensifies the bunchgrass dominance of the habitat, ultimately rendering the area unsuitable for lichens.

The continued demand for housing and recreational areas also suggests that most lowland areas will be overrun in the foreseeable future. The increasing concerns to conserve watersheds, however, could counteract this potential loss of habitat.

(*d*) *Upper rainforest* Bryophytes, particularly *Bazzania* spp., dominate corticolous rainforest habitats. Typical species include *Cladonia scabriuscula*, *Erioderma pulchella*, *Hypotrachyna gigas*, *H. microblasta*, *Megalospora sulphurata*, *Pannaria lurida*, *Phaeophyscia hispidula*, and *Pertusaria isidiophora*. *Placopsis ?parellina* and *Stereocaulon ramulosum* inhabit the few rock surfaces along stream-banks protected from the raging torrents. *Cladina skottsbergii* is often dominant in open boggy areas, together with *Siphula* spp.. Species diversity and biomass in this habitat are low. Endemism is less than 10 per cent.

Many of these ecosystems are relatively intact though feral animals, particularly pigs, are opening up the understorey. Many bogs have been destroyed. Alien plants, especially vines such as *Passiflora mollissima*, are another serious disruptive influence (Smith 1985).

(*e*) *Inversion layer* The inversion layer is a region where the cloud layer moves up the mountain around midday and then retreats by evening, creating ideal growing conditions for lichens. Yet, pendent fruticose species, such as *Usnea australis*, are only common in *Acacia koa* forest with its open canopy. In the denser, canopied *Metrosideros polymorpha* forest they are rare. Typical species include: *Bryoria smithii*, *Hypotrachyna imbricatula*, *H. microblasta*, *H. sinuosa*, *Lecanora pallida*, *Pannaria rubiginosa*, *Parmotrema reticulatum*, *Pseudocyphellaria crocata*, *P. flavicans*, *Sticta fuliginosa*, *St. tomentosa*, *St. wiegelii*, and *Xanthoparmelia taractica*. Overall species diversity is high, with over 80 species recorded so far, but the biomass is quite low. Endemism is estimated to be less than 15 per cent.

These relatively intact ecosystems were severely impacted in the past by feral cattle and are now being damaged by feral pigs. A number of alien grasses (e.g. *Dactylis glomerata*, *Holcus lanatus*) now threaten these forests (Smith 1985). Some large areas are now protected in National Parks and The Nature Conservancy reserves.

## 2. Above the inversion layer

(*a*) *Alpine shrubland* This open community on unweathered lava has been severely damaged by feral animals, particularly goats. Native bunchgrasses have been replaced by more invasive alien species that create denser shade. Common saxicolous lichens include: *Lecanora muralis*, *Placopsis cribellans*, and *Xanthoparmelia taractica*. Corticolous species include: *Lecanora pallida*, *Hypotrachyna sinuosa*, *Parmotrema dominicanum*, and *Ramalina celastri*. Over 40 species are known from this habitat, mostly crustose species with little biomass. The level of endemism is difficult to establish because so many species appear to be sterile. The threats to this habitat are the same as those for the inversion layer.

(*b*) *High elevation grassland* This ecosystem is characterized by an almost monotypic community of the bunchgrass *Deschampsia australis*. Only three lichens, *Peltigera dolichorhiza*, *Pseudocyphellaria crocata*, and *Cladonia scabriuscula*, are present in deep shade between the closely packed grass tussocks. The species, though common, are very 'etiolated' with thin thalli, minimal cortex, and few soredia, pseudocyphellae or squamules.

The few remnants of this ecosystem have been devastated by feral pigs but one large area is now protected in Haleakala National Park. Fire is a potential threat and was used to 'improve the keep' at the turn of the century.

(*c*) *Alpine 'desert'* These areas above 3300 m are generally devoid of vegetation though some protected areas may be covered by the moss *Rhacomitrium lanuginosum*. The common lichens are: *Lecanora polytropa*, *Pseudephebe minuscula*, *Rinodina hawaiiensis*, *Umbilicaria decussata*, and *U. hirsuta*. Less than 30 species are present, with little biomass, though over half the species are endemic. These areas are threatened by telescope, communication, and military installations, skiers, and off-road vehicles. It is hard to convince managers and politicians to take precautions to preserve this environment because there is very little indication of life. There is a large Natural Area Reserve on Mauna Kea, however, set up to conserve the glaciated landscape which includes an excellent sample of this lichen flora.

## 3. Leeward communities

(*a*) *Upper dry forest* Rock (1974) described this forest type as '. . . richest in species . . .'. The open canopy structure of many of the trees coupled with frequent inundation in cloud make this an ideal lichen habitat. Common species include: *Chiodecton perplexum*, *Graphis anfractuosa*, *Haematomma punicea*, *Lecanora flavovirens*, *Lecidea granulosa*, *Pannaria lurida*, *Parmotrema reticulatum*, *Phaeophyscia laciniata*, *Ramalina cerinella*, *R. farinacea*, *R. sandwicensis*, *R. soraligera*, *Teloschistes flavicans*, *Usnea osseoleuca*, and *U. entoviolata*. *Stereocaulon octomerellum* is common on exposed rocks. The diversity of lichens in this area is very high, with over 150 species present. The biomass is also high, promoted by the frequent cloud inundation. Approximately 40 of the species are endemic.

Many of these areas have been logged and replaced by ranchland. In fact, state land leases encourage their conversion to grassland, generally *Pennisetum clandestinum*, which prevents regeneration of the forest (Smith 1985) and smothers exposed rocks as well. Grassland is also assessed at a lower tax value than forest increasing the incentive for conversion. Cattle destroy whatever arborescent vegetation remains. The long-term prognosis for this ecosystem is very poor, though there is a growing awareness of the impact of watershed degradation.

(*b*) *Low elevation scrub* The lowland scrub between sea level and 1000 m was one of the most heavily impacted by Hawaiians who preferred dwelling in these areas. Very few native flowering plant species remain. Rocks are inhabited by *Caloplaca inconstans*, *Dirinaria aegialita*, and *Xanthoparmelia subramigera*. Corticolous species are few, generally *Dirinaria applanata*, *Chrysothrix candelaris*, and *Candellariella concolor*. Diversity is moderate, with over 30 species recorded, but many more need identification. Though apothecia are present they are nearly always without spores. The biomass is very low. The degree of endemism is unknown due to the lack of determination of many species.

Most areas were abandoned after 1778 and rapidly invaded by alien shrubs. Recently introduced fire-enhanced bunchgrasses, particularly *Pennisetum setaceum*, brought in as a garden ornamental and still available from seed catalogues, interacting with more frequent and increasingly widespread fires, are destroying the alien scrub (Smith and Tunison 1990). These areas have recently become a favoured area for resort development.

### 4. *Lava flows*

Lava flows of various ages and types are widespread at all elevations on Hawaii, but confined to Haleakala on Maui, and absent on all other islands. *Stereocaulon vulcani* is the dominant species up to 3000 m. Above that elevation, colonization is extremely sparse involving other species such as *Placopsis cribellans* and *Lecanora polytropa*. Stature and biomass are directly related to annual rainfall. Above 600 m *Stereocaulon ramulosum*, and *Psilolechia lucida* are common in areas where the rainfall exceeds 2000 mm per annum. Lichen diversity in any area is low, but because this habitat occurs at every available elevation and rainfall regime, the overall diversity is high. There are very few endemic species in this habitat. In wet and mesic areas the biomass is high.

Such a severe environment might be thought to be safe from invasive species. The evergreen, alien, nitrogen-fixing tree *Myrica faya*, however, invades lava flows where the rainfall exceeds 1500 mm per annum, short-circuiting the slow colonization dominated in the early periods by *Stereocaulon vulcani*. Many lava flows have been subdivided though the majority of the lots have not yet been developed. One subdivision has since been overrun by a subsequent lava flow.

## 5. Coral and compacted sand dunes

These are the only calcareous habitats in the Islands. *Endocarpon pusillum* and an unidentified crustose cyanobacterial lichen are the commonest species. *Diplotomma alboatra* and the endemic *Caloplaca perminuta* and *Lecania oahuensis* are common, but inconspicuous, clements in the flora. This is the poorest habitat for lichens in Hawaii with fewer than 10 species present and with very low cover. Over half of the species, however, are endemic.

This habitat is among the most threatened in Hawaii. Many areas have already been devastated by development or agriculture. One of the largest areas of compacted dunes was destroyed by a housing development on Maui, for example, even though it was the principal habitat for an endangered plant (*Scaevola coriacea*). Off-road vehicle use has destroyed many compacted dune habitats. Most remaining areas are overrun by weeds, particularly *Leucaena leucocephala* and *Prosopis pallida*. One extensive dune system has recently been acquired by The Nature Conservancy.

## Discussion

The conservation of Hawaii's lichen flora will always be a secondary consideration of the Island's conservation programmes. The angiosperm, bird, fern, and insect biota are well-known for their remarkable adaptive radiation. They also have a much greater appreciation by the general public which translates into powerful lobbying groups. Lichens have a very small public and virtually no lobbying potential. They can be used as additional elements, however, when trying to preserve areas proposed to conserve the more appreciated groups.

Disturbance is the common thread that runs through the threat to lichen conservation in Hawaii. Natural disturbance events, principally lava flows and tropical storms, are common in Hawaii. However, they are normally confined to relatively small areas and recovery is fast. Human-mediated disturbance, on the other hand, has been diverse and widespread.

The impact of the early Hawaiians was confined principally to dry and mesic areas below 1000 m (Cuddihy and Stone 1990). Yet pockets of native vegetation remained in gullies, rock outcroppings, and other areas. The rate of degradation of Hawaiian ecosystems accelerated after the arrival of Cook in 1778 and has since affected every lichen habitat in the Islands. Goats and cattle, protected by royal decree, were released, became feral, reproduced rapidly, and spread through all ecosystems. In the mid-nineteenth century, sugar and pineapple plantations were established. Hawaiian agricultural areas were consolidated and extended. Western engineering techniques were used to clear extensive areas of rocks, gullies were filled in, etc. A number of alien plants, for example, *Lantana camara*, *Psidium guajava*, and *Opuntia megacantha* escaped cultivation and formed dense monotypic stands, invaded native forests and became serious pests in agricultural areas. Lowland lichen habitats were destroyed except for very rocky areas. The burgeoning populations of feral animals were destroying the forests and their understorey at an alarming rate and changing the integrity of the watershed.

The turn of the century brought a realization that the forests had to be restored to preserve the watershed and major agricultural weeds controlled. Biological control organisms were introduced and some remarkable successes were recorded. The establishment of forest reserves, reafforestation programmes and removal of feral cattle resulted in a significant rehabilitation of watersheds. Other equally important programmes controlling alien influences were not implemented. Feral goat populations expanded unchecked. Alien plants continued to naturalize at an essentially constant rate (Wester 1990). After statehood, the forest reserves were converted into 'sustained-yield' hunting areas and pigs were introduced into many areas. The explosive growth of the tourist industry in the last quarter-century has had enormous impacts.

*1. Alien plants*

Alien plants are a mixed blessing for lichens. Certain groups, such as the conifers and Myrtaceae, are poor hosts for Hawaiian lichens. Yet, some lichens are best developed on some of these species, for example, *Dimerella zonata* on *Eugenia cuminii*. Other groups, particularly Fabaceae, are excellent hosts. Unfortunately, the lichen communities on alien species are generally confined to a few species such as *Coccocarpia eryth-*

*roxyli*, *Parmotrema cristiferum*, and *Ramalina exiguella* on *Schinus terebinthifolius*, particularly in monotypic stands. In mixed stands, the lichen communities are more diverse. A more significant problem with alien plant communities is their recency. There are no old-growth alien plant communities. It is hard to find a community much over 50 years old and even then there are indications of recent disturbance events in those communities. Consequently, even if these communities did provide suitable habitats for lichens, the diversity of species normally associated with stable conditions are not present.

## 2. Fire

Fire is the most immediately destructive influence on the Hawaiian lichen flora. Fire was not an important evolutionary determinant in Hawaiian ecosystems (Smith and Tunison 1990). Fires are now common below 1000 m due to the invasion of alien bunchgrasses. Each fire favours the growth and spread of the grasses resulting in hotter, more extensive fires. Woodlands are converted to monotypic grassland where very few lichens habitats remain.

## 3. Alien animals

Feral ungulates have been responsible for long-term, large scale changes in the character of Hawaiian ecosystems. Feral cattle destroy all but the highest strata by browsing and grazing. Goats and deer curtail regeneration by browsing the undergrowth. Feral pigs can also influence all strata by their rooting activity. They also introduce seeds from the fruit of alien plants that they have eaten. The activity of these animals opens the forest and accelerates erosion. Corticolous, foliicolous, and terrestrial lichens are affected. Forested areas that included diverse lichen floras with over 60 species can be reduced to a few of the more hardy species after only three years of pig activity.

Dry forests are equally affected. Feral goat and cattle activity open up the habitat, drying it out. This is followed by a dramatic drop in the number of lichens present. The endemic *Ramalinopsis mannii*, which used to be common in native dryland forests on West Maui (Magnusson 1954), now appears to be confined to the Polynesian introduction *Aleurites moluccana* along semi-permanent streams.

## 4. Direct human impacts

(*a*) *Development*  Housing was generally confined to lowland habitats until 25 years ago. It was usually low density, trees were used for shading, and concrete was rarely used. In the last twenty years, construction has focused on large subdivisions of closely-packed houses or high-rises, clearing large areas with little regard to conserving vegetation. Dust from construction activities has become a temporary negative influence beyond the construction site. Housing is no longer confined to the lowlands. Consequently, large areas of vegetation are disturbed for long periods of time. Trees and shrubs, if planted at all, are generally maintained for a few years before being replaced due to modifications by the householder. Pioneer lichen communities are the most common and representative of more xeric environments due to the city-desert climate. Lichen communities are often best-developed in poorer neighbourhoods where both house and garden maintenance is minimal. The larger gardens with trees are also good habitats because they are cooler and generally watered.

Arterial road construction through forests also has a major negative impact on lichen floras except, perhaps, in the rainforest habitats. Dust from road construction decreases adjacent lichen communities. The reduction in relative humidity due to black-body radiation from the asphalt, results in decreased lichen cover for some distance from the roadway. Rainforest communities do not appear to be severely impacted because of cloud cover and higher rainfall.

(*b*) *Agriculture* Agriculture has had three major impacts on lichen habitat, (1) habitat destruction, (2) dust, and (3) chemical contamination. Habitat destruction continues even though pineapple and sugarcane plantations are in decline. Property tax laws encourage the conversion of forest into grassland because timber is rated at a higher value than ranching.

Tilling sugarcane and pineapple lands creates large dust clouds. Lichens in forests immediately downwind are smothered and killed. The impact is still observable up to about 3 km from the field. The impact of the chemicals used in agriculture on lichens are not fully appreciated. Corticolous habitats immediately adjacent to fields are devoid of lichens even when there is no evidence of dust. The lichen desert reaches at least 50 m from the fields except in high-rainfall areas.

(*c*) *Recreation* Recreational activities have profound impacts, some direct, others indirect. Off-road vehicles, particularly cross-country motorbikes, are a significant problem. Not only do they open up the vegetation, they create large clouds of dust and accelerate erosion. Lichen diversity decreases over large areas. Skiers disturb the very fragile high-elevation saxicolous communities. Opening new trails in remote areas provides avenues to feral animals, particularly pigs, invading these ecosystems. Popular scenic sites often suffer from fire damage.

## 4. Pollution

There is little evidence of man-made air pollutants affecting lichens in Hawaii. Smoke from sugarcane harvest was a problem until recently when the health authorities required that cane could be burned only on windless days so that the smoke ascended from the field. Previously, the smoke was often blown into forested areas the consequences of which are unknown. Sulphur dioxide from fumaroles and from eruptions occasionally has a major impact on the lichen flora when the volcanic fumes, locally called 'vog', hang over the Islands. Most of the time, however, the trade winds blow the fumes out to sea.

In conclusion, all is not lost in Hawaii. There are excellent pockets of native vegetation and lichen habitats remaining. However, these areas are extremely vulnerable to the relentless march of 'development'.

## References

Cuddihy, L. and Stone, C.P. (1990). *Alteration of native Hawaiian vegetation: effects of humans, their activities and introductions*. Co-operative National Park Resources Studies Unit, University of Hawaii, Honolulu. (in press).

Magnusson, A.H. (1954). A catalogue of the Hawaiian lichens. *Archiv för Botanik, ser. 2*, **3**, 223–402.

Rock, J.F. (1974). *The indigenous trees of the Hawaiian Islands*. Pacific Tropical Botanical Garden, Lawai, Kauai, Hawaii, and Charles E. Tuttle Co., Rutland, Vermont, and Tokyo, Japan. (Reprint of 1913 edition.)

Smith, C.W. (1985). Impact of alien plants on Hawai'i's native biota. In *Hawai'i's terrestrial ecosystems: preservation and management* (ed. C.P. Stone and J.M. Scott). Co-operative National Park Resources Studies Unit, University of Hawaii, Honolulu.

Smith, C.W. and Tunison, J.T. (1990). Fire and alien plants in Hawai'i: research and management implications for native ecosystems. In *Alien plant invasions in native ecosystems of Hawai'i: management and research* (ed. C.P. Stone, C.W. Smith, and J.T. Tunison). Co-operative National Park Resources Studies Unit, University of Hawaii, Honolulu.

Wagner, W.L., Herbst, D.R., and Sohmer, S.H. (1990). *Manual of the flowering plants of Hawaii*. University of Hawaii Press, Honolulu. (in press).

Wester, L. (1990). Origin and distribution of adventive alien flowering plants in Hawai'i. In *Alien plant invasions in native ecosystems of Hawai'i: management and research* (ed. C.P. Stone, C.W. Smith, and J.T. Tunison). Co-operative National Park Resources Studies Unit, University of Hawaii, Honolulu.

# 4. The tropical Pacific species of *Usnea* and *Ramalina* and their relationship to species in other parts of the world

G.N. STEVENS

*Botany Department, The University of Queensland, St. Lucia, Brisbane, Queensland 4072, Australia*

### Abstract

Many of the species of *Usnea* and *Ramalina* which occur in tropical Australia and the western Pacific region have been found to extend much farther afield in their distribution. These taxa can be divided into maritime species (occurring on mangroves or lowland forest trees) and montane forest species (occurring in upland open-forest and rainforests). Species of *Ramalina* are mostly in the former category, whereas species of *Usnea* mainly occupy montane forests. The altitude of tablelands in the tropics of Australia and the high mountains of the Pacific islands tends to nullify the effects of high temperatures found at sea level in lower latitudes, so that species found in elevated tropical regions are not necessarily of tropical origin. Several species of *Usnea* which occur in montane rainforests at high elevations appear to be temperate taxa which tolerate tropical conditions at the elevation at which they grow. By tracing the palaeobiogeography of the phorophytes for both maritime and montane species of *Usnea* and *Ramalina*, the tropical or temperate origins of the lichens can be hypothesized.

### Introduction

In the western Pacific region, the largest island is Australia, with nearly half of its area in the tropics. This position makes Australia well-suited for research into tropical lichen taxonomy and this, in turn, is of immense value in interpreting the taxonomy, distribution, and biogeography of lichens on the tropical islands to the north, for example, New Guinea, Indonesia, Borneo, and the Philippines and to the east, New Caledonia, Vanuatu, Fiji, Cook Islands, and Tahiti.

## Geography

The eastern coastline of Australia borders the Pacific and the eastern highlands extend in a north–south alignment following this coastline. The landmass of Australia is generally low, exceeding 1000 m in relatively few places. The tableland form of the highlands allows species dispersal over a wide geographical area and the distribution patterns of the species can be easily plotted to their limits. Contrary to this, many islands of the Pacific have quite high mountains. In Papua New Guinea (PNG), Borneo, Indonesia, and the Philippines, the mountains rise to 3700 m in height and taxonomic interpretation of *Usnea* and *Ramalina* on these islands is difficult because of altitudinal zonation which allows tropical and temperate species to coexist in the same area.

By determining distribution patterns of these species in Australia it becomes easier to establish which species have tropical origins and which originally may have migrated to the islands from more temperate regions.

The majority of Australian corticolous lichens occur in the region from the eastern shoreline on mangroves to the mountain tops in rainforest. The rain-bearing south-east trade winds drop their moisture along this strip of land so that this part of Australia is densely forested. To the west of the mountains where the land is much drier, open woodland is extensive but fruticose lichens appear almost totally restricted to the area between the Great Divide and the coast.

## Habitats

In the tropics species of *Ramalina* and *Usnea* are usually corticolous, and are found in maritime habitats (mostly on mangroves) and in the mountains in open forests or rainforests.

### 1. Mangroves

Optimal development of mangroves occurs in the tropics and the greatest number of mangrove species in Australia is on the tropical north-eastern coastline. Mangroves grow in sheltered bays and estuaries and line the river banks of coastal streams. The mangrove shore is rarely exposed to the open ocean; in such conditions lowland rainforest replaces mangrove communities. Nearly 30 species from 15 plant

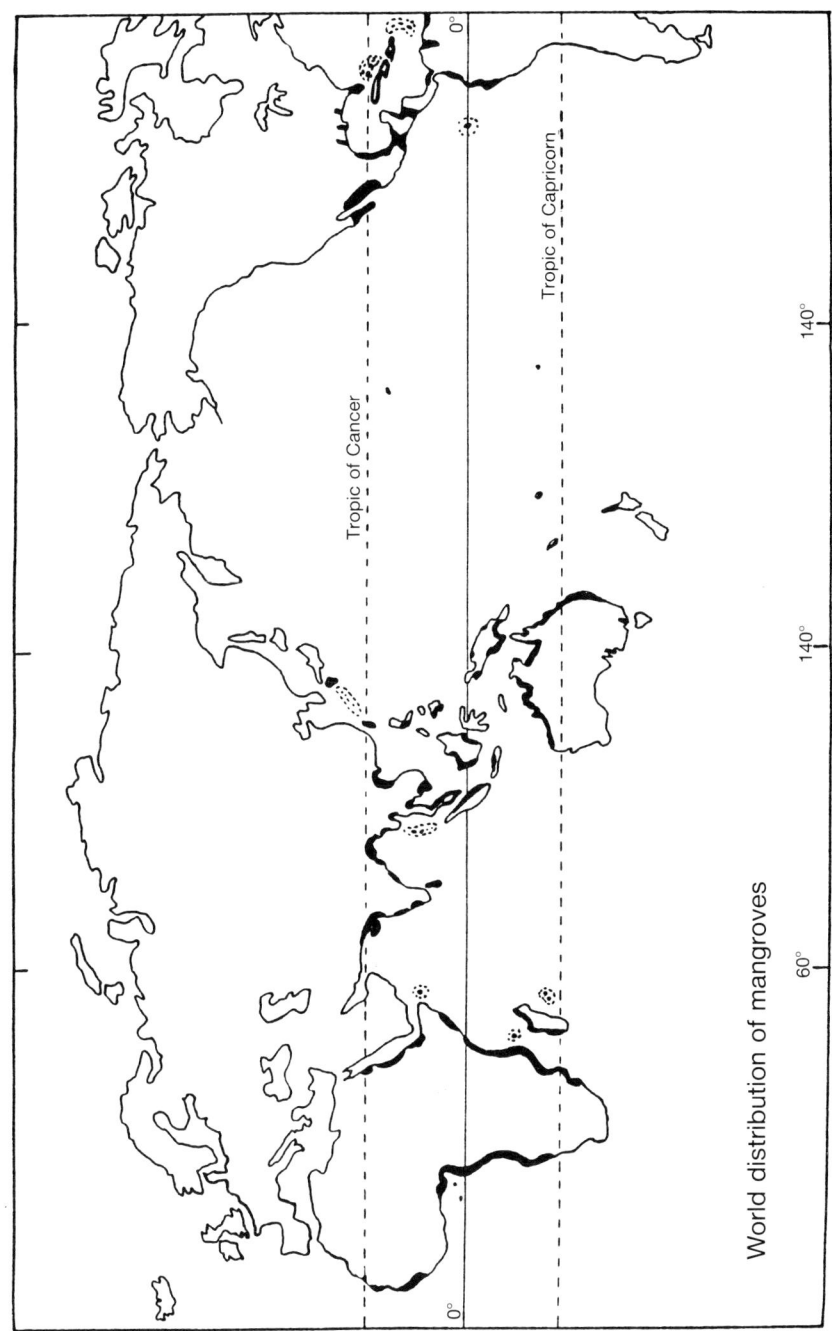

**Fig. 4.1.** World distribution of mangroves (modified after Chapman 1977).

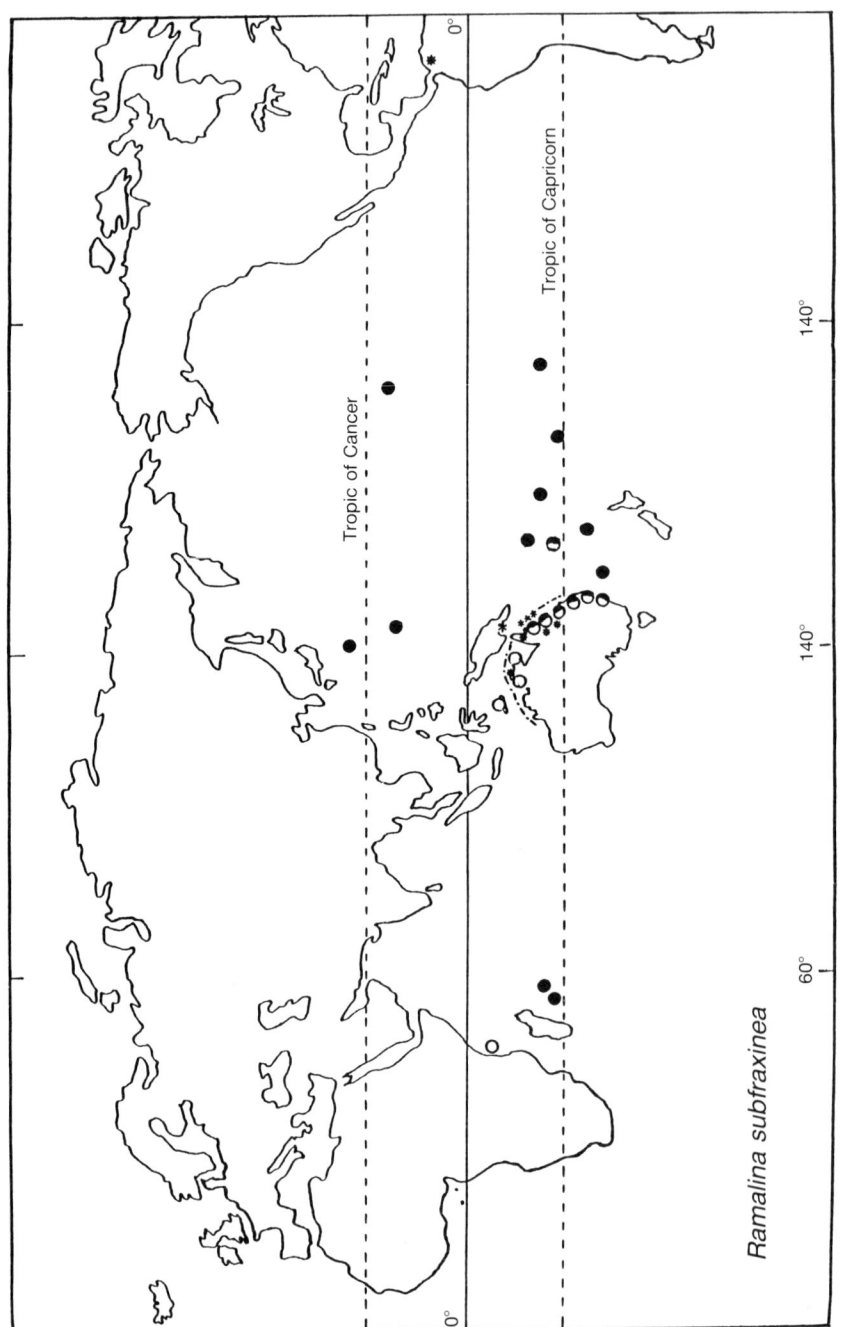

**Fig. 4.2.** Distribution of the five chemical taxa in the *Ramalina subfraxinea* complex. (○) = *R. subfraxinea* var. *confirmata*, (●) = *R. subfraxinea* var. *leiodea*, (◐) = the occurrence of both var. *confirmata* and *leiodea*, (–·–·) = *R. subfraxinea* var. *norstictica*, (✳) = *R. subfraxinea* var. *subfraxinea*, (∗) = *R. tropica*.

families occur as mangroves but the family Rhizophoraceae provides species most favoured by lichens.

Species of *Ramalina* are more numerous than *Usnea* on mangroves. The phorophyte mangroves for *Usneas* are *Rhizophora stylosa* and *Ceriops tagal*, whereas species of *Ramalina* grow on *Avicennia marina*, *Excoecaria agallocha*, and *Lumnitzera racemosa* as well as on *Rhizophora* and *Ceriops*.

To the north of Australia, mangroves occur on portions of the shoreline of most islands (Fig. 4.1). Most of the Polynesian islands are devoid of mangroves and here rainforest trees, *Hibiscus tiliaceus*, and coconut palms replace the mangrove community.

(*a*) *Maritime species of* Ramalina  The most widely distributed maritime *Ramalina* taxa in the tropics are grouped into species complexes by Stevens (1987), for example, the fertile *R. subfraxinea* complex and the sorediate *R. nervulosa* group (which belongs to the tropical section of the *R. farinacea* complex). Taxa within each complex produce similar morphologies with different acids and have different but overlapping distribution patterns.

The *Ramalina subfraxinea* complex is composed of six chemical taxa. Each of these have distinct but overlapping distributions (Fig. 4.2). Although mainly confined to the tropics, two taxa, *R. subfraxinea* var. *confirmata* and var. *leiodea* extend into the subtropics. An Australian endemic is *R. subfraxinea* var. *norstictica* which grows on mangroves around the whole of the tropical coastline of Australia from east to west. *Ramalina tropica*, a salazinic acid-producing taxon within the *R. subfraxinea* group, was given species status because it was distinguished both morphologically and chemically from other chemical races in the group. So far, it is only known from Australia and PNG, but further investigation may prove that this salazinic acid species occurs in Java under the name *R. zollingeri* Szat..

Another tropical *Ramalina* complex is part of the *R. farinacea* group. It is composed of *R. nervulosa* and its varieties var. *lucie* and var. *dumeticola*. It comprises three acid races, divaricatic acid in the type and several acids in the sekikaic acid aggregate in the two varieties (Stevens 1983). Kashiwadani (1986) placed all three races under the name *R. nervulosa* but there are differences, both ecological and in distribution, between both of these sekikaic races (Fig. 4.3).

(*b*) *Maritime species of* Usnea  The most common maritime *Usnea* is *Usnea nidifica*, which shows great morphological variation but produces mainly only one set of acids over the whole of its Pacific/Indian Ocean range (Fig. 4.4). (Two minor strains are also represented but neither

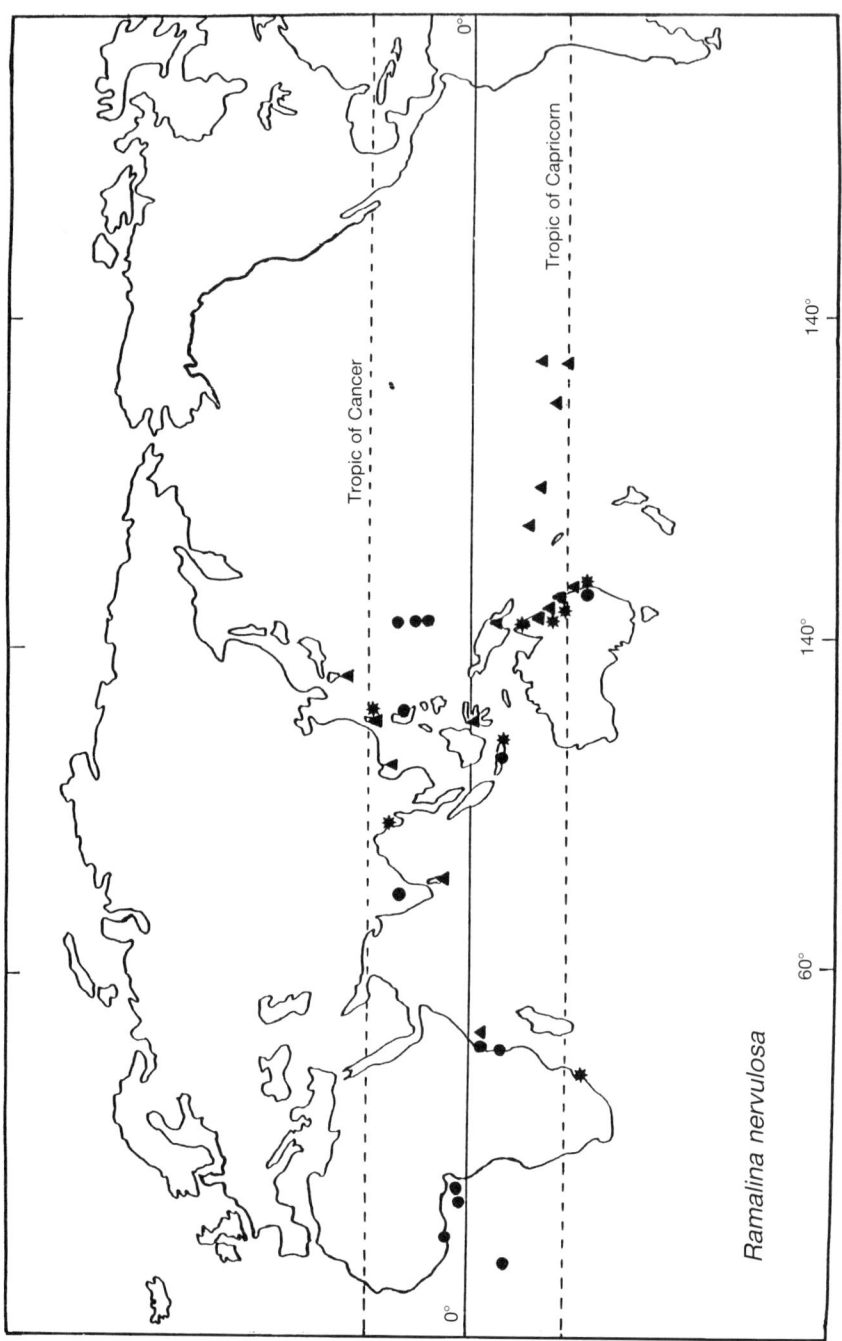

**Fig. 4.3.** Distribution of the three chemical taxa in the *Ramalina nervulosa* group. (✱) = *R. nervulosa* var. *nervulosa*, (▲) = *R. nervulosa* var. *lucie*, (●) = *R. nervulosa* var. *dumeticola*.

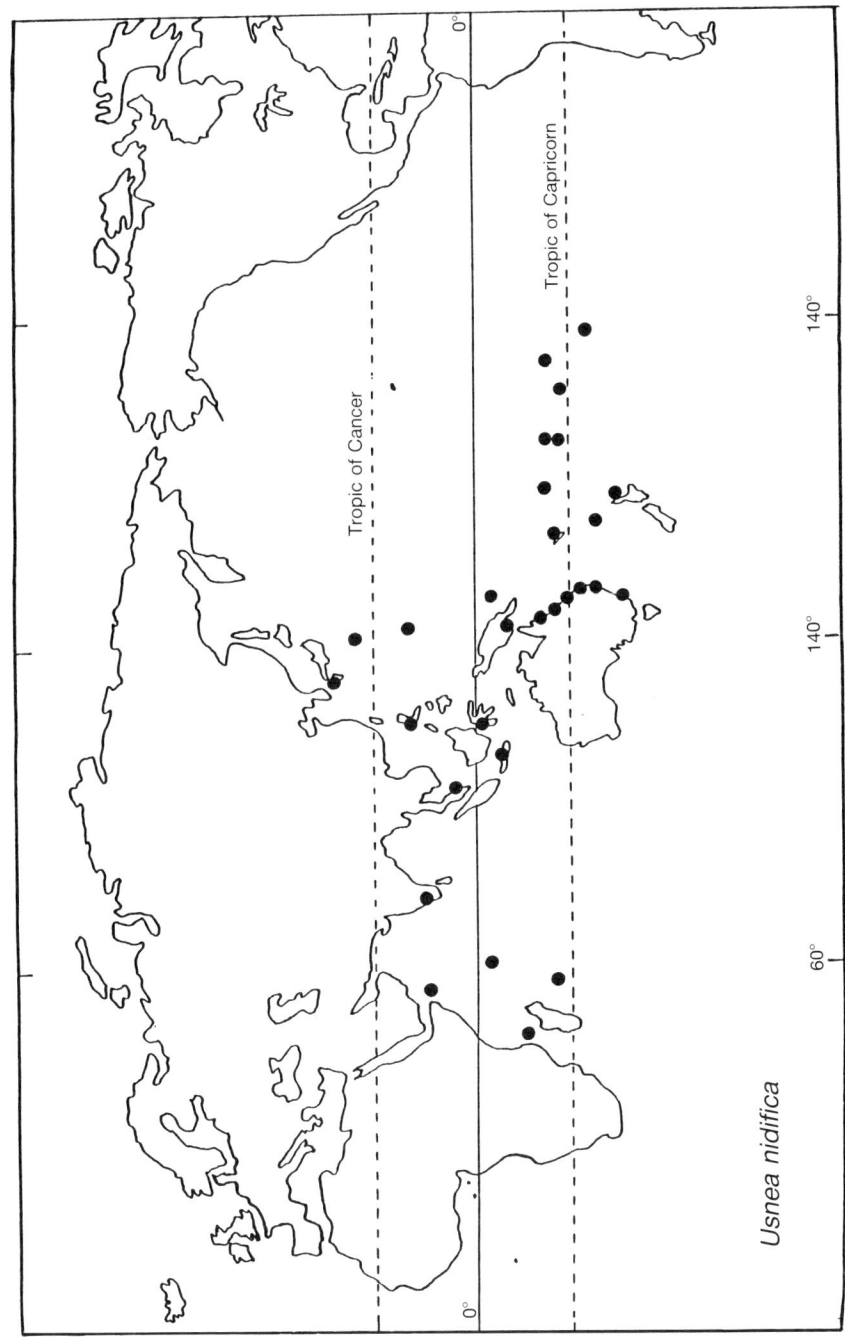

**Fig. 4.4.** Distribution of the polymorphic *Usnea nidifica*.

have any distinct distribution pattern; one contains psoromic acid, and the other strain produces protocetraric acid.)

*Usnea nidifica* (type material from Norfolk Island) has a wide morphological variation and, in the past, many names have been applied to it when it was collected from different islands of the Pacific such as *U. neocaledonia* (from New Caledonia), *U. intercalaris* (from Fiji), *U. societatas* (from Tahiti), *U. grandis* (from Java), *U. japonica* (from Japan), *U. nexilis* (from Norfolk Island), and *U. straminea* (from Mauritius); in Australia it is known as *U. propinqua*. Because of the intermediate forms which have been found to exist between all of these type specimens, it was considered that the rank of subspecies should be applied to taxa which occupy different geographical zones if they had sufficiently different morphological traits to warrant this arbitrary division (ssp. *propinqua*, ssp. *japonica*, ssp. *straminea*).

The taxon from Fiji named *U. intercalaris*, is distinguished by rings of white material at intervals along the branches, giving it a distinct morphology. As this is merely an excretion of calcium oxalate, purely environmentally induced, it has no taxonomic merit.

The usual maritime, tropical *Usnea* and *Ramalina* taxa are often absent from the shoreline in low equatorial latitudes, but are present inland at slightly higher altitudes.

From distribution patterns of maritime species of *Usnea* and *Ramalina*, it is evident that they are not restricted to the Pacific but extend also to the islands of the Indian Ocean and to the east coast of Africa.

*2. Two ubiquitous tropical taxa*

Both *Ramalina peruviana* and *Usnea baileyi* have a very wide distribution which includes the tropics, both occur in PNG, Java, and the Philippines. In Australia, they occur from temperate to tropical regions, in both open forest and rainforest environments or on fence posts, and in the subtropics they occur on mangroves. These two species are plainly not tropical taxa but they apparently tolerate a tropical climate when growing in the mountains under cool, mild conditions.

In the Pacific region *R. peruviana* extends from Japan to New Zealand, to Hawaii, Tahiti, and to South America (Fig. 4.5). Until recently the name *R. intermediella* was applied to specimens of *R. peruviana* found in Japan but this name has now been reduced to synonymy (Stevens and Kashiwadani 1987).

The section *Eumitria* species, *Usnea baileyi* also has a very wide distribution range as many *Eumitria* species names from different regions have now been reduced to synonymy with *U. baileyi* (Rogers and Stevens 1988) (Fig. 4.6). All of these taxa produce similar acids or only

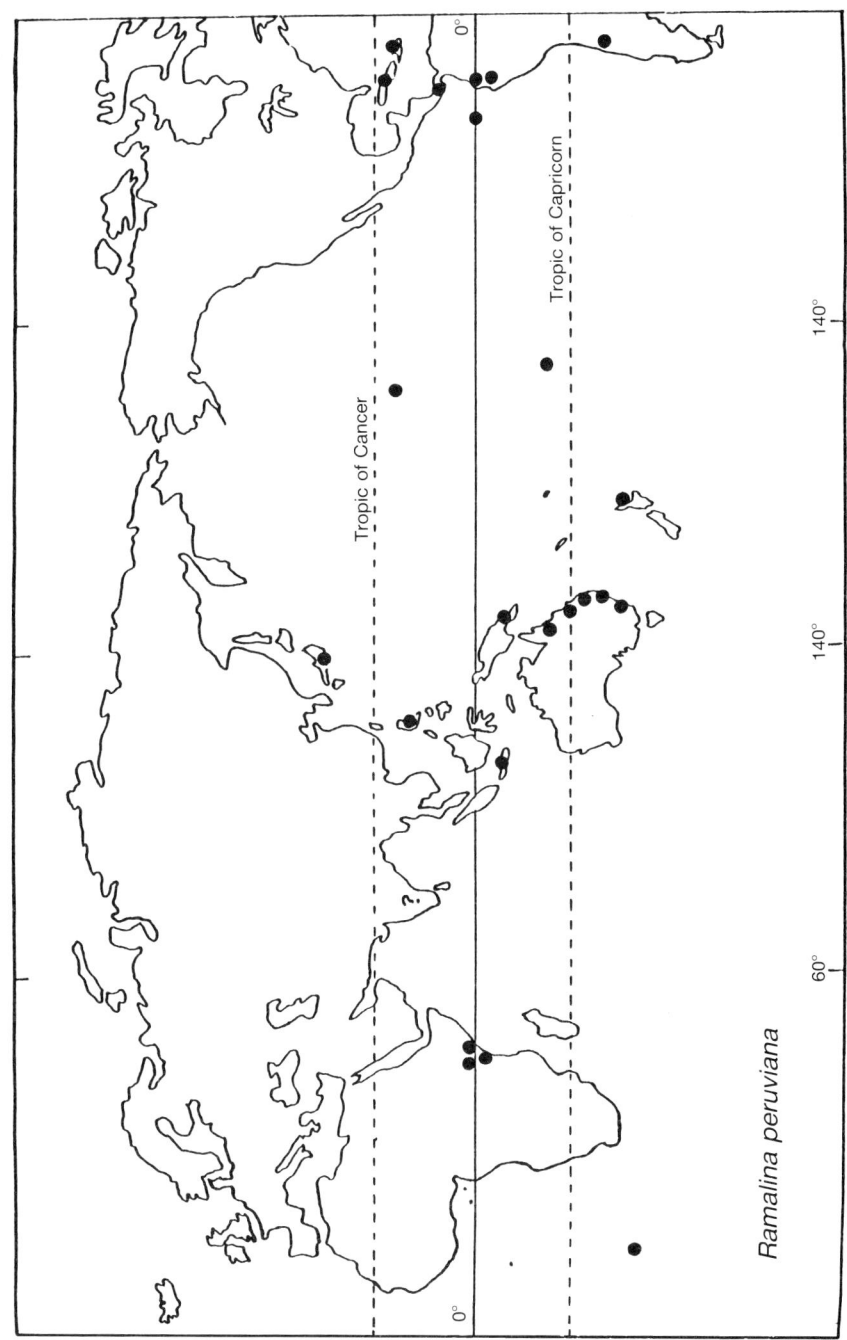

**Fig. 4.5.** Distribution of *Ramalina peruviana*.

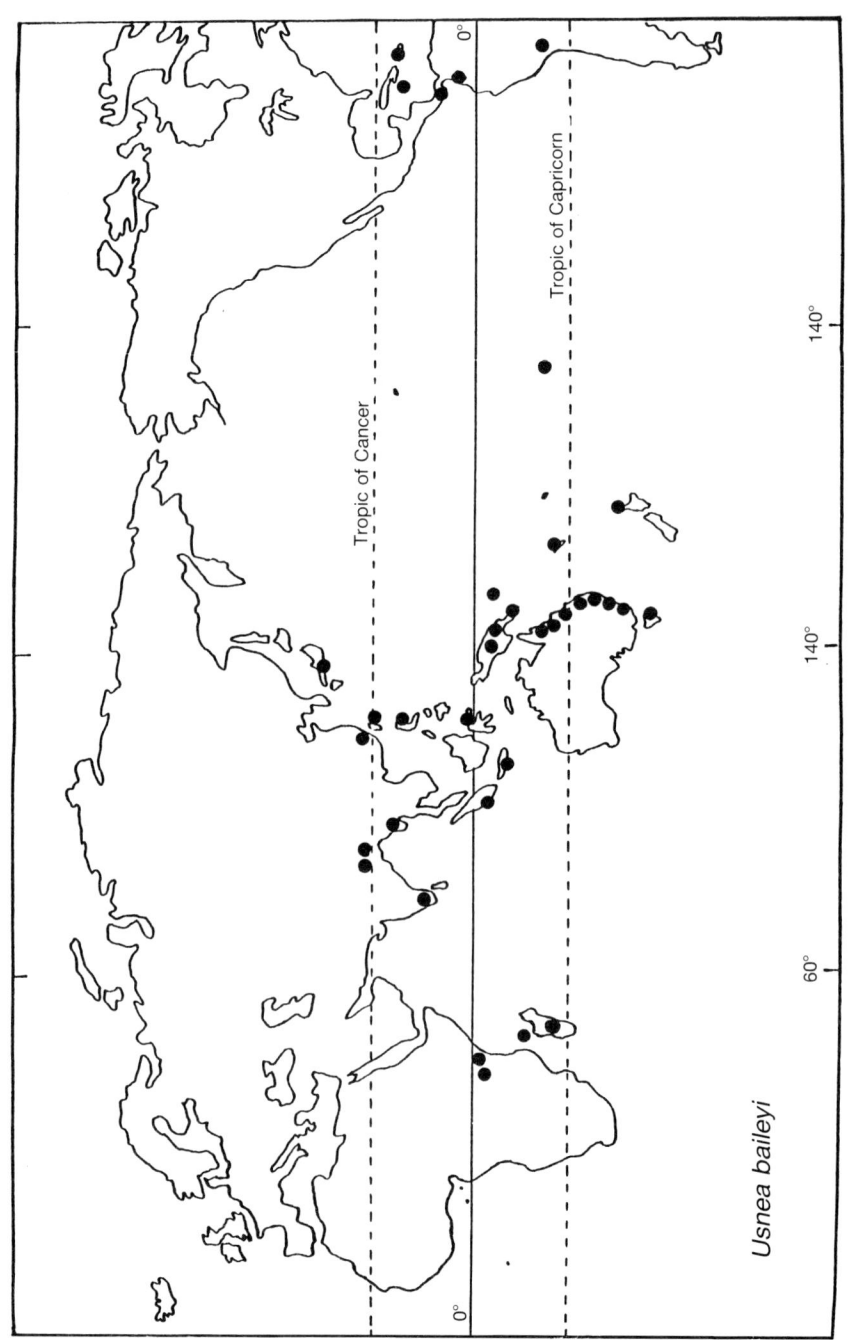

**Fig. 4.6.** Distribution of *Usnea baileyi*.

slight variations to those of *U. baileyi* and although medullary pigment colour shows great variation (yellow, orange, pink, rust, maroon or chocolate brown) this proves to be of no taxonomic importance. In PNG, *Usnea baileyi* occurs over a wide altitudinal range (600 m–2300 m) and grows on *Araucaria*, *Nothofagus*, and other forest trees.

The morphological variation found in both *U. baileyi* and *R. peruviana* throughout their wide distribution ranges is not greater than the variation found within a single population of the taxon concerned.

*3. Montane rainforest*

(*a*) *Rainforest species of* Ramalina  Species of *Ramalina* are neither prominent nor numerous in the tropical rainforests of Australia. *Ramalina inflata* ssp. *perpusilla*, the tropical subspecies of the widely distributed *R. inflata* complex, is the usual *Ramalina* found in rainforests of north Queensland. In PNG a taxon occurs at high altitudes which morphologically resembles the temperate region taxon in this complex. *Ramalina pumila* Mont., an Asian taxon which resembles *R. inflata* ssp. *perpusilla*, but with different chemistry, probably lies within the *R. inflata* group, but more investigation is needed into this in the Pacific region. *Ramalina javanica* Nyl., an inflated, sorediate taxon found at very high altitudes in PNG and Java, does not occur in Australia as the mountains are not high enough to allow its distribution to continue southwards.

(*b*) *Rainforest species of* Usnea  Species of *Usnea*, however, are common in rainforest communities of the region and are often prominent in the canopy of *Araucaria* and other emergent trees in Australia and PNG and in south-east Asia they occur on dipterocarps. Some pendent species are also found in open-forest abutting rainforest where there is maximum light and where humidity is constant and mists common. Such species as *U. misamisensis*, *U. trichodeoides*, *U. hossei*, and *U. himantodes* (form *neoguineensis*) are commonly found in this environment in tropical Australia, PNG, and the islands of south-east Asia.

The pendent *Usnea misamisensis* Motyka, contains acids in the stictic acid aggregate. It was first named *U. longissima* var. *misamisensis* by Vainio from a specimen collected in Thailand. It now seems that the original varietal status was nearer the correct relationship.

*Usnea longissima* is a temperate taxon in Europe and America; it grows in Japan where Asahina (1956) recorded five chemical races (barbatic, evernic, diffractaic, salazinic, and protocetraric acids); the evernic acid race was collected recently by A. Aptroot in PNG at an elevation of 2800 m; and *U. misamisensis* also occurs in PNG and is morphologically similar. Thus *U. misamisensis* could be regarded as another member of the *U. longissima* group.

*Usnea trichodeoides* occurs in Australia in both subtropical and tropical montane open-forests, abutting rainforests, or on *Araucaria* within rainforest at altitudes of 700 to 1200 m. It produces protocetraric acid or a second chemistry of salazinic ± norstictic acids. Both chemical races are represented in the four 'type' specimens held at H. In East Africa, *U. trichodeoides* occurs at high elevations to 3000 m (Swinscow and Krog 1978). Although variable in morphology, its erose main branches, terete or flattened, and pectate arrangement of fibrils, clearly distinguish it from other pendent *Usneas*, except *U. longissima* and *U. misamisensis*, and it may be that *U. trichodeoides* should be placed in the *U. longissima* complex (Fig. 4.7).

*Usnea himantodes* Stirton, in its coarsest form was called *U. neoguineensis* by Asahina. It occurs in open eucalypt forests and rainforest canopies in Australia at elevations of 700 to 1100 m and in PNG it festoons *Araucaria* forests at 850 m altitude. It appears to be closely related to the East African taxon *U. gigas*. Apart from a difference in chemistry [*U. neoguineensis* produces stictic acid aggregate and *U. gigas* has several acid strains, e.g. protocetraric or salazinic or constictic acids (Swinscow and Krog 1978)], they have a similar morphology and a distinct, coloured axis which indicate they are vicariant species. If *U. gigas* is the East African equivalent of the Pacific region *U. himantodes*, then this group also has a wide distribution (Fig. 4.8).

*Usnea hossei* is a pendent species with a morphology similar to *U. neoguineensis* var. *gracilor* Asahina, but it does not produce the yellow-orange to brown axis that is so characteristic of the *U. gigas–U. himantodes* group. Three chemical races occur in *U. hossei*: (1) the type chemistry, stictic acid aggregate; (2) protocetraric acid; and (3) salazinic acid, in which the taxon *U. squarrosa* is included. Its distribution is restricted to Australia and south-east Asia where it grows in dipterocarp forests at elevations of 100 to 1800 m (Fig. 4.9).

The pendent species *U. flexilis* occurs only in montane rainforests in PNG, Sabah, and the Philippines at very high altitudes, and in Japan and India. It does not occur in Australia but the Australasian taxon *U. contexta* (syn. *U. capillacea*) which it closely resembles morphologically but not chemically, is a temperate taxon found in Tasmania and New Zealand and, rarely, in southern Australia (Fig. 4.10).

It appears that species of *Usnea* and *Ramalina* found in the tropical Pacific region are elements from both tropical and temperate ancestral stock. The altitude provided by the tablelands in the tropics of Australia and the high mountains of the Pacific islands offsets the high temperatures found at sea-level in these low latitudes. When species are found in tropical countries, it does not necessarily mean that they are of tropical origin (Fig. 4.11).

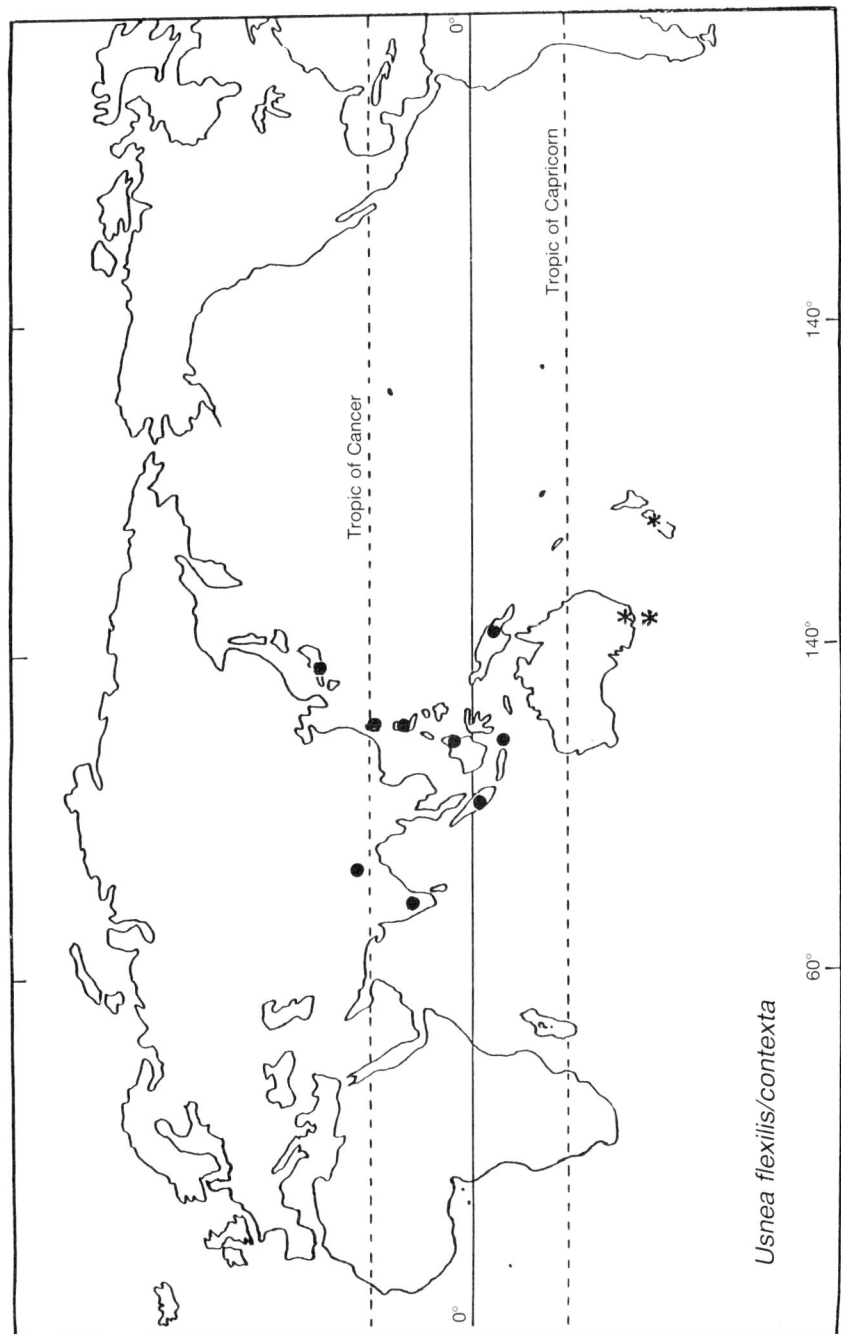

**Fig. 4.7.** Distribution of: (∗) *Usnea contexta* (fumarprotocetraric acid) and (●) *Usnea flexilis* (salazinic acid).

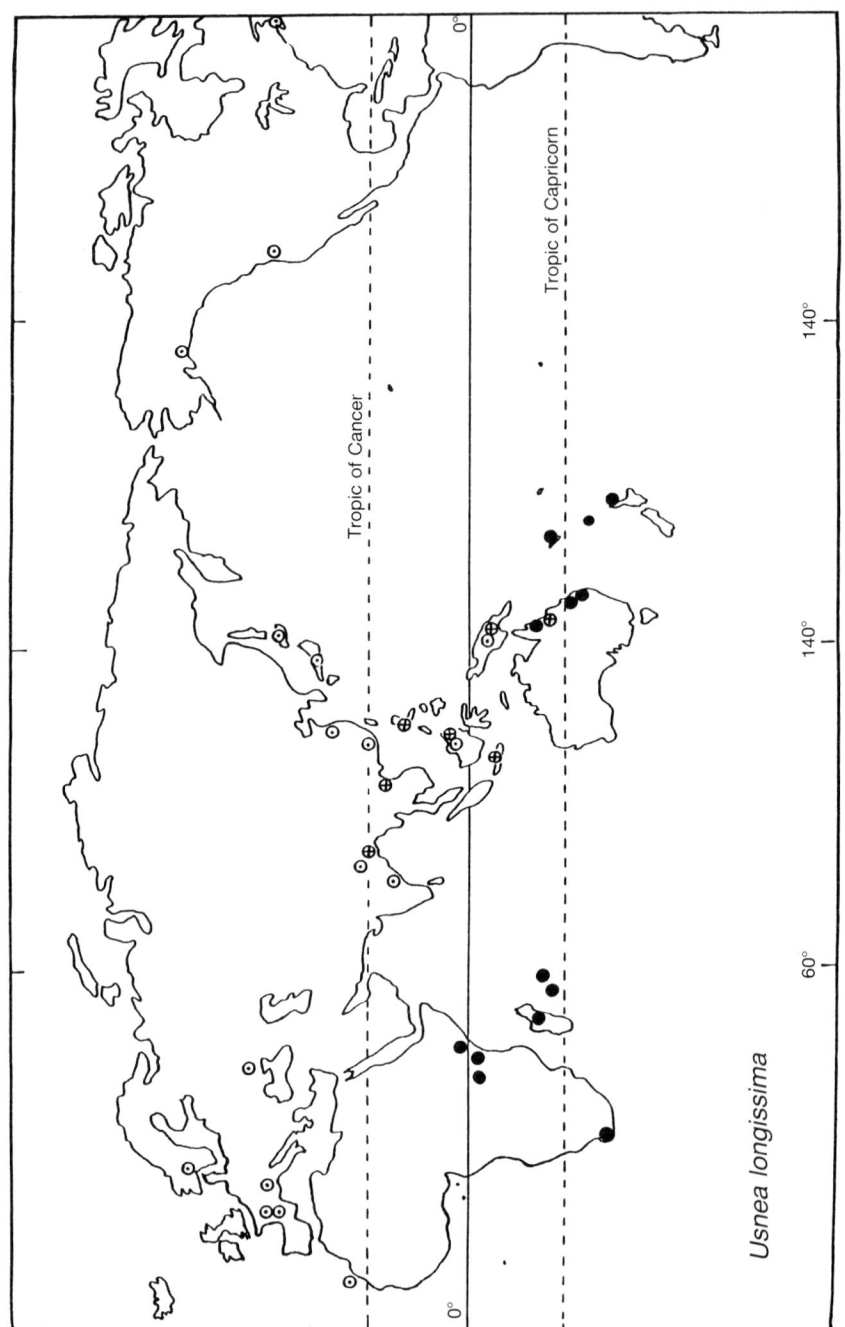

**Fig. 4.8.** Distribution of taxa belonging to the *Usnea longissima* complex. (☉) = *U. longissima* (diffractaic, barbatic, evernic, salazinic), (⊕) = *U. misamisensis* (stictic acid aggregate), (●) = *U. trichodeoides* (protocetraric, salazinic, norstictic taxa).

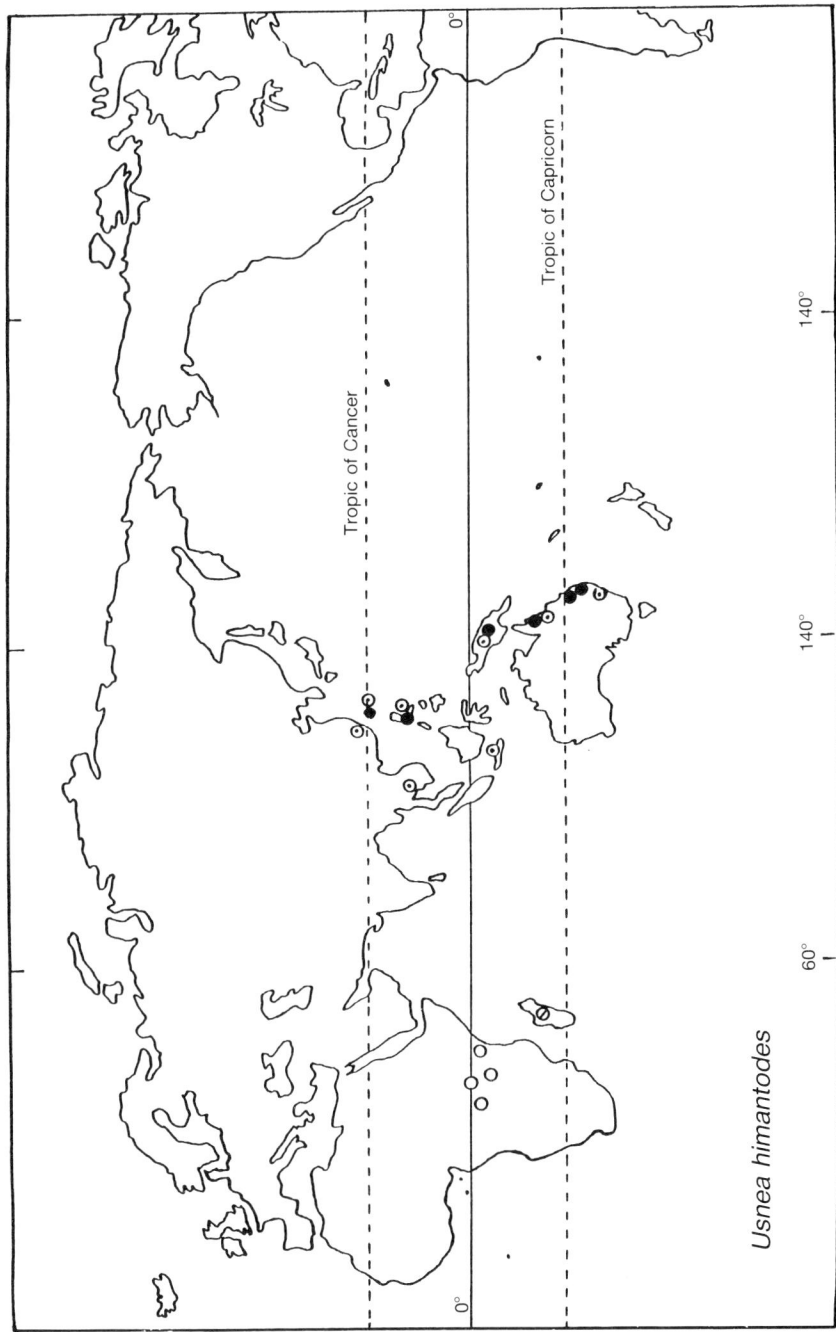

**Fig. 4.9.** Distribution of taxa in *U. himantodes* and *U. gigas* group: (☉) = *U. himantodes*, (●) = *U. himantodes* f. *neoguineensis*, (○) = *U. gigas*, (⊘) = *U. eburnea*.

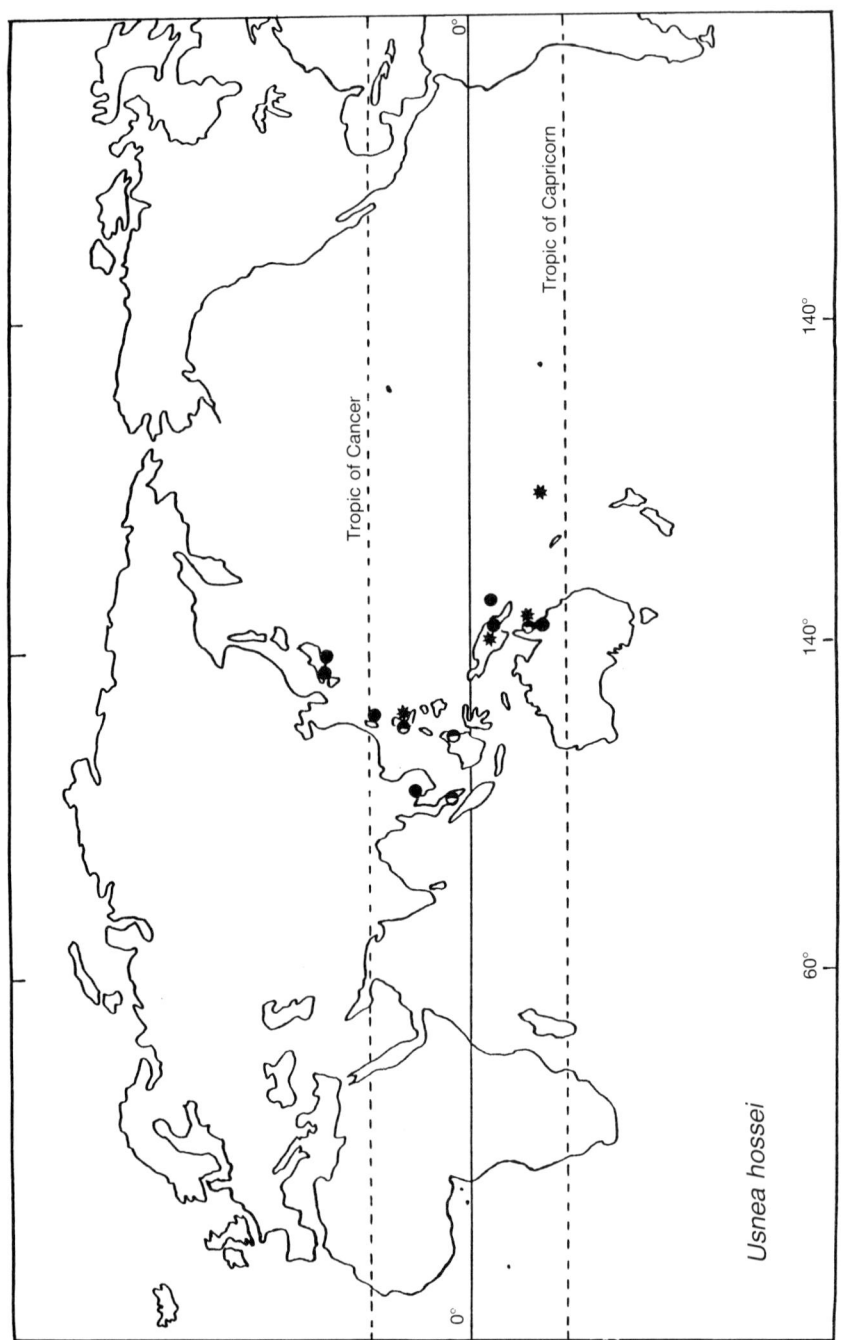

**Fig. 4.10.** Distribution of the three chemical taxa in *Usnea hossei* group: (●) = *U. hossei* (stictic acid aggregate), (✷) = *U. hossei* (protocetraric acid), (○) = *U. hossei* (syn. *U. squarrosa*—salazinic acid), (◐) = occurrence of both salazinic and stictic acid taxa.

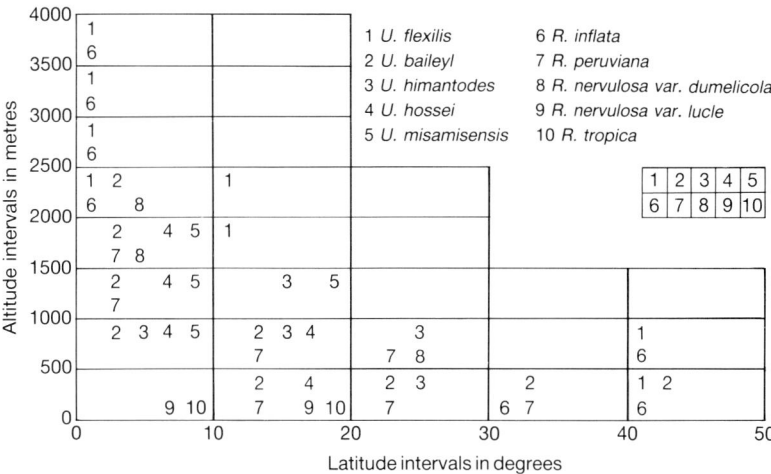

**Fig. 4.11.** Distribution of ten tropical lichens (five *Usnea* and five *Ramalina*) showing altitudinal segregation with latitudinal gradation. For each interval of altitude and latitude, the position of each species number is related to its arrangement in the legend box.

## Palaeobiogeography

By examining the palaeobiogeography of the phanerogams found in the tropics with reference to those which, in present times, act as phorophytes to species of *Usnea* and *Ramalina*, we can gain some idea of, and hypothesize on, the direction from which the lichen taxa might have migrated. It would appear they were derived from two sources, one tropical in origin and the other temperate.

### 1. Tropical origins

Fossil evidence shows that early Rhizophoraceae mangroves first evolved in the upper Eocene in the Caribbean area and in Brazil (Muller 1981). They are known in the late Oligocene from Queensland, Australia, and north-west Borneo. Muller also reports that, in the Miocene, differentiation of species of *Rhizophora* was occurring around the Pacific. This early establishment of phanerogams on the islands would have ensured that a suitable substrate was available for lichen propagules being dispersed across the Pacific.

The Dipterocarpaceae is an exclusively tropical family which evolved in the Oligocene; Muller (1981) reported fossil records from Belgium at this time and dipterocarp pollen from north-west Borneo in the

Miocene. The three taxa in the *U. hossei* group all occur in present day dipterocarp forests in the Philippines, Malaya, and Sabah, which would indicate that the group can be regarded as tropical in origin as is its present day distribution.

2. *Temperate origins*

The migration of plant communities with their epiphytes enables those plants with similar habitat requirements to expand as a whole in suitable climates and so migrate when conditions are favourable and, conversely, to disappear from areas when environmental conditions have deteriorated. Lichens growing within these communities would migrate at the same rate and in the same direction.

(*a*) *Gymnosperms* The gymnosperm families Araucariaceae and Podocarpaceae were well established on both sides of the Tethys Sea by the Jurassic. Fossil evidence shows that the family Araucariaceae occurred in Great Britain, Europe, India, and North America, as well as in Africa, Australia, and South America at this time (Stockey 1982). It is also known that *Araucarias* occurred in the Cretaceous of Japan (Stockey *in litt.*). As a present-day phorophyte for taxa of both *Usnea* and *Ramalina*, it is feasible that *Araucaria* may have been a phorophyte for ancestral lichen distribution.

Present-day species of *Araucaria* occur in tropical and subtropical areas of the Western Pacific. They are emergent and sometimes the dominant element in rainforests and fringing communities. Their migration in the Southern Hemisphere from cooler latitudes to the tropics (as determined by fossil evidence, Dettmann 1989), would have enabled species of *Usnea* to move into the tropics via the same route, for example *U. trichodeoides* and *U. himantodes*, as both of the species occur on *Araucaria* today.

Another phorophyte may have been *Dacrycarpus* (Podocarpaceae), a canopy tree in present-day rainforests. Dettmann (1989) reports the earliest occurrence of pollen from *Dacrycarpus* in the Late Cretaceous of southern Australia, with megafossils known from the late Miocene of PNG. *Dacrycarpus* also occurs in the islands of the western Pacific extending to PNG, Malaysia, and south-east Asia.

(*b*) *Angiosperms* The first angiosperms appeared in the Early Cretaceous in southern Laurasia and northern Gondwana (i.e. North Africa and northern South America) and then migrated by individualistic routes to Antarctica and thence to Australia and New Zealand. Angiosperm families at this time include *Nothofagus*, Proteaceae, Winteraceae, Epacridaceae, Ericaceae, and Trimeniaceae (Dettmann 1989). In the

Cretaceous, Gondwanan climates are thought to have been cool-to-warm temperate with little or no seasonality and high moisture levels (Dettmann and Jarzen 1990). Today there is a high concentration of the primitive angiosperm families in the mountains on the islands to the north of Australia and in the highland rainforests of north-eastern Queensland, in areas with high humidities and mild temperatures (Webb, Tracey, and Jessup 1986).

In the Late Cretaceous and early Tertiary, *Nothofagus* spread across Antarctica from South America and entered southern Australia and New Zealand in the Eocene, with one group, *N. brassii* migrating to New Caledonia and PNG. *Usnea contexta* (syn. *U. capillacea*) contains fumarprotocetraric acid and occurs on species of *Nothofagus* today. It occurs in New Zealand, southern Australia, and Tasmania. *Usnea flexilis* (containing salazinic acid) closely resembles *U. contexta* morphologically. Since *Nothofagus brassii* migrated north as far as PNG, it could have acted as the carrier phorophyte for the salazinic race which now occurs in PNG, south-east Asia, Japan, and India at high altitudes.

How did these plant communities migrate to south-east Asia? Recent geological findings indicate that parts of south-east Asia formed the north-eastern rim of Gondwanaland, which separated from the Australian/New Guinea continental margin during the Jurassic (160 Ma). Audley-Charles (1987) suggests that Burma, Thailand, Malaya, and Sumatra comprise continental fragments which became isolated in the Tethys Ocean. From the late Cretaceous (100 Ma) onwards these provided an archipelago of islands between the Asian mainland and Australia/New Guinea, which could have permitted land plant dispersal to occur in both directions throughout that time.

## Conclusion

It would appear that most species of *Usnea* and *Ramalina* occurring in the western Pacific region are not confined to this region but form part of species groups which have much wider distributions around the world. Those species growing at sea level or at low altitudes in the tropics are true tropical species which follow the mangrove distribution patterns or occur on phorophytes whose origins are tropical. Those species which are found in the mountains in cool and misty conditions originally came from Antarctica with such phanerogams as *Araucaria*, *Nothofagus*, etc. and should be regarded as temperate species rather than tropical.

## Acknowledgements

I wish to thank the curators and keepers of the herbaria from which material was borrowed, e.g. BM, CHR, H, O, TNS, TUR. I also want to thank Dr M.E. Hale for allowing me to examine his collections of *Usnea* from Sabah, Malaya, the Philippines, and New Guinea. I thank Dr J.A. Elix, Mr P. Lambley and Mr A. Aptroot for loaning me their personal collections of New Guinea material. Financial support for this study was provided by an Australian Biological Resources Study Grant. The author acknowledges the monetary contribution made by The Royal Society which helped towards the cost of travel to the Tropical Lichen Symposium.

## References

Asahina, Y. (1956). *Lichens of Japan*, Vol. 3. Genus *Usnea*. Research Institute for Natural Resources, Shinjuku, Tokyo.

Audley-Charles, M.G. (1987). Dispersal of Gondwanaland: relevance to evolution of the Angiosperms. In *Biogeographical evolution of the Malay Archipelago*. Oxford monographs on Biogeography No. 4 (ed. T.C. Whitmore), pp. 5–25. Clarendon Press, Oxford.

Chapman, V.J. (1977). Africa B. The remainder of Africa. In *Ecosystems of the world. I. Wet coastal ecosystems* (ed. V.J. Chapman), pp. 233–40. Elsevier, Amsterdam, Oxford, New York.

Dettmann, M.E. (1989). Antarctica: Cretaceous cradle of austral temperate rainforests? In *Origins and evolution of the Antarctic biota* (ed. J.A. Crame), pp. 89–105. Geological Society London, Special Publication.

Dettmann, M.E. and Jarzen, D.M. (1990). The Antarctic/Australian rift valley late Cretaceous cradle of north-eastern Australasian relects. *Review of Palaeobotany and Palynology*, **65**. (in press).

Kashiwadani, H. (1986). Genus *Ramalina* (Lichens) in Japan (2). On *Ramalina pacifica* Asah. and its allies. *Bulletin of the National Science Museum* Series B (Botany), **12**, 117–25.

Müller, J. (1981). Fossil pollen records of extant angiosperms. *The Botanical Review*, **47**, 1–145.

Rogers, R.W. and Stevens, G.N. (1988). The *Usnea baileyi* complex (Parmeliaceae, lichenized Ascomycetes) in Australia. *Australian Systematic Botany*, **1**, 355–61.

Stevens, G.N. (1983). Tropical–subtropical *Ramalinae* in the *Ramalina farinacea* complex. *Lichenologist*, **15**, 213–29.

Stevens, G.N. (1987). The lichen genus *Ramalina* in Australia. *Bulletin of the British Museum (Natural History)*, Botany, **16**, 107–223.

Stevens, G.N. and Kashiwadani, H. (1987). Synonymy of *Ramalina intermediella* Vain. with *R. peruviana* Ach. *Journal of Japanese Botany*, **62**, 373–6.

Stockey, R.A. (1982). The Araucariaceae: An evolutionary perspective. *Review of Palaeobotany and Palynology*, **37**, 133–54.

Swinscow, T.D.V. and Krog, H. (1978). Pendulous species of *Usnea* in East Africa. *Norwegian Journal of Botany*, **37**, 221–41.

Webb, L.J., Tracey, J.G. and Jessup, L.W. (1986). Recent evidence for autochthony of Australian tropical and subtropical rainforest floristic elements. *Telopea*, **2**, 575–89.

# 5. Lichens of Papua New Guinea

P.W. LAMBLEY
*The Cottage, Elsing Road, Lyng, Norwich, NR9 5RR, UK*

## Abstract

An attempt is made to provide an overview of the macrolichen flora of Papua New Guinea and to relate this to the island's complex geological history and to the range of habitats now present. The lichen flora is shown to be composed of a number of different elements, including an interesting Gondwanaland element. The mountainous nature of the island provides opportunities for studying the altitudinal zonation of lichen species and their communities. The conservation of the flora is discussed in relation to the increasing destruction of the forests through logging, mining developments, and agricultural activities.

## Introduction

Papua New Guinea is a country occupying the eastern half of the world's second largest island (Fig. 5.1), together with the associated islands of the Bismark archipelago, Bougainville (North Solomons), Admiralty Islands, the Louisiade and D'Entrecasteaux archipelagos. The land surface is about 800 000 km², i.e. about twice that of the United Kingdom. The island is highly mountainous with about 30 per cent of Papua New Guinea lying above 1000 m. The Central Ranges form a cordillera that runs unbroken from Milne Bay in the east to the isthmus of the Vogelkop in Irian Jaya to the west. Along this length there are only a few passes below 1500 m. In several regions the cordillera broadens into a series of parallel ranges separated by high, flat, inter-montane valleys. The mountains are highest in Irian Jaya with Mt. Carstenz (Mt. Jaya) reaching 4884 m, other high peaks in the Snow Mountains include Mt. Idenburg (4820 m), and Mt. Wilhelmina (4720 m). In Papua New Guinea the highest peaks are Mt. Wilhelm

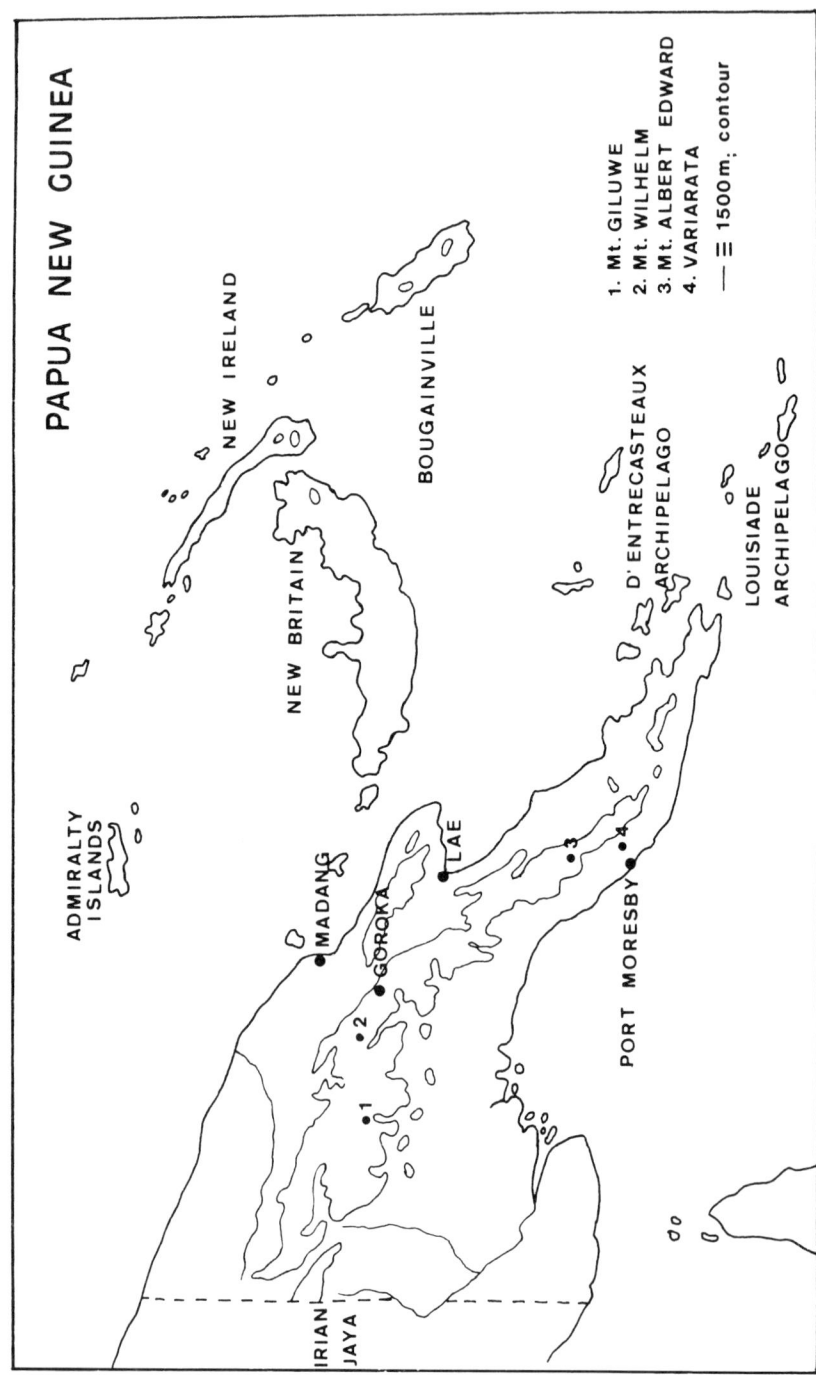

**Fig. 5.1.** Locality map Papua New Guinea showing major towns, mountains and the 1500 m contour.

(4509 m) and Mt. Giluwe (4367 m), both in the Central Ranges. In the west, Mt. Capella reaches 3993 m in the Star Mountains close to the Irian Jaya border, while Mt. Victoria (4035 m) and Mt. Albert Edward (3990 m) are the highest peaks in the Owen Stanley Range in the east. There are a number of outlying ranges to the north but only those on the Huon Peninsula reach any great height (4160 m). Many of the satellite islands are also mountainous. New Britain has peaks reaching 2440 m (Mt. Sinewit), New Ireland 2400 m (Hans Meyer Range) and Bougainville 2743 m (Mt. Balbi). These mountains are the source of a number of major rivers including the Sepik and the Fly which flow into extensive lowland plains which are still largely forested.

## Geology

New Guinea in its present form is geologically young and probably first became land in the Miocene. However, the main mountain-building phase took place during the Pliocene, about 4 to 5 Ma. New Guinea is a product of the interaction between the northward moving Australian plate and the Pacific plate, a movement that began in the middle Mesozoic. It lies on the boundary between the two plates and is still tectonically active with a continuing history of earthquakes and volcanism. Geologically, New Guinea can be divided into three distinct provinces. The first is the youthful, uplifting north coastal area which extends through the Huon peninsula to the Bismark archipelago and Bougainville. The second is a 250 km wide mountain belt including the entire central cordillera, a region of deformed and uplifted sediments and volcanics. This region is older than the north coastal area and volcanism has now ceased, though it is still subject to earthquakes. The third area lies to the south of the main range and is low-lying. It is composed of the original continental crust of the Australian plate extensively buried by recent sediments. This part is geologically continuous with the Australian continent.

## Climate

The climate of New Guinea is largely determined by its equatorial and oceanic position and by the mountainous terrain. It is generally warm and humid with little seasonal range in temperature and has abundant rainfall.

Temperatures range from hot in the lowlands, to frequent frosts on the higher mountains and permanent ice-caps on the highest mountains in Irian Jaya. Daily mean maxima on the coast are about 30–32 °C with mean minima around 23 °C. The months of the Austral winter

(June to September) are generally slightly cooler. Temperatures decrease by about 2 °C for every 300 m altitude. At 1500–2000 m in the Highland valleys, where many of the population live, average daily maxima of 22–25 °C and minima of 11–15 °C are experienced. Frosts are rare below 2000 m, but increase with altitude so that on the slopes of Mt. Wilhelm at 3450 m frosts can be expected on 30–50 per cent of nights. Readings at this altitude give mean maxima of 11 °C and mean minima of 4 °C.

The climate is influenced by two major wind systems; the north-west monsoon which blows from December to April, and the south-east trades which blow from May to October. Most parts of the country receive most rain during the period of the north-west winds, but coasts and mountains facing south-east are generally wetter during the period of the south-easterlies. Gulf Province, the south coast of New Britain, parts of the Huon, and southern New Ireland are wettest at this time. Rainfall is generally not markedly seasonal except in the areas of low rainfall such as Port Moresby which receives 75 per cent of its rain, on average, during the period December to the end of April. Rainfall totals vary from about 1100 mm in the driest areas around Port Moresby to a general average of 2500–3500 mm over the rest of the country. The wettest regions such as Gulf, the Star Mountains, and the south coast of New Britain receive totals in excess of 6000 mm. Snow falls on Mt. Wilhelm and perhaps some of the other highest peaks and may lie for several days. The permanent snowline lies at about 4800 m and is only reached in the Snow Mountains of Irian Jaya. Fog and cloud are important features of the montane climate. In the mountains, mornings are frequently clear and bright, with clouds building up by midday and rain showers falling during the afternoon and evening. Clouds often settle on ridges and humidity rises to 100 per cent most nights. A distinct cloud-line often forms at a particular altitude which varies from place to place. This cloudline usually forms at 2200 m or above in the central ranges but may be much lower on isolated ranges or those near the coast (Massenerhebung effect). Cloud frequently develops, for instance, on the Variarata escarpment behind Port Moresby at an altitude of only 800–1000 m.

During the Pleistocene, ice-caps formed on many of the mountains above 3500 m. In Papua New Guinea the largest ice-cap developed on the volcanic dome of Mt. Giluwe covering 188 sq. km and extending down to 3200–3500 m with valley glaciers reaching 400 m lower down in places. Present evidence indicates that the mountains were free of ice by 9000 BP though permanent snowfields may have been re-established on occasions.

## Vegetation

The warm, humid climate supports the development of forests which extend from the mangrove formations of the coastal waters almost up to the summits of the highest peaks clothing an estimated 75 per cent of the land surface. There have been a number of attempts to describe the wide variety of forest types which result from differences in altitude, rainfall, past history, and other environmental influences. Johns (1982) proposed a classification based on a series of nodal forest types and this appears to fit most closely the altitudinal zonation shown by lichens in New Guinea. The lowland zone is considered by Johns to extend from sea level to 700 m, though lower in places. Within this zone there is a wide variety of vegetation types. Mangrove communities form extensive stands in Gulf Province, parts of Central Province, and the lower reaches of the Sepik and Ramu Rivers. The natural vegetation of much of the lowlands is rainforest. Lowland rainforest is structurally and taxonomically complex with a high species richness. Mixed, lowland rainforest is comprised of trees with large buttresses which reach a canopy height of 45 m. Light levels are low in undisturbed forest and, as a result, undergrowth is relatively sparse. Branches in the canopy are often covered in epiphytic orchids and ferns. In contrast, disturbed forest is often a tangle of vines and secondary shrubs and trees. In many of the flood plains, drier forest grades into swamp-forest of various kinds. In areas with a long dry season, the vegetation resembles that of parts of northern Australia, with savanna or open woodlands dominated by species of *Eucalyptus* and with gallery forest fringing the watercourses. Monsoon forests with seasonally deciduous trees such as Kapok (*Bombax*) form a transitional community to rainforest in these areas. These savanna communities are best developed in the Trans Fly and on the coastal plain of Central Province around Port Moresby. With increasing altitude the structure and species composition of the forest gradually changes. The foothills and low mountains are clothed in mixed evergreen forest which is less luxuriant than those of the plains as conditions are generally less favourable for tree growth on the steep slopes and unstable soils.

The montane zone as defined by Johns, occupies a zone from about 700 m to 3000 m. Mixed, broad-leaved forest occurs throughout but shows considerable variation. Johns recognized considerable differences between the forests in the lower part of this zone and those above. This change occurs between 1500 and 2000 m.

The lower montane forests are dominated by mixed evergreen forest which differs from lowland rainforest in having fewer palms, lianes, and trees with buttress roots. Tree ferns and bryophytes are uncommon. In places, especially on ridge-tops, oak forests dominated by *Castanopsis acuminatissima* and *Lithocarpus* occur. The canopy of these forests only reaches about 25 m and is rather thin. Despite the relatively high illumination, there are few secondary stage trees. These lower montane forests have a rather drier aspect than the mid-montane forests, with less bryophyte growth. *Araucaria* forests grow in this zone in the Lower Watut valley around Bulolo and elsewhere.

The mid-montane forest replaces the lower montane at altitudes between 1500 m and 2000 m. The forest is dominated by Southern Beech (*Nothofagus* spp.) or by a mixed forest. These forests have a rich growth of epiphytes with orchids, ferns, bryophytes, and lichens prominent in the canopy. Members of the Ericaceae especially rhododendrons and *Vaccinium* are also frequent. Climbing rattans and other palms disappear, but are replaced by species of the pandan, *Freycinetia* and scrambling bamboo (*Nastus* spp.). Stilt-rooted *Pandanus*, tree ferns, and woody shrubs are commonly present in the undergrowth.

There is a change to upper montane forest between about 2700 and 3000 m. This forest is dominated by conifers notably *Podocarpus*, *Dacrycarpus* and *Libocedrus* (Fig. 5.2; 5.3). With increasing altitude, this forest becomes more stunted and the structure simplified into one storey. Eventually, near the tree-line it may only be a few metres tall. The forest also becomes increasingly fragmented with grasslands separating islands of forest (Fig. 5.4). There are often well-developed shrub communities on the forest edge with species of *Rhododendron*, *Vaccinium*, *Coprosma*, *Olearia*, and *Pittosporum* present. A feature of the New Guinea montane vegetation is the development of areas of tree fern savanna (Fig. 5.5), especially on slopes between the forest and poorly drained valley floors. There is some debate about whether this community is natural or maintained by man through fire.

The tree-line is reached at about 3900 m on Mt. Wilhelm, but is probably lower on more isolated peaks and ranges. At the highest altitudes, the forest is replaced by alpine grassland broken by rock outcrops and occasional patches of scrub.

## The lichen flora

The history of the exploration of the lichen flora is detailed in the accounts of Mattick (1942) and Streimann (1986). Streimann in his 'Checklist' (1986) enumerates 495 species in 126 genera. This list has since been extended by collecting expeditions by Aptroot, Sipman, and Lambley. Since publication of the Checklist, the genus *Cladonia* has also been revised in a series of papers by Stenroos (1986a, b, 1987, 1988).

**Fig. 5.2.** Shrub communities and *Dacrycarpus*, upper montane forest 3000 m Mt. Kenevi, Northern Province.

Forests are the major habitat type in New Guinea, the lichen flora is, therefore, predominantly corticolous and foliicolous although, particularly at high altitudes, rock outcrops and alpine grasslands provide habitats for saxicolous and terricolous species.

The lichen flora of the extensive mangrove systems of New Guinea has not been well studied in contrast to that of Australia (Stevens, 1978). Limited studies in the vicinity of Port Moresby suggest that tall, mature mangrove forest dominated by species of *Rhizophora* and *Bruguiera* has a lichen flora similar to that of closed, mature, lowland rainforest, with crustose lichens predominant and macrolichens rare. A richer macrolichen flora develops in *Avicennia* scrub which often forms a zone on the landward side of mangrove systems in low-rainfall areas. Mangrove species associated with *Avicennia* include *Lumnitzera* and

**Fig. 5.3.** Interior of upper montane forest 3000 m Mt. Kenevi, Northern Province.

**Fig. 5.4.** Alpine grassland and upper montane forest 3000 m Mt. Kenevi, Northern Province.

*Ceriops tagal* in the vicinity of Port Moresby. Lichens present in this community include species of *Ramalina*, such as *R. tropica* and *R. nervulosa* var. *luciae*, and *Roccella* spp. on twigs and small branches. Larger branches and dead wood, especially low down, often have communities dominated by species of *Dirinaria* and *Physciaceae* including *Pyxine* spp. and *Physcia* spp., e.g. *P. erumpens*. A similar community grows on *Bougainvillea* and other shrubs in the grounds of the University of Papua New Guinea, indicating that the community is not restricted to mangroves. On well-lit boughs higher up in the canopy, species of *Parmelia* s.l., are found, including *Parmotrema saccatilobum*, *Relicina* spp., and *Bulbothrix* spp.. In the interior of the *Avicennia* scrub, the higher humidity and lower light levels favour species with blue-green photobionts such as *Coccocarpia palmicola*, *C. erythroxyli*, *Collema rugosum*, *Leptogium* spp., *Pannaria mariana* s.l., and *Physma* spp..

Lichens are generally scarce in the *Eucalyptus* savanna around Port Moresby. This is, in part, due to the nature of *Eucalyptus* bark and the frequent fires which are lit every dry season. In gardens, where fires are not a regular occurrence and where there are a greater variety of tree species, a few macrolichens such as *Parmelia* s.l., and *Dirinaria* occur. A crustose lichen community does, however, grow on chert

**Fig. 5.5.** Tree fern savannah 2800 m Myola, Northern Province.

boulders scattered in the savanna, the community includes *Caloplaca* cf. *cinnabarina* and a species of *Lecidea*.

In lowland rainforest, foliicolous and other crustose lichens are prominent but macrolichens are uncommon. A few species grow on boughs in the canopy notably *Dirinaria* spp., *Pyxine* spp., *Pannaria mariana*, *Parmotrema* spp., e.g. *P. sulphurata*, and *Coccocarpia* spp.. Around Lae, and in other humid areas, these communities are developed on trees in the open.

Hill forest is richer in macrolichens. Species of *Usnea* (e.g. *U. baileyi* and *U. nidifica*) start appearing in the canopy of trees in exposed, ridge-top situations above about 500 m. Macrolichens are best developed on stream-side trees, in the canopy and other situations where light levels are higher than within the forest. *Lobaria insularis*, *Coccocarpia erythroxylii*, *C. pellita*, *C. glaucina*, *Relicina amphithrix*, *R. butleri*, *R. connivens*, *Leptogium* spp., *Coenogonium* spp., *Pannaria* spp., and *Parmotrema cristiferum* are all characteristic of this lower, hill forest.

The macrolichen flora of the lower montane zone is richer than the lowland forests, with members of the Stictaceae and Lobariaceae becoming prominent. As Brass (1964) observed, the lower part of this zone appears to coincide with the level at which clouds form. The effects of this regular cloud zone are evident at 800 m on the crest of the escarpment at Variarata near Port Moresby where species such as *Lobaria insularis*, *Pseudocyphellaria argyracea*, *P. aurata*, *Psoroma* spp., and *Sticta* spp., grow on trees along the ridge. Species of *Usnea* are prominent in the canopy. Variarata is also one of the few places at low altitude where there are extensive rock outcrops. These rocks and boulders composed of a volcanic agglomerate, provide a substrate for a lichen community with *Parmotrema cristiferum*, *Dirinaria aegialita*, *Physma byrsaenum*, *Pyxine sorediata*, *Relicina abstrusa*, and *Usnea baileyi*. The very local *Thysanothecium scutellatum* is abundant on burnt wood and, occasionally, other substrates in areas of the park dominated by *Eucalyptus terticornis* and *Casuarina*. The oak forests are apparently a suitable habitat for many lichens especially members of the Lobariaceae and Stictaceae. In contrast to forests at higher altitudes, there are fewer mosses and other epiphytes and light levels are higher. *Lobaria* spp., *Sticta* spp., *Pseudocyphellaria* spp., and *Psoroma* spp., are attractive features of this forest, growing on branches and, occasionally, on the main trunks. These forests deserve more study as they appear to contain communities having affinities with those of southern oceanic temperate regions.

The mid-montane zone lichen flora is characterized by an increase in both abundance and species numbers of *Lobaria*, *Menegazzia*, *Sphaerophorus*, and *Pseudocyphellaria*. A number of species of *Sticta* and *Psoroma*, and *Lobaria insularis* become scarce or disappear at this altitude. Preliminary observations suggest that there are no noticeable differences between the lichen floras of the *Nothofagus* and mixed forests; of greater importance is the structure of the forest and the presence of secondary habitats like those of the forest edge. Macrolichen communities are rarely developed on tree trunks unlike in the lower montane zone, but are largely restricted to 'edge' situations including the canopy. A few species grow regularly within the forest interior, for example, *Lobaria pseudopulmonaria*, *L. subscrobiculata*, *Pseudocyphellaria* cf. *lombokensis*, *Dictyonema* spp., *Sticta* aff. *filix*, and species of *Leptogium*. In places, the humidity is sufficiently high for species of *Pannaria* and *Leptogium* to grow on the leathery leaves of shrubs growing in the understorey. *Menegazzia propagulifera* is of interest as the only member of the genus which regularly penetrates the interior of the forests and then mainly in relatively well-lit situations on ridge tops. The twigs and small branches of *Olearia*, *Vaccinium*, *Saurauia*, and other small trees and shrubs growing on the forest edge are covered in lichens including *Nephroma helveticum*, *Erioderma* spp., *Leioderma* spp. (rarely), *Heterodermia leucomelos*, *H. podocarpa*, *Pseudocyphellaria crocata*, *P. pickeringii*, *Hypotrachyna* spp., and *Menegazzia* spp.. The rich lichen flora of these forests also includes other species of *Pseudocyphellaria* like *P. intricata*, *P. argyracea*, *P.* cf. *rufovirescens*, *P. multifida*, *Sticta* spp. (particularly those with stalks), *Lobaria dendrophora*, *L. isidiosa*, *L. isidia*, *L. clemensiae*, and *L. pseudopulmonaria*.

At altitudes above about 2800–3000 m, the mid-montane forest gives way to upper montane forest in which conifers become more prominent and there are sometimes breaks in the forest with grasslands having tree ferns of the genus *Cyathea*. Tree ferns are resistant to fire and, therefore, usually survive the regular burning of the grassland which takes place in many of these localities. These fires and the fibrous texture of the bark result in rather specialized lichen communities in which species of *Peltigera* and *Cladonia* are important. Other lichens growing in this community include *Pannaria pezizoides*, *Psoroma hypnorum*, *Pseudocyphellaria crocata*, *Lobaria isidiosa*, *Heterodermia leucomelos*, and species of *Usnea*.

The conifer forests are heavily mossed, but as the canopy lowers, more light penetrates into the interior. With increasing altitude the forest tends to become more fragmented and grasslands are increasingly a feature of the landscape. *Cladia aggregata* is occasional in such habitats. These high altitude forests, with their attendant shrub communities, provide a suitable habitat for many lichen species, particularly those which grow on twigs, such as *Nephroma* spp., *Erioderma* spp., *Bryoria* spp., *Anzia* spp., *Hypotrachyna* (especially *H. sinuosa*), and *Cetrelia* spp.. Species of *Menegazzia* are also very well-represented in this community. The moss-covered poles and small branches are the habitat for the local and spectacular *Nephromopsis stracheyi*. Two endemic monotypic genera, *Calathaspis* and *Compsocladium*, are probably also characteristic of this habitat, though more collections are needed to confirm this. Species of *Sphaerophorus*, *Lobaria isidiosa*, *L. pseudopulmonaria*, and *Pseudocyphellaria crocata* are frequent in the forest interior.

At high altitudes (above c. 3800 m) alpine grasslands predominate. The peaty alpine soils support a lichen community of *Thamnolia vermicularis*, *Siphula*, *Cetrelia*, *Pannaria* spp., *Cladonia* spp., and *Hypogymnia lugubris*. The commonest species at the highest altitudes are *Thamnolia vermicularis* and *Hypogymnia lugubris*. *Cetraria islandica* ssp. *antarctica* is recorded from Mt. Wilhelm. *Arthrorhaphis alpina* is recorded from Mt. Wilhelm (Galloway and Bartlett 1986) but was observed on a number of other peaks (Mt. Giluwe and Mt. Albert Edward) at elevations of c. 4000 m. Species of *Stereocaulon* are often abundant on rock outcrops and stream-sides. *Hypotrachyna costaricensis* is another common macrolichen of rock outcrops. Species of *Placopsis* are rather uncommon but occur at higher altitudes where water runs over rocks. The alpine nature of the flora is further indicated by *Xanthoria elegans* which grows on rocks near the summit of Mt. Wilhelm.

Secondary habitats created by man increase the diversity of the lichen flora. Rubber plantations, which are usually sited below 500 m, provide an interesting contrast to the rainforest they replace. The well-spaced uniform stands provide a light, but still humid environment which is favourable for many lichens including *Parmotrema cristiferum*, *Usnea* spp., *Physma* spp., *Relicina* spp., *Leptogium* spp., *Coccocarpia* spp., and *Pannaria*. In the highland valleys where man has cleared most of the forest between 1400–2500 m, lichen habitats are often rather scarce and confined to isolated trees and fences. However, hedges of *Cordyline* which are grown as boundaries and also for clothing (arse-grass) support a surprisingly rich lichen community in which *Pseudocyphellaria argyracea*, *P. crocata*, *Lobaria isidiosa*, *Collema* spp., and *Pannaria* feature. Man-made road cuttings create bare soil and rock which are colonised by species of *Stereocaulon* and *Baeomyces*. Isolated trees in towns and in cultivated land are often covered in lichens especially of the *Physiaceae*; *Heterodermia diademata*, *H. barbifera*, *H. leucomelos*, *Pyxine subcinerea*, and also *Hypotrachyna formosana* and *Parmotrema reticulatum*. Many of these species also grow in forest or the edge, but a number, such as *Pseudocyphellaria clathrata*, may be confined to disturbed habitats.

## Biogeography

Gressitt (1982) states

'The biogeography of New Guinea is an intriguing problem, and one difficult of resolution. Not only did the biota come from different directions, such as from south-east Asia, the Phillipine islands and from Australia, but even from S. America, or Gondwanaland, via Australia.'

It is the only place in the world where oaks (*Castanopsis* and *Lithocarpus*) grow with southern beech (*Nothofagus*), while gymnosperms with southern affinities grow with rhododendrons and other Ericaceae from Asia.

The lichen flora also reflects this diversity of origins though it is perhaps too early to attempt a detailed analysis at species level of these various affinities. It is clear that at generic level a number of primarily Southern Hemisphere genera are well represented. Species of *Pseudocyphellaria* and *Menegazzia* are conspicuous and attractive members of the flora, especially in the mid- and upper montane forests and both genera appear to have speciated considerably within New Guinea. Other Southern Hemisphere genera such as *Placopsis*, *Psoroma*, *Siphula*, and *Cladia*, though represented in the flora, do not appear to have speciated to the same extent. *Cetrelia* and *Nephromopsis* are examples of a number of genera which have distributions centred on East Asia and which are rare or absent from the rest of Australasia. Others, like *Thysanothecium*, have distributions extending around the western side of the Pacific Basin. Not unexpectedly, widespread tropical genera such as *Coccocarpia*, *Erioderma*, *Physma*, *Pyxine*, and *Hypotrachyna* (tropical montane) are well represented in the flora. There is a smaller bipolar or high montane element present on the higher peaks including *Arthrorhaphis*, *Thamnolia*, and *Xanthoria*. So far, species of *Umbilicaria* have not yet been recorded from New Guinea. Two monotypic endemic genera have been described (Lamb, 1956, Lamb *et al.*, 1972); *Calathaspis*, a member of the *Cladoniaceae* and *Compsocladium*, which is considered close to *Bacidia*. Both species grow mainly in high altitude forests dominated by conifers, though *Calathaspis* has been collected as low as 2600 m.

Altitude is the major determinant in influencing the distribution of lichens. This zonation is presumably a response to temperature and humidity changes with altitude. There are however, also some distribution patterns which apparently reflect differences in rainfall totals. Species of *Roccella*, on present evidence, appear to be restricted to the low rainfall area on the coast north and south of Port Moresby. *Thysanothecium scutellatum* may also be a species restricted by rainfall, as its normal substrate, charred wood, is very common in most parts of the country. So far it is known only from the slopes of Mt. Albert Edward, a small population at about 1400 m in Milne Bay Province and at Variarata where it is abundant. From observations at Variarata it would seem to be primarily a species of the savanna–rainforest transition, where annual rainfall is less than 2000 mm. Other distributions are less easily explained, for instance, the rarity of *Teleoschistes flavicans*, known from only two sites, one on the Huon peninsula and the other in the highlands near Goroka.

## Conservation

The lichens of New Guinea are beautiful, conspicuous, and diverse, but they are not of any utilitarian or economic value to the people of Papua New Guinea. Therefore, any attempts to conserve them must be developed with a more general framework of measures to protect habitats. Wildlife does have a utilitarian, economic, aesthetic, and spiritual value to the people of New Guinea. This is reflected in the National constitution which has the conservation of its wildlife as one of its aims and a Bird of Paradise is depicted on the National flag. In rural Melanesian society there have always been traditional methods of environmental management and conservation. A mixture of taboos and customary rules have helped to protect certain species or sites. However, with the coming of Europeans many of these practices are dying out and new economic and population pressures are increasingly damaging the environment. The demands for tropical hardwoods are increasing, especially now that there are restrictions on logging in Queensland and the more accessible timber reserves in the rest of south-east Asia are now so depleted. Once logged, these areas may revert to secondary forest, may be gardened, or be replanted with cash-crops such as coconut, oil palm, rubber, coffee, or cocoa. In the highlands coffee growing is important and, to a lesser extent, tea and European type vegetables, in addition to traditional gardening. Exploitation of timber resources is not restricted to the lowlands, the *Araucaria* forests in the Lower Watut valley have been exploited for many years and in more recent times the fine *Nothofagus* forests of Mt. Giluwe are being logged. The country is rich in gold and copper and these deposits are increasingly exploited with large mining operations at Ok Tedi in the Star Mountains, Porgera, and Bougainville and many more are planned. These operations, besides affecting the site, have implications for the surrounding area with increased population and exploitation of the forests for timber and slash and burn agriculture. Oil has also been found in commercial quantities in the Southern Highlands and these reserves are likely to be developed in the near future.

It was appreciated that traditional laws would not be sufficient to protect the environment against these new pressures and there is now a body of legislation designed to protect wildlife and the environment.

The three most important statutes for the conservation of wildlife and the environment are the National Parks, Conservation Areas, and Fauna (Protection and Control) Acts. Two full National Parks have been established under this Act—Varirata and McAdam, both having an interesting lichen flora. Three others have been declared under different designations, these are smaller in area and are not significant lichenologically. Four others await final gazetting, of which Mt. Wilhelm and Mt. Gahvisuki Provincial Park are of considerable importance for lichens. An alternative method of establishing sanctuaries is provided under the Fauna (Protection and Control) Act which allows for the setting up of Wildlife Management Areas on customary land after consultation with landowners and local government authorities. These have been described by Eaton (1986). There are 12 at present covering 870 000 ha with varying degrees of effectiveness. In some cases they are principally to protect individual species such as turtles and megapodes. At Siwe Utame south of Mt. Giluwe, measures to protect Birds of Paradise have also protected a fine area of *Nothofagus* forest. Even where there is no formal protection, an income through tourism as at Myola on the borders of Central and Northern provinces can provide an additional incentive not to destroy the forests.

In a country with a vascular plant flora of over 9000 species and only a few professional botanists, the initiative for research on the lichen flora must come from outside the country, though it is important that every effort is made to stimulate local interest by depositing collections at the National Herbarium at Lae, or at the University of Papua New Guinea. This is stipulated before Research Visas are normally issued. Wherever possible local botanists should accompany collectors, this is generally of value to both parties. Further, by producing simple illustrated keys we might at least increase awareness of the group within the local scientific community.

## Acknowledgements

I am indebted to the Research Committee of the University of Papua New Guinea for providing funds for travel within Papua New Guinea; to Dr P.L. Osborne, Chairman of the Biology Department for help and encouragement during these studies, to Kipling Naoni and Tamari Mala for providing valuable assistance in the field and to P.W. James, Dr D.J. Galloway and Dr J.A. Elix for their encouragement and help in identifications.

## References

Brass, L.J. (1964). Results of the Archbold Expeditions No. 86. Summary of the sixth Archbold Expedition to New Guinea (1959). *Bulletin of the American Museum of Natural History*, **127** (Art. 4), 149–215.

Eaton, P. (1986). Grassroots conservation: Wildlife management areas in Papua New Guinea. *Land Studies Centre Report* **86/1** University of Papua New Guinea.

Galloway, D.J. and Bartlett, J.K. (1980). *Arthrorhaphis* Th. Fr. (lichenised Ascomycotina) in New Zealand. *New Zealand Journal of Botany*, **24**, 393–402.

Gressitt, J.L. (1982). General introduction. In *Biogeography and ecology of New Guinea*, Vol. 1 (ed. J.L. Gressitt), pp. 3–13. Dr W. Junk, The Hague.

Johns, R. (1982). Plant zonation. In *Biogeography and ecology of New Guinea*, Vol. 1 (ed. J.L. Gressitt), pp. 309–30. Dr W. Junk, The Hague.

Lamb, I.M. (1956). *Compsocladium*, a new genus of lichenised Ascomycetes. *Lloydia*, **19**, 157–62.

Lamb, I.M., Weber, W.A., Jahns, H.M., and Huneck, S. (1972). *Calathaspis*, a new genus of the lichen family Cladoniaceae. *Occasional Papers from the Farlow Herbarium of Cryptogamic Botany*, **4**, 1–12.

Mattick, F. (1942). Beitrage zur Flora von Papuasien 26. Die Fletchen von NeuGuinea. 1. Allgemeines. Die Gattung *Cladonia*. *Botanische Jahrbücher für Systematik, Pflanzengeschichte und Pflanzengeographie*, **72**, 151–8.

Stenroos, S. (1986a). The family Cladoniaceae in Melanesia. 1. *Cladonia* sect. *Unciales*. *Annales Botanici Fennici*, **23**, 161–4.

Stenroos, S. (1986b). The family Cladoniaceae in Melanesia. 2. *Cladonia* sect. *Cocciferae*. *Annales Botanici Fennici*, **23**, 239–50.

Stenroos, S. (1987). Studies on the genus *Cladonia* sect. *Cocciferae* in Papua New Guinea and the adjacent regions. *Bibliotheca lichenologica*, **25**, 421–2.

Stenroos, S. (1988). The family *Cladoniaceae* in Melanesia. 3. *Cladonia* sections *Helopodium*, *Perviae*, and *Cladonia*. *Annales Botanici Fennici*, **25**, 117–48.

Stevens, G.N. (1978). Lichens on mangroves along the east coast of Australia. M.Sc. Thesis (unpublished) University of Queensland.

Streimann, H. (1986). Catalogue of the lichens of Papua New Guinea and Irian Jaya. *Bibliotheca Lichenologica*, **22**, 1–145.

# 6. Lichenological observations in low montane rainforests of eastern Tanzania

H. KROG

*Botanical Museum, University of Oslo, Trondheimsveien 23B, N-0562 Oslo 5, Norway*

### Abstract

Three genera, *Cladia*, *Menegazzia*, and *Siphula*, are additions to the East African macrolichen flora. The new species *Parmotrema fragilescens* Krog and *P. laciniatulum* Krog are described. *Erioderma leylandii* and *Relicina planiuscula* are reported for the first time from Africa, while *Menegazzia terebrata*, *Phyllopsora mauritiana*, and *Stereocaulon fibrillosum* are new to continental Africa. New to East Africa are *Cladia aggregata*, *Hypotrachyna pseudosinuosa*, and *Siphula decumbens*. Few macrolichens occur in the lowland rainforest below *c.* 1000 m, although a variety of species are found in drier forest types, on roadside trees, and on riverine rocks. The dominant elements in some montane rainforests below 2500 m are discussed.

### Introduction

A special topographic feature of eastern Tanzania is a series of isolated mountainous areas from the Usambara and Pare Mts in the northeast to the Poroto and Livingstone Mts near Lake Malawi in the south, including the Nguru and Kanga Mts west of Turiani and the Uluguru Mts south of Morogoro. Parts of the Usambara Mts are only 40 km from the coast while the Nguru and Uluguru Mts are 130–140 km. A warm ocean current along the East African coast south of the equator generates high air humidity which is carried inland by easterly winds, causing high precipitation and more or less permanent mist and cloud formation on the eastern slopes and summits of these mountains. The Poroto and Rungwe Mts are farther from the coast, but the nearness of Lake Malawi and local climatic conditions lead to a high annual rainfall there also, between 3000 and 4000 mm on the wetter side of Rungwe Mountain.

The mountains are surrounded by woodland and savanna, and the forest flora of each ridge has developed independently over a long time. Thus, a large number of endemic species are known among vascular plants; among woody plant species there are more than 40 endemics in the Uluguru Mts alone (Pócs 1976). Another feature of the flora of these mountains is a definite affinity to that of the Mascarenes and Madagascar.

The chief aim of the present project is to report on some aspects of the macrolichen flora of lowland and low montane rainforests below c. 2500 m in eastern Tanzania. Some preliminary results are presented here.

## Materials and methods

This report is based mainly on personal field observations and specimens collected during a three-week visit to Tanzania in October 1988, and a five-week stay in March–April 1989. Some records based on specimens kindly made available by Dr T. Pócs and Ms E. Farkas are also included. Parts of my own collections are as yet undetermined; there are a number of specimens which require critical study, including several species which may prove to be new to science.

Chemical data were obtained by thin-layer chromatography by means of standard techniques (White and James 1985). Sections of thalli and apothecia were cut by microtome. Spores and conidia were studied in squash preparations in Melzer's reagent and water, respectively. My own collections are deposited in O, those of Pócs and Farkas in VBI, some of them with duplicates in O.

The nomenclature conforms to that of Swinscow and Krog (1988). The geographical divisions (provinces, districts) conform to those applied in Flora of Tropical East Africa (Polhill 1988).

## Results and discussion

Not surprisingly, a number of genera and species which were not previously known from East Africa have now been discovered in Tanzania. Two species of *Parmotrema* new to science are described here. Both of them belong in the low montane rainforest element.

1. Parmotrema fragilescens *Krog sp. nov.* (Fig. 6.1)

Thallus corticola, membranaceus, griseo-viridis. Lobi 1–1.5 cm lati, superne laevigati vel foveolati vel rimulosi, marginibus crenatis vel laciniatis, laciniis simplicibus vel repetite ramosis, 1–2(–4) mm latis. Cilia tenuia, 4–6 mm longa, ad margines loborum laciniarumque. Soredia isidiaque nulla. Apothecia ignota. Conidia filiformia, 10–16 μm longa. Acidum lecanoricum et atranorinum continens.

Type: Tanzania. Tanga Province, Lushoto District, West Usambara Mts, Kwagoroto Summit WNW of Mazumbai village, 04°48′S, 38°29′E, in montane rainforest, 1850–1950 m coll. X 1988, *H. Krog* 2T 10/64 (O-holotype).

*Thallus* corticolous, thin, fragile, loosely attached, pale grey-green. *Lobes* 1–1.5 cm broad, crenate, or with simple or repeatedly branched laciniae of uneven width, mostly 1–2(–4) mm wide. *Cilia* numerous, slender, simple or bifurcate, 4–6 mm long, at margins of lobes and laciniae. *Upper side* even or minutely pitted peripherally, appearing coarsely maculate centrally, with the cortex cracking and often flaking. *Medulla* white. *Underside* black, with a brown or mottled marginal zone; lower cortex cracking and often flaking; rhizines simple, short, sparse. Soredia and isidia absent. *Apothecia* not seen. *Conidia* filiform, 10–16 μm long. T.l.c.: lecanoric acid, atranorin.

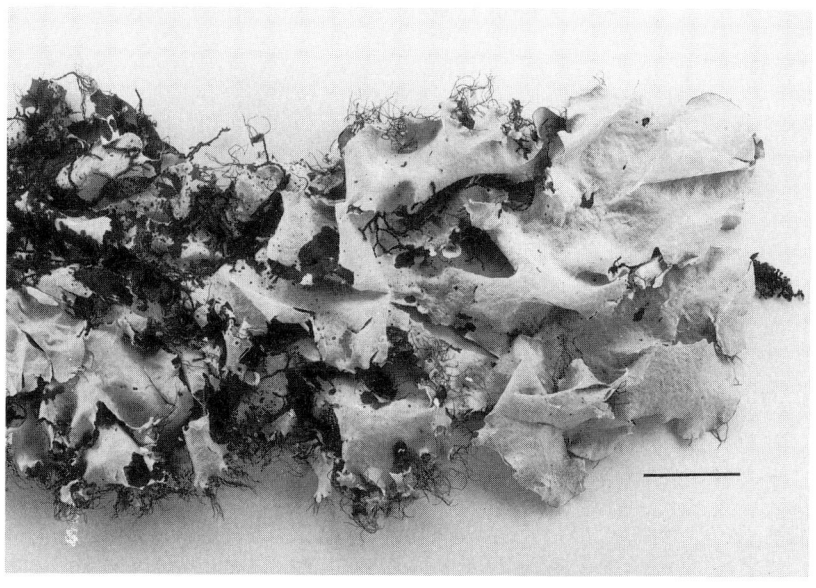

**Fig. 6.1.** *Parmotrema fragilescens* Krog, holotype (O). Rule = 1 cm.

The fragile laciniae are easily shed and are believed to function as propagules.

For differences from *P. planatilobatum* and *P. laciniatulum*, see below.

*Specimens examined*   Tanzania. Tanga Province, Lushoto District, West Usambara Mts, Kwagoroto Summit WNW of Mazumbai village, in montane rainforest, 1850–1950 m, *Krog* 2T 10/64 (O-type collection); West Usambara Mts, in montane rainforest between Mazumbai and Kambi Falls, 1600–1700 m, *Krog* 2T 7/62 (O); West Usambara Mts, Mazumbai University Forest Reserve, Sagara Ridge, in mossy elfin forest and *Philippia* heath, 1850–1980 m, *Krog* 2T 9/69 (O).

2. Parmotrema laciniatulum *Krog sp. nov.* (Fig. 6.2)

Thallus corticola, membranaceus, griseo-viridis. Lobi 0.5–0.8 cm lati, superne emaculati, marginibus laciniis simplicibus vel repetite ramosis, 0.2–0.3 mm latis. Cilia tenuia, 3–4 mm longa, ad margines loborum laciniarumque. Soredia isidiaque nulla. Apothecia ignota. Conidia sublageniformia, 7–7.5 μm longa. Acidum alectoronicum, acidum α-collatolicum, et atranorinum continens.

**Fig. 6.2.** *Parmotrema laciniatulum* Krog, holotype (O). Rule = 1 cm.

Type: Tanzania. Tanga Province, Lushoto District, West Usambara Mts, Mazumbai University Forest Reserve, Sagara Ridge, 04°49′S, 38°30′E, in mossy elfin forest and *Philippia* heath, 1850–1980 m, coll. X 1988, *H. Krog* 2T 9/75 (O-holotype; UPS-isotype).

*Thallus* corticolous, thin, fragile, loosely attached, pale grey-green. *Lobes* 0.5–0.8 cm broad, slightly concave, the margins with simple or repeatedly branched laciniae up to 9 mm long and 0.3 mm wide. *Cilia* slender, 3–4 mm long, at margins of lobes and laciniae. *Upper side* emaculate, with a continuous cortex. *Medulla* white. *Underside* black, with a brown or mottled marginal zone; rhizines simple, short, sparse. Soredia and isidia absent. *Apothecia* not seen. *Conidia* sublageniform, 7–7.5 μm long. T.l.c.: alectoronic acid, α-collatolic acid, atranorin.

As in *P. fragilescens* the laciniae are easily shed and are believed to function as propagules.

Although superficially similar, *P. fragilescens* and *P. laciniatulum* differ from each other in lobe and lacinia width, properties of the upper cortex, shape of conidia, and secondary medullary products. *Parmotrema planatilobatum* differs from both in being basically an isidiate–lobulate species, with mainly laminal isidia and lobules; furthermore, it produces gyrophoric acid in the medulla. While *P. fragilescens* and *P. laciniatulum* are species of the low montane rainforest, where they have been collected between 1600 and 2000 m, *P. planatilobatum* seems to prefer more open habitats, including riverine rocks; in the Usambara Mts it has been collected between 800 and 1000 m.

*Specimens examined* Tanzania. Tanga Province, Lushoto District, West Usambara Mts, Mazumbai University Forest Reserve, Sagara Ridge, in mossy elfin forest and *Philippia* heath, 1850–1980 m, *Krog* 2T 9/75 (O, UPS-type collection), 2T 9/76 (O); West Usambara Mts, Kwagoroto Summit WNW of Mazumbai village, montane rainforest, 1850–1950 m, *Krog* 2T 10/63 (O); West Usambara Mts, on W side of Gonja Mt., 5 km E of Mgwashi village, montane evergreen mossy forest, 1600–1700 m, *Krog* 2T 6/46–48 (O), Pócs 8505/s (VBI, O).

*3. Genera and species new to East Africa*

Three genera, namely *Cladia*, *Menegazzia*, and *Siphula*, are additions to the East African lichen flora. The corresponding species will be commented upon below.

New to Africa are *Erioderma leylandii* and *Relicina planiuscula*. *Erioderma leylandii* grows on tree trunks and branches in low montane rainforest in the Usambara Mts at *c.* 1900 m. Galloway (1985) reported the species from New Zealand under the name *Erioderma* cf. *glaucescens*, its world distribution was given as pantropical.

*Relicina planiuscula* occurs in *Philippia* heath in the Usambara Mts at 1800 m and in elfin forest in the Nguru Mts at 1900 m. The genus *Relicina* is very rare in Africa. When Hale's world monograph appeared (Hale 1975*b*), the genus was not yet known to occur in continental Africa, although one species, *R. subabstrusa*, with a wide distribution in the tropics, was known from a single collection in the Comore Islands. A record from continental Africa of a second species of *Relicina*, *R. abstrusa*, was added by Swinscow and Krog (1988). The species was collected in mangroves and low coastal hills between sea level and 300 m in Kenya; its world distribution includes south-east Asia and South America. The new addition to the African lichen flora, *R. planiuscula*, was previously known only from south-east Asia.

An additional three species appear to be new to continental Africa, namely *Menegazzia terebrata*, *Phyllopsora mauritiana*, and *Stereocaulon fibrillosum*. *Menegazzia terebrata*, a widespread species in temperate regions of the Northern Hemisphere, has not yet been reported from continental Africa, although it would be expected to occur there since it is known from Madeira (Tavares 1952 and personal observations) and was reported from Madagascar (des Abbayes 1961). In Tanzania it grows in mossy elfin forest and *Philippia* heath at 1700–1900 m in the Usambara Mts, and in a very wet montane rainforest at 2000 m in the Poroto Mts.

*Phyllopsora mauritiana* occurs in submontane forest at 1300–1500 m in the Usambara Mts. It was previously known only from Mauritius.

*Stereocaulon fibrillosum* was collected by Dr Pócs in 1984 on a rocky summit in the West Usambara Mts just below 2000 m. The species was described by Lamb (1977) on the basis of material collected by des Abbayes in Madagascar, and has not been reported on since.

New to East Africa are *Cladia aggregata*, *Hypotrachyna pseudosinuosa*, and *Siphula decumbens*. *Cladia aggregata* was collected on the Lukwangule Plateau in the Uluguru Mts at 2450 m by Pócs *et al.*, in 1988. The species is known to occur in southern Africa and islands of the Indian Ocean (des Abbayes 1948); it is widely distributed, especially in the Southern Hemisphere.

*Hypotrachyna pseudosinuosa* occurs in submontane and low montane rainforests in the Usambara Mts between 1500 and 1900 m. It is known to occur in southern Africa (Hale 1975*a*), and Macaronesia (Østhagen and Krog 1976), and has a wide distribution in tropical and warm temperate regions.

*Siphula decumbens* grows in mossy elfin forest and *Philippia* heath close to 2000 m in the Usambara and Nguru Mts. It is widely distributed in tropical regions, and has been recorded for southern Africa and Madagascar (Mathey 1974). All Tanzanian specimens belong to the thamnolic acid strain.

Close to 40 species so far are new to Tanzania; they have all been reported previously from one or more of the other East African countries.

## 4. The rainforest

Pócs (1976) provided vegetation maps for most of the northern Uluguru Mts and for the Lukwangule Plateau in the central part of the southern Uluguru Mts. These mountains have much in common with the Usambara and Nguru Mts and in the following account the terminology of Pócs (1976) is largely adopted.

Lowland, submontane, and montane rainforests dominate the eastern slopes of the isolated mountains in eastern Tanzania. At altitudes above c. 2000 m and on the main ridges, mossy forest, elfin forest, and *Philippia* heath are encountered.

In the Nguru and Kanga Mts lowland evergreen rainforest gradually merges with the submontane rainforest. This forest type was studied between c. 500 and 1200 m. In the dense forest with canopy trees of 40–50 m or more, the lack of light prevents most macrolichens from growing there, at least at ground level. Fallen canopy branches were sometimes found to bear common and widespread species such as *Parmotrema austrosinense*, *P. reticulatum*, and *P. tinctorum*, but no macrolichens which could be said to be characteristic for lowland rainforest were observed. However, in drier forest types at the same altitudes, in artificial habitats, and on rocks in streams and rivers, a number of species occurred which are not part of the rainforest flora. On riverine rocks were noted species of, for example, *Cladonia*, *Dirinaria*, *Hypotrachyna*, *Leptogium*, and *Parmotrema*, and on roadside trees *Bulbothrix*, *Leptogium*, *Physma*, and *Pyxine*. A dry, semi-evergreen forest at 400–450 m had few but interesting lichens, for example, *Leptogium austroamericanum*, *Phyllopsora martinii*, and an undescribed species of *Pyxine*.

In the West Usambara Mts *Lobaria holstiana* is locally common in the submontane rainforest, extending into the low montane forest to c. 2000 m. This conspicuous species was not found in any of the other mountain areas visited, and it may prove to be endemic to the Usambaras. Another rare species collected in the submontane forest was *Phyllopsora mauritiana*.

Low montane rainforests were studied between 1500 and 2000 m in the West Usambara Mts. In the Nguru and Kanga Mts I did not go much above 1200 m, but a few lichens were brought back by other expedition members from altitudes up to 1900 m, including mossy and elfin forest. In the Poroto Mountains I visited a very wet montane rainforest at 2000 m, and a short collecting trip was made to the ericaceous heath on the wet north slope of Rungwe Mountain.

Although some species are more or less widespread in most rainforests, there is considerable variation in species composition on mountains and ridges located in the same general area. This was the case with Gonja Mountain, Sagara Ridge, and Kwagoroto Summit in the West Usambara Mts, all within walking distance from Mazumbai.

Gonja Mountain is a mist mountain reaching 1700 m, probably subjected to permanent high humidity. Its summit is covered with evergreen microphyllous forest rich in bryophytes. On the west slope of the mountain one of the dominating lichens was *Physcidia wrightii*; the only East African records of this species to date consist of two small, sterile scraps from Tanzania. *Pseudocyphellaria argyracea* was extremely common, as were *Pseudoparmelia sphaerospora*, *Sphaerophorus melanocarpus*, and *Sticta papyracea*. Noteworthy, too, are at least nine different species of *Phyllopsora*. Present, but uncommon, were *Parmotrema hicksii* and *P. laciniatulum*. Previously known from a single locality in Tanzania, *Ramalina roesleri* was collected both on Gonja Mountain and in several other localities in the montane forest.

Sagara Ridge culminates at 1980 m on a summit with rocky outcrops, mossy elfin forest, and *Philippia* heath. Although insignificantly higher than Gonja Mountain, the open vegetation at the summit allows different elements of lichens to grow there. Worth mentioning are *Anzia afromontana*, *Cladonia diplotypa*, *Hypotrachyna leiophylla*, *Leprocaulon arbuscula*, *Leptogium furfuraceum*, *L. hibernicum*, *Menegazzia terebrata*, *Pannaria santessonii*, a number of *Parmotrema* species including *P. degelianum*, *P. fragilescens*, *P. hensseniae*, *P. hicksii*, and *P. laciniatulum*, five *Phyllopsora* species, *Relicina planiuscula*, *Siphula decumbens*, and a number of *Stictas* and *Usneas*.

Kwagoroto Summit, at about the same altitude as Sagara Ridge, likewise has *Philippia* heath at, or near, the top. Lichenologically, it has much in common with Sagara Ridge, but the following species may be noted: *Erioderma leylandii*, *E. sorediatum*, *Hypotrachyna pseudosinuosa*, *Pannaria fulvescens*, *P. lurida*, and *Pseudocyphellaria intricata*. Among the uncommon species also found on Sagara Ridge are *Pannaria santessonii*, *Parmotrema degelianum*, *P. fragilescens*, *P. hensseniae*, and *P. laciniatulum*.

In the Nguru Mts some specimens were collected by Edit Farkas in montane forest, mossy forest, and elfin forest between 1400 and 1900 m. They include *Bulbothrix meizospora*, *Cetrariastrum africanum*, *Parmotrema degelianum*, *P. hicksii*, *Ramalina pocsii*, *Recilina planiuscula*, *Siphula decumbens*, and *Usnea sorediosula*.

The wet montane rainforest at 2000 m in the Poroto Mts farther to the south had the following noteworthy species: *Anzia afromontana*, *Cetrariastrum vexans*, *Leptogium burnetiae*, *L. caespitosum*, *L. marginellum*, *L. phyllocarpum*, *L. sessile*, *Lobaria retigera*, *Menegazzia terebrata*, *Nephroma tropicum*, *Sticta ambavillaria*, and *S. weigelii*.

On the north slope of Rungwe Mountain the natural forest line, consisting of ericaceous heath, has been depressed to about 2400 m owing to a rainfall close to 4000 mm annually. Lichens encountered here include, among others, *Anzia afromontana*, *Cetrariastrum vexans*, *Cladonia diplotypa*, *Erioderma meiocarpum*, *Hypotrachyna ducalis*, *Leptogium adpressum*, *Parmeliopsis aleurites*, *Punctelia neutralis*, *Sticta ambavillaria*, *S. weigelii*, and *Usnea bicolorata*.

On the basis of the great variation in the lichen flora even between adjacent hills and ridges in the montane rainforest, one may presume that there are a great many species yet to be discovered there. However, the rainforest in Tanzania is a threatened vegetation type, as it is all over the world's tropics. Deforestation and erosion have destroyed numerous former natural forest habitats and agriculture encroaches steadily on forest reserves. There is reason to fear that a number of lichen species may already be in danger of disappearing, some of them probably before they have ever been known to science. Unfortunately, there is at present little evidence that this development will change in the foreseeable future.

## Acknowledgements

I am indebted to the Tanzania Commission for Science and Technology who granted permission to undertake research in Tanzania, and to Professor T. Pócs, Sokoine University of Agriculture, Morogoro, who acted as my local contact. I thank Professor T. Pócs and Ms E. Farkas for placing their collections of Tanzanian macrolichens at my disposal; Professor P.M. Jørgensen for nomenclatural advice on *Erioderma leylandii*; and Professor R. Santesson for the determination of *Siphula decumbens*. The photographs were taken by Mr P.E. Aas, the Natural History Museums, Oslo. Field work in Tanzania was supported by travel grants from the Norwegian Research Council for Science and the Humanities.

## References

Abbayes, H. des. (1948). Caracteres et affinites de la flore des *Cladonia* (lichens) de la region Malgache. *Mémoires de l'Institut Scientifique de Madagascar, Série B*, **1**, 57–63.

Abbayes, H. des. (1961). Lichens récoltés a Madagascar et a la Réunion (Mission H. des Abbayes, 1956). *Mémoires de l'Institut Scientifique de Madagascar, Série B*, **10**, 81–121.

Galloway, D.J. (1985). *Flora of New Zealand Lichens*. P.D. Hasselberg, Government Printer, Wellington.

Hale, M.E., Jr. (1975a). A revision of the lichen genus *Hypotrachyna* (Parmeliaceae) in tropical America. *Smithsonian Contributions to Botany*, **25**, 1–73.

Hale, M.E., Jr. (1975b). A monograph of the lichen genus *Relicina* (Parmeliaceae). *Smithsonian Contributions to Botany*, **26**, 1–32.

Lamb, I.M. (1977). A conspectus of the lichen genus *Stereocaulon* (Schreb.) Hoffm. *Journal of the Hattori Botanical Laboratory*, **43**, 191–355.

Mathey, A. (1974). Contribution à l'étude du genre *Siphula* (lichens) en Afrique. *Nova Hedwigia*, **22**, 795–878.

Østhagen, H. and Krog, H. (1976). Contribution to the lichen flora of the Canary Islands. *Norwegian Journal of Botany*, **23**, 221–42.

Pócs, T. (1976). Vegetation mapping in the Uluguru Mountains (Tanzania, East Africa). *Boissiera*, **24**, 477–98.

Polhill, D. (1988). *Flora of tropical East Africa. Index of collecting localities*. Royal Botanic Gardens, Kew.

Swinscow, T.D.V. and Krog, H. (1988). *Macrolichens of East Africa*. British Museum (Natural History), London.

Tavares, C.N. (1952). Contributions to the lichen flora of Macaronesia. I. Lichens from Madeira. *Portugaliae Acta Biologica (B)*, **3**, 308–91.

White, F.J. and James, P.W. (1985). A new guide to microchemical techniques for the identification of lichen substances. *British Lichen Society Bulletin*, **57** (suppl.), 1–41.

# 7. New and interesting records of Tanzanian foliicolous lichens

E. FARKAS

*Institute of Ecology and Botany, Hungarian Academy of Sciences, Vácrátót, H-2163, Hungary*

## Abstract

A total of 98 species of obligately foliicolous lichens is recorded to date from submontane and montane rainforests of the Usambara Mountains, Tanzania, due mainly to collections made during the Usambara Rain Forest Project expeditions. 35 species are new to the Usambaras including 15 species new also to Tanzania and 6 to Africa. Seven species are described by Farkas and Vězda as new to science: *Porina sphaerocephaloides* Farkas; *Macentina borhidii* Farkas and Vězda; *Dimerella flavicans* Vězda and Farkas; *D. pocsii* Vězda and Farkas; *D. tanzanica* Vězda and Farkas; *D. usambarensis* Vězda and Farkas, and *Byssoloma usambarense* Vězda. 36 species belong to Arthoniaceae, Opegraphaceae, and genera with pyrenocarpous ascomata. 62 species belong to genera with discocarpous fruiting bodies and *Lichenes imperfecti*. The most common species of the two main taxonomic groups mentioned above are *Porina epiphylla* Fée, and *Byssoloma leucoblepharum* Nyl. Most species are pantropical. Species recently described from the Usambaras probably have a wider distribution. Our results indicate that the foliicolous lichen flora of the tropical rainforests is very little known and emphasizes the necessity for further research.

## Introduction

The only area of Tanzania studied in detail for its foliicolous lichen flora is the submontane and montane rainforest belt of the Usambara Mountains (Farkas 1987*a*, *b*).

The first collections of foliicolous lichens from the Usambara mountains were published by Santesson (1952). In his world-wide monograph on obligately foliicolous lichens, 44 species were reported for the Usambara Mountains.

More than twenty years later Vězda (1975) published 43 species from the material of T. Pócs collected between 1969 and 1972. In this paper, 24 species were added to the 44 species of the Usambara mountains known previously from the Engler, Brunnthaler, and Hedberg collections. Two species (*Porina longispora*, *Bacidia subsimilis*) and two varieties (*Porina multipunctata* var. *schizospora*, *Porina papillifera* var. *rubrofusca*) were described from the Usambara mountains as new to science by Vězda (1975).

As a result of the Integrated Usambara Rain Forest Project expeditions between 1982 and 1986, 98 foliicolous lichen species were collected in the Usambara Mountains in collaboration with T. Pócs, A. Borhidi, E. Farkas, M. Hedrén, S. Iversen, I. Krisai, W. Mziray, M. Steiner, and R.P.C. Temu (Farkas 1988). The herbarium specimens are housed in Vácrátót (VBI), Uppsala (UPS), Morogoro (Herbarium of the Sokoine University of Agriculture) and Brno (Herbarium of A. Vězda).

## General characterization of the foliicolous lichen flora

Foliicolous lichens from the Usambara Mountains were collected in submontane and montane rainforest from living trees, shrubs, tree ferns, and ferns. In the forest reserve areas visited, 20–30 species were usually found. In the richest areas for foliicolous lichens there were about 40 species, for example, 44 species in Balangai East F. R., 40 species in Baga II F. R., and 32 species in Mazumbai University F. R.

Sometimes, similar species numbers were detected beyond the reserve areas, for example, at Kambi Falls, 37 species; at Gonja Hill, 31 species; and in submontane rainforests SW of Ambangulu Tea Estates, 31 species. It was suggested that these areas should be as forest reserves as well because they are not only rich in foliicolous lichens but are the habitats of rare species. In submontane rainforests SW of the Ambangulu Tea Estates a recently described species, *Byssoloma usambarense* (Vězda 1987), was collected for the second time. Among the species found in these rainforests one species, *Arthonia calamicola*, is new to Africa, two species, *Linhartia patellarioides* and *Echinoplaca intercedens*, are new to Tanzania and one, *Byssoloma tricholomum*, is new to the Usambara Mountains.

The common species (collected in 15 or more localities) of the Usambara Mountains (the number of localities in parentheses) are: *Byssoloma leucoblepharum* (34), *Porina epiphylla* (28), *Calopadia puiggarii* (20), *Bacidia dimerelloides* (19), *Porina sphaerocephaloides* (18), *Tricharia vainioi* (17), *Bacidia palmularis* (16), *Phylloporis phyllogena* (15), *Byssolecania fumosonigricans* (15), *Dimerella epiphylla* (15), and *Echinoplaca pellicula* (15).

## Geographical distribution

The majority of the common species are pantropical. Some of them also occur in subtropical or temperate regions, such as *Opegrapha filicina*, *Strigula elegans*, *S. nitidula*, *Gyalectidium caucasicum*, *G. filicinum*, *Byssoloma leucoblepharum*, *B. subdiscordans*, and *Fellhanera bouteillei*.

Some species (e.g. *Arthonia accolens*, *Porina subpilosa*, *Aulaxina submuralis*, and *Tricharia dilatata*) are known from tropical America and Africa, but they probably occur in other parts of the tropical world, too.

*Trichothelium alboatrum*, *Aulaxina epiphylla*, and *Tricharia vainioi* have been recorded from the palaeotropics and Australia. Data on *Arthonia calamicola* are insufficient to judge its world distribution, but it is so far known only from palaeotropical regions. *Porina epiphylloides* is known from Tanzania and Vietnam.

There are many new records in the foliicolous collections of the Usambara project expeditions. Thirty-five species are considered to be new to the flora of the Usambara Mountains, for example, *Chroodiscus mirificus*, *Aulaxina epiphylla*, and *Dimerella hypophylla*. Fifteen species are new to Tanzania and 6 to Africa. New species for Tanzania include *Arthonia cyanea*, the species of the *Opegrapha lambinonii* group (Sérusiaux 1985), *Raciborskiella prasina*, and *Strigula concreta*. New species for the continent are *Arthonia accolens*, *A. calamicola*, *Porina subpilosa*, *Aulaxina submuralis*, *Dimerella fallaciosa*, and *Fellhanera cateilea* (Vainio) Farkas **comb. nov.** [Basionym: *Pilocarpon cateileum* Vainio, *Univ. Calif. Pubs Bot.* **12**, 11 (1924).]

Species described recently from the Usambara Mountains are known to have a restricted distribution. *Porina sphaerocephaloides* is known from tropical Africa, *Dimerella tanzanica* is recorded from the Kilimanjaro, Ukaguru, Uluguru, and the Usambara Mountains, and Mafi Hill of Tanzania. *Macentina borhidii*, *Dimerella flavicans*, *D. pocsii*, *D. usambarensis*, and *Byssoloma usambarense* are recorded only from different localities of the Usambara Mountains (Farkas 1987a; Farkas and Vězda 1987; Vězda 1987; Vězda and Farkas 1988). Further research is necessary to establish whether they are really endemic taxa or not.

## Species recently described from the Usambara Mountains

Species described from the Usambara Mountains in 1987 and 1988 are

*Porina sphaerocephaloides*, *Macentina borhidii*, *Dimerella flavicans*, *D. pocsii*, *D. tanzanica*, *D. usambarensis*, and *Byssoloma usambarense*. They are presented in taxonomic sequence:

### Trichotheliaceae

Porina sphaerocephaloides *Farkas*   While *P. sphaerocephala* Vainio has radiate ridges on the thallus, this species is obviously verrucose. There are also large differences between the ascus and ascospore sizes of the two species. *P. sphaerocephala* has ascospores of $30-38 \times 3-4.5$ μm and asci of $50-60 \times 8-12$ μm (Santesson 1952), the corresponding data for *P. sphaerocephaloides* are $42-60 \times 6-9$ μm and $100-110 \times 10-20$ μm (Farkas 1987a). Described from Tanzania and the Republic of Guinea. Very common in the Usambara Mountains.

### Verrucariaceae

Macentina borhidii *Farkas and Vězda*   Specimens with isidia of $50-70$ μm in diameter are frequent in the Usambara Mountains, but the very small ascomata ($0.15-0.20$ mm wide) occurred in only one location, which allowed description of this species (Farkas and Vězda 1987). Asci are 8 spored. Spores are more or less elliptical, $18-23$ μm long, $4.5-5$ μm wide, 3-septate. It is thought to be a pioneer species on large leaves due to its rapid colonizing ability by special isidia.

### Gyalectaceae

Species of *Dimerella* usually have orange apothecia and are very similar to each other. Their ascospores are almost of the same size and shape, so that conidia (pycnospores) can serve to distinguish them (Fig. 7.1). With a little practice, sterile *Dimerella* thalli can be identified on the basis of the conidia.

Dimerella flavicans *Vězda and Farkas*   Described from the Usambara Mountains (Vězda and Farkas 1988). The apothecia are cadmium yellow, relatively large ($0.4-1$ mm in diameter), often found also at the edge of the leaves where the thallus turns to grow downwards. The conidia are 1-septate, relatively small ($3 \times 2$ μm).

Dimerella pocsii *Vězda and Farkas*   Known only from the Usambara Mountains (Vězda and Farkas 1988), its characteristic feature is the white, yellowish-white prothallus and the thallus growing over the leaf edge to the underside of the leaf. Apothecia are yellow to reddish-orange and are usually found at the edge of the leaf. Their size ($0.5-0.7(-1)$ mm in diameter) is more or less the same as that of *Dimerella flavicans*, but the conidia are simple, and much longer ($13-16 \times 1.2$ μm).

|  | Ascospores [μm] | 10 μm | Conidia [μm] | 10 μm |
|---|---|---|---|---|
| D. usambarensis Vězda & Farkas | 9–12×2.5–3 | 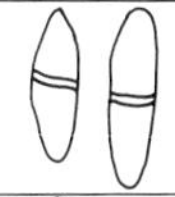  | 3×0.8 |   |
| D. flavicans Vězda & Farkas | 6.5–7×2 | 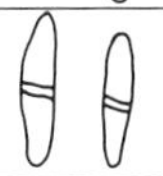  | 3×1 |   |
| D. tanzanica Vězda & Farkas | 11–13.5×2–2.5 | 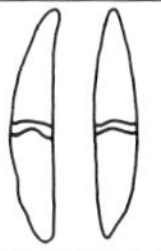  | 5–5.5×1–1.5 |   |
| D. pocsii Vězda & Farkas | 10–13×1.8–2 | 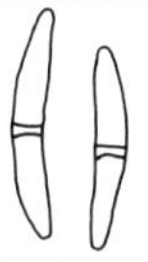  | 13–16×1.2 | 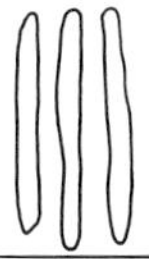  |

**Fig. 7.1.** Ascospores and conidia of some recently described species of *Dimerella*.

Dimerella tanzanica *Vězda and Farkas* Known from the Usambara, Uluguru, Ukaguru, and the Kilimanjaro Mountains of Tanzania (Vězda and Farkas 1988), often on *Marattia* leaves. Apothecia are frequent, yellow, orange-red, sometimes reddish-brown, 0.25–0.35 mm in diameter. Conidia are simple, 5–5.5 × 1–1.5 μm.

Dimerella usambarensis *Vězda and Farkas* Described from the Usambara Mountains (Vězda and Farkas 1988). The colour of the apothecia is variable, from orange-brown to reddish-brown, various shades of orange, and brown. Its size is small (0.2–0.5 mm). The conidia are very slim (3 × 0.8 μm). This species is most similar to *D. epiphylla*, which has much bigger, one-septate conidia (14–18 × 1.5–2.5 μm).

*Pilocarpaceae*
Byssoloma usambarense *Vězda*  B. *usambarense* is most similar to *Bacidia palmularis*. The most important differences are the pale coffee-brown, 0.45–0.8 mm apothecia sitting on the epiphyllous thallus and the 7-septate, 23–40 × 2.5–3.5 μm ascospores. Described from Mazumbai University Forest Reserve of West Usambara Mountains (Vězda 1987). It seems to be very rare even in the Usambara Mountains. A second specimen was collected in rainforest near the Ambangulu Tea Estates.

## Some interesting records

*Gomphillaceae*

The foliicolous lichen flora of the Usambara Mountains is rich in thalli with hyphophores. Hyphophores characterize the Gomphillaceae (Vězda and Poelt 1987). These organs are of assistance in the delimitation of genera. Hyphophores of *Actinoplaca*, *Bullatina*, *Calenia*, *Echinoplaca*, *Gyalectidium*, *Gyalideopsis*, and *Tricharia* develop on the alga-bearing thallus, whereas those of *Aulaxina* and *Caleniopsis* grow on the prothallus.

Bacidia apiahica *(Müll. Arg.) Zahlbr.* A pantropical species. Frequent in the Usambara Mountains. Very similar to *B. scutellifera*, but it has no isidia.

Bacidia scutellifera *Vězda* Probably a pantropical species. Frequent in the Usambara Mountains. An excellent example of a special type of isidium is that of *Bacidia scutellifera* described by Vězda (1975) from Tanzania. Even sterile thalli of this species, often found in our material, can be identified on the basis of the isidia as no other species is known with the same type of isidium. Its shape is similar to that of a drawing-pin whose peak is held by a small cup on the surface of the thallus. On touching it the isidium springs out of the cup and its top falls on to the surface of the leaf where it can produce a new thallus (Fig. 7.2).

*Ectolechiaceae*

In our material we have often found thalli with campylidia. Campylidia are known in all genera of the Ectolechiaceae as recognized by Vězda (1986). This family is divided into genera on the basis of numerous features, for example, ascospore septation, the nature of the paraphyses and asci, or the colour of the apothecia. In addition to these classically-used features, the type of campylidium-produced conidia also characterizes each of the genera. *Barubria*, *Loflammia*, *Logilvia* (in part), and *Sporopodium* have one-celled more-or-less ellipsoidal or clavate conidia. *Badimia*, *Calopadia*, and *Tapellaria* have long, transversely-septate conidia curved in a different way. Those of *Lasioloma* are branched and septate (Fig. 7.3).

Leptopelis uluguruensis *Barbour and Loveridge* *Leptopelis uluguruensis* is not a foliicolous lichen, a bryophyte, nor a fungus but an endemic tree frog species of Tanzania (Fig. 7.4). Why do we mention it among our foliicolous lichen records? Collecting foliicolous lichens and liverworts in the East Usambara Mountains we found this tree frog which looked

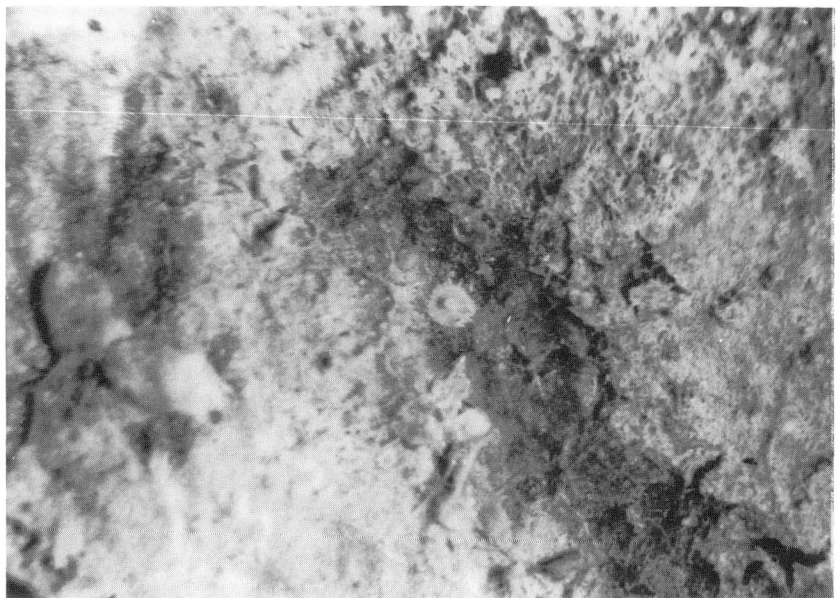

**Fig. 7.2.** An isidium of *Bacidia scutellifera* just beginning its development.

as if its back was covered in foliicolous lichen thalli (Farkas and Pócs 1989). There were, in fact, no lichens on its skin, but the tree frog seemed to be mimicking them. We could isolate neither ascospores, nor conidia, so it was not possible to identify any species of lichens on it. Nevertheless, to us it meant, perhaps, the most interesting record of the Tanzanian foliicolous lichens.

## Summary

The relatively short life-cycle of foliicolous lichens growing on this special, short-lived substrate increases the speed of evolutionary processes through the quick succession of generations. This is the theoretical importance of research into foliicolous lichens. Phylogenetic questions of general lichenology may be solved studying foliicolous communities (Vězda personal communication). For the sake of the continuation of studies like ours, one should never forget that, primarily, the natural state of the rainforests must be maintained.

Our systematic and floristic results have revealed the luxuriance and scientific value of the investigated area. They indicate that the foliicolous lichen flora of tropical rainforests is still relatively unknown and

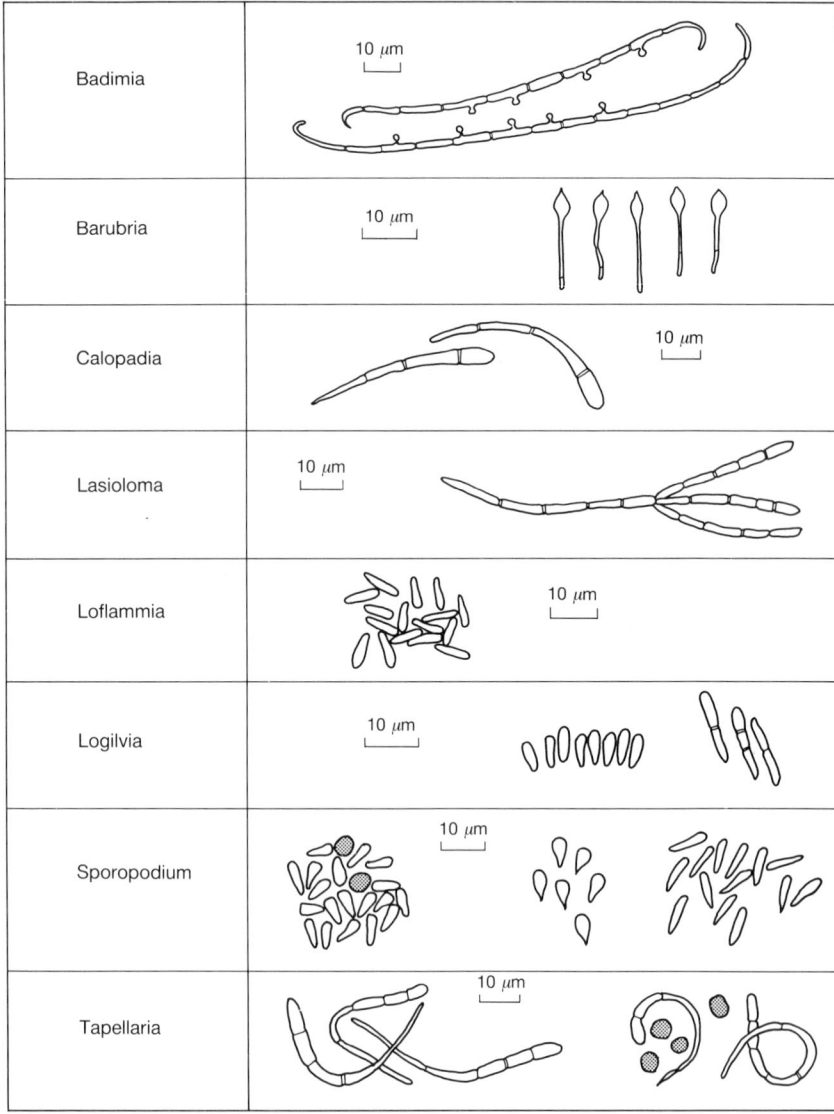

**Fig. 7.3.** Conidia of genera from the family Ectolechiaceae (based on Vězda (1986)).

emphasize the necessity of further research. Not only species new to different areas, but several species new to science were recognized recently, as a result of the Usambara Rain Forest Project expeditions.

# New and interesting records of Tanzanian foliicolous lichens 103

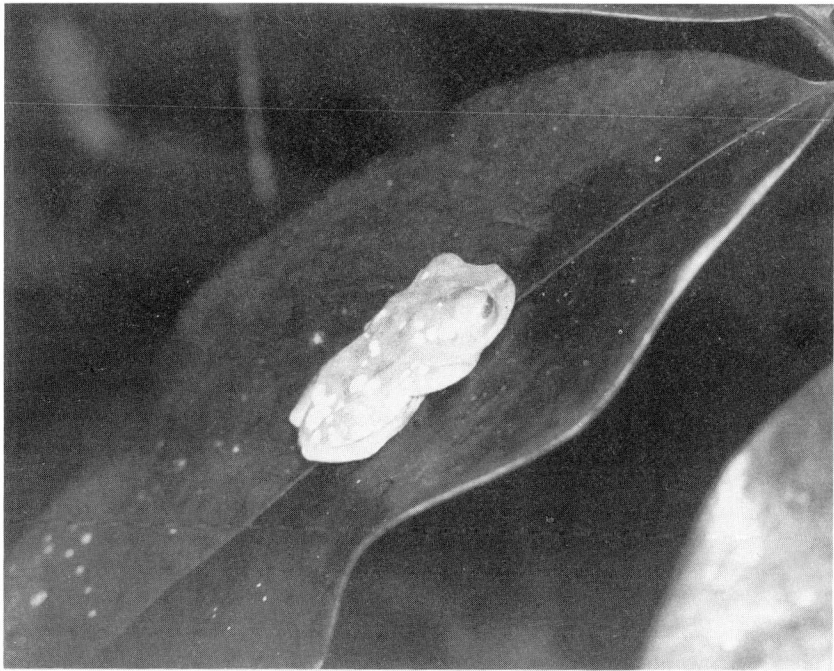

**Fig. 7.4.** *Leptopelis ulugurensis* with foliicolous lichen-like patches (photograph by T. Pócs).

In March 1989 work was continued in the Nguru Mountains of Tanzania. This resulted in rich and promising collections from rainforests where no, or only small collections were made previously. For a thousand years, the superstitions, beliefs, and traditions of the black people prevented both damage to and scientific recognition of the natural vegetation; for example, on the Kanga mountains. Analysis of collections from this area will probably increase our knowledge of the foliicolous lichen flora of Tanzania in the near future.

Sensitive bioindicators direct our attention to potentially harmful environmental changes before they are dangerous for human life. Since foliicolous lichens are also excellent indicators of environmental pollution they will vanish before their discovery and description, if we do not take good care to preserve their habitats.

## Acknowledgements

I would like to express my gratitude to Professors A. Borhidi (Vácrátót/

Pécs), D.L. Hawksworth (Kew), Hildur Krog (Oslo), T. Pócs (Vácrátót/ Morogoro), Dr A. Vězda (Průhonice/Brno) and L. Lökös (Budapest) for their comments on the manuscript.

## References

Farkas, E. (1987a). Foliicolous lichens of the Usambara Mountains, Tanzania, I. *Lichenologist*, **19**, 43–59.

Farkas, E. (1987b). Importance of the asexual reproductive organs in foliicolous lichen taxonomy. In *Abstracts of the general lectures, symposium papers and posters* (ed. W. Greuter, B. Zimmer, and H.-D. Behnke), XIV International Botanical Congress. pp. 252. Berlin.

Farkas, E. (1988). Foliicolous lichens; Checklist: Obligately foliicolous lichens of the Usambara Mountains. In *The SAREC supported Integrated Usambara Rain Forest Project Tanzania. Report for the period 1983–1987.* (Introduction written by I. Hedberg and O. Hedberg) p. 30; Appendix 3. (5 unnumbered pages) Department of Systematic Botany, Uppsala.

Farkas, E. and Pócs, T. (1989). Foliicolous lichen mimicry of a rainforest tree frog? *Acta Botanica Hungarica*, **35**, 73–6.

Farkas, E. and Vězda, A. (1987). *Macentina borhidii*, eine neue foliicole Flechte aus Tansania. *Acta Botanica Hungarica*, **33**, 295–300.

Santesson, R. (1952). Foliicolous lichens. I. A revision of the taxonomy of the obligately foliicolous lichenized fungi. *Symbolae Botanicae Upsalienses*, **12**(1), 1–599.

Sérusiaux, E. (1985). Goniocysts, goniocystangia and *Opegrapha lambinonii* and related species. *Lichenologist*, **17**, 1–25.

Vainio, E.A. (1924). Lichenes a W.E. Setchell et H.E. Parks in insula Tahiti a 1922 collecti. *University of California Publications in Botany*, **12**. Berkeley, California, 1924–31.

Vězda, A. (1975). Foliikole Flechten aus Tanzania (Ost-Afrika). *Folia geobotanica et phytotaxonomica, Praha*, **10**, 383–432.

Vězda, A. (1986). New genera of the family Lecideaceae s. lat. (Lichenes). *Folia geobotanica et phytotaxonomica, Praha*, **21**, 199–219.

Vězda, A. (1987). Foliicole Flechten aus Zaire. (III). Die Gattung *Byssoloma* Trevisan. *Folia geobotanica et phytotaxonomica, Praha*, **22**, 71–83.

Vězda, A. and Farkas, E. (1988). Neue foliicole Arten der Flechtengattung *Dimerella* Trevisan (Gyalectaceae) aus Tansania. *Folia geobotanica et phytotaxonomica, Praha*, **23**, 183–97.

Vězda, A. and Poelt, J. (1987). Flechtensystematische Studien. XII. Die Familie Gomphillaceae und ihre Gliederung. *Folia geobotanica et phytotaxonomica, Praha*, **22**, 179–98.

# 8. Evolutionary rates in the Teloschistaceae

I. KÄRNEFELT

*Department of Systematic Botany, University of Lund, Östra Vallgatan 18–20, S-223 61 Lund, Sweden*

### Abstract

The Teloschistaceae is generally recognized as a rather natural unit in the lichenized Ascomycotina. However, the same family has long been known to include an assemblage of rather unnatural genera and, in addition, there are numerous problems concerning delimitation of species among those genera. Problems related particularly to gradual geographical speciation are discussed in three of the most important genera in the Teloschistaceae and illustrated mainly from the African flora in the genera *Xanthoria*, *Teloschistes*, and *Caloplaca*. In one species in particular, *Xanthoria mendozae*, the known distributional pattern in the Southern Hemisphere supported by geological events and with only few opportunities for long distance dispersal, indicates an ancient history for this lichen. Two cladograms or strict consensus trees are presented, supporting the presumed unnatural affinities among accepted genera in the family.

### Introduction

Compared to other plant groups, not much attention has been paid to the evolutionary process and its impact on speciation or the emergence of higher taxonomic categories among the lichenized Ascomycotina. It is natural because of the difficulties presently encountered in carrying out genetical experimental work on lichens which could form a basis for evolutionary theories. The mechanism of the evolutionary process is occasionally interpreted differently (Lövtrup 1987; Stebbins 1987) but many circumstances argue for the neo-Darwinian theory to be the most likely, i.e. the evolutionary process is viewed as a continuous process and divergence between related evolutionary lines goes through a series of stages with time (Grant 1977; Futuyma 1986; Scharloo 1987). Concerning structural changes at the population level, this process is recognized as geographical or allopatric speciation, i.e. an ancestral population gives rise to two or more local races, geographical races, species, and species groups (Fig. 8.1) (Dillon 1978; Futuyma 1986). Important changes from lower to higher levels of branching related to speciation mainly concern increasing differentiation of the genotype, increasing differentiation in structural characters, and ecological behaviour followed by increasing isolation (Grant 1977). Divergence can then continue to higher levels such as genus and family which mainly concern the process of macroevolution and which is an entirely different historical process, related to the geological time-scale (Fig. 8.1).

Since this process is presumably continuous, it should be possible to observe the process at any stage and not only on fixed points or categories such as species and genus. Why do most of us recognize only species or genera in our efforts to describe the evolutionary process? The minor chemical and structural differences recognized within populations of many different lichen groups may not always necessarily have involved the evolution of species and genera. Leuckert and Poelt

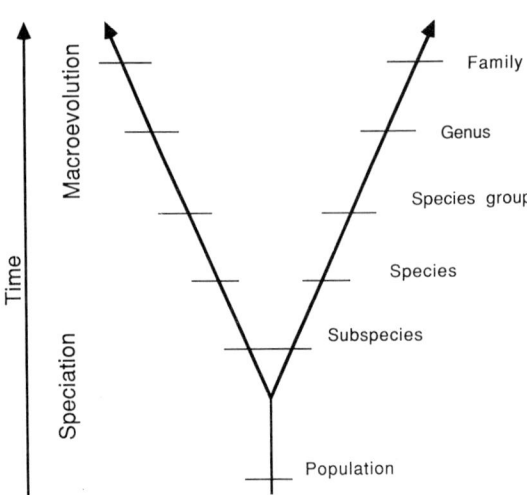

**Fig. 8.1.** Amount of differentiation in geographical speciation and macroevolution. Adapted from Dillon (1978) and Grant (1977).

(1989) in the *Lecanora rupicola* group and Sipman (1983) in the Megalosporaceae take a different view from Elix *et al.* (1986) and Hale (1986, 1989) in the genus *Xanthoparmelia* from southern Africa and Australasia, and instead choose to distinguish minor chemical and structural differences in subspecific entities.

The evidence of geographical speciation is extensive and recognized mainly from studies in geographical variation of structural and chemical characteristics in many groups among the lichenized Ascomycotina. There is nothing which actually confirms a rapid mode of speciation among lichenized fungi. On the contrary, many biogeographical patterns from the various groups studied infer extremely slow evolutionary rates and many recent entities may have remained unchanged through time and space over millions of years (Arvidsson 1983; Galloway 1988*a*; Hawksworth 1982; Sipman 1983; Tehler 1983).

## Problems related to speciation

There are numerous problems regarding difficulty-delimited species from all major groups in the Teloschistaceae and it is naturally also difficult to select a few representative examples. However, as examples from the subtropical and tropical regions, mainly related to the African flora, but also concerning closely-related entities in other floristic regions and problems related to generic interrelationship, I will discuss briefly a few cases in the genera *Xanthoria*, *Teloschistes*, and *Caloplaca*.

### 1. Xanthoria

The foliose genus *Xanthoria* presumably belongs to the most advanced group in the family. Many of the species which are widespread in the Northern Hemisphere, such as *X. candelaria*, *X. elegans*, *X. fallax*, *X. parietina*, and *X. polycarpa*, are apparently structurally unchanged and occur on the African continent south of the Sahara. Of special interest are a few species with very limited ranges. The isolated East African species *X. africana* is thoroughly described by Almborn (1963), and Swinscow and Krog (1988), and will not be discussed further here. It is still, however, remarkable that a species which mainly differs from *X. parietina* in the presence of marginal soredia, is known only at a few isolated localities at higher elevations in Kenya and Uganda. Sorediate taxa, in general, appear to be much more widely distributed.

*Xanthoria capensis* Almb. ined., which is characterized by rather narrow, convex, bright yellow lobes having small pseudocyphellae, and by the relatively thick upper cortical layer (Figs 8.2, 8.3) occurs locally in abundance on the shores of southernmost Africa. The lobes are occasionally provided with small marginal lobules, which is a

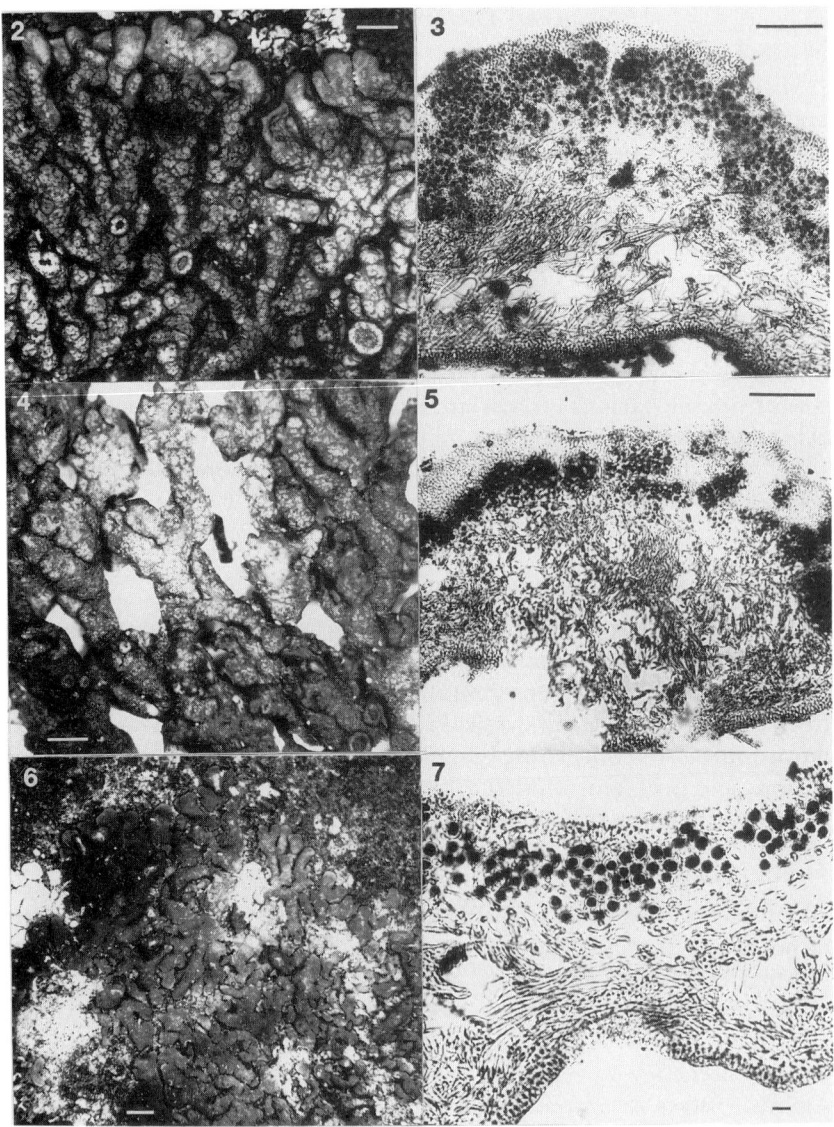

**Figs 8.2–7.** Morphology and anatomy in species aggregates in the genus *Xanthoria*. (Fig. 8.2) *X. capensis*, structure of thallus showing pseudocyphellae, *Almborn and Kärnefelt* 8409-23 (LD). (Fig. 8.3) cross-section of thallus of *X. capensis*, *Almborn and Kärnefelt* 8409-22 (LD). (Fig. 8.4) *X. mandschurica*, thallus portion, 1928 *Asahina* 554, isotype (LD). (Fig. 8.5) cross-section of thallus of *X. mandschurica*, *Asahina* 554 (LD). (Fig. 8.6) *X. marlothii*, thallus portion, *Kärnefelt* 8610-30 (LD). (Fig. 8.7) cross-section of thallus of *X. marlothii*, *Kärnefelt* 8605-45 (LD). Bar in Figs 8.2, 4, and 6 = 1 mm, in Figs 8.3 and 8.5 = 100 μm, and in Fig. 8.7 = 10 μm.

characteristic of *X. ligulata* from Australasia (Galloway 1985). On sea shores of East-Asia a similar species, characterized by the same type of pseudocyphellae in addition to thick cortical layers (Figs 8.4, 8.5), was separated as *X. mandschurica* (Asahina 1954). Because of the relatively weak structural differences it would presumably be more suitable to treat these entities as infraspecific taxa. Since there are no known localities between these widely separated populations, how is this remarkably disjunct range explained? The ancestral species presumably had a more continuous range which, during the course of time, must have been separated in now slightly morphological different populations. Biogeographical patterns involving entities occupying areas in southern Africa are normally more related to other regions in the Southern Hemisphere (Galloway 1988*b*). Furthermore, affinities with the Australasian *X. ligulata*, also characterized by rather convex lobes with scattered marginal lobules and growing on maritime rocks, are not clear.

*Xanthoria marlothii* is another locally separated taxon, known today mainly from Namibia, but described on material from Namaqualand in the Cape Province (Zahlbruckner 1926). This species is characterized by rather narrow, deeply pigmented lobes with relatively thin cortical layers (Figs 8.6, 8.7). It is presumably a locally developed form within the *X. parietina* group, but with more firmly attached lobes adapted to semi-desert habitats. I think that as this taxon is not yet sufficiently genotypically isolated from *X. parietina* to be regarded as an independent species, an infraspecific status would be more appropriate for it.

A remarkable species, *X. mendozae*, occurring at high elevations in the Drakensberg mountains of southern Africa is characterized by densely aggregated, subumbilicate lobe portions which are sorediate on their lower surface. However, *X. mendozae* was originally described from higher elevations in western Argentina (Räsänen 1939) and since then has been discovered in the Andean regions of Peru. A remarkable fact about a few other major disjunct populations in southern Africa and in western South America is that they seem to have remained virtually unchanged through time and space (Kärnefelt 1990). With very few possibilities for dispersal and new establishment in the specialised, high altitudes where *X. mendozae* occurs today, in my opinion, the most likely explanation must be that existing disjunct populations are remnants of an ancient Cretaceous Southern Hemisphere range. A consequence of this assumption is that we must also estimate the age of this lichen to be extremely great. A plausible, but in my opinion, less likely explanation is that these populations have developed along parallel lines of evolution.

Another difficultly-delimited species, known from shrubs along the coast of south-west Africa, has been recognized as *X. turbinata* (Vainio 1900). It is characterized mainly on the turbinate form of the apothecia, in addition to rather loosely attached lobe portions, and on the relatively thin cortical layers (Figs 8.8, 8.9). Certain forms of this species in a way approach the monotypic genus *Xanthodactylon*, which is endemic to roughly the same area in south-west Africa where it occurs in similar types of habitats on shrubs along the sea shores. At least in mature stages, *X. flammeum* is characterized by pulvinate lobes where the lower cortical layer has become reduced (Figs 8.10, 8.11).

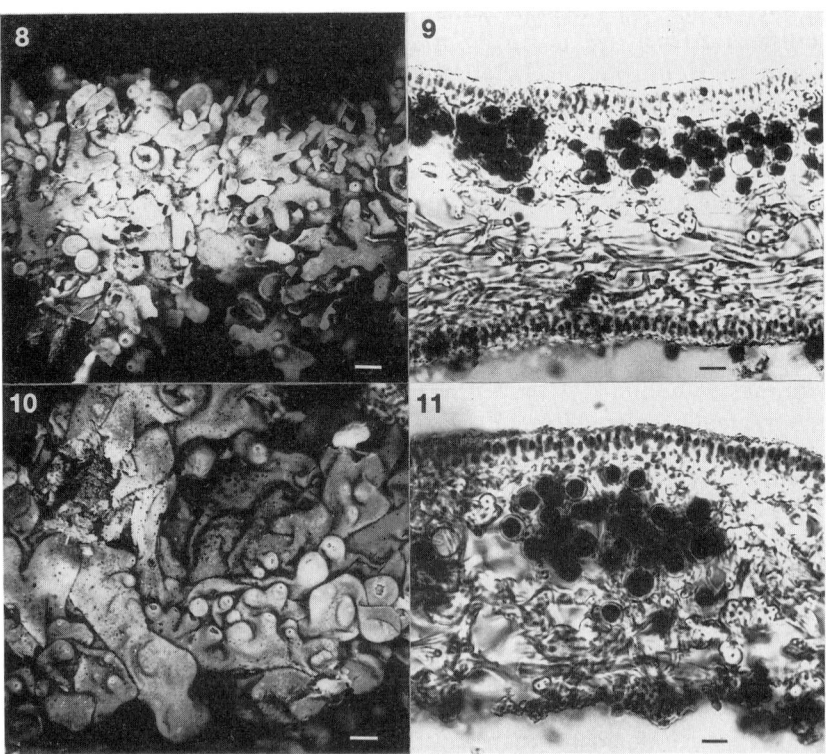

**Figs 8.8–11.** Morphology and anatomy in *Xanthoria turbinata* and in the genus *Xanthodactylon*. (Fig. 8.8) *X. turbinata*, thallus with foliose lobe portions and turbinate apothecia, *Almborn and Kärnefelt* 8412-22 (LD). (Fig. 8.9) cross-section of thallus of *X. turbinata*, *Almborn and Kärnefelt* 8412-22 (LD). (Fig. 8.10) *Xanthodactylon flammeum*, thallus with depressed pulvinate lobe portions and apothecia, *Almborn and Kärnefelt* 8412-13 (LD). (Fig. 8.11) cross-section of thallus of *X. flammeum* showing reduced lower cortex, *Almborn and Kärnefelt* 8412-12 (LD). Bar in Figs 8.8 and 8.10 = 1 mm, in Figs 8.9 and 8.11 = 10 μm.

## 2. Teloschistes

Within the genus *Teloschistes* some major problems in the African lichen flora concern affinities between the obviously closely-related species, *T. capensis*, *T. perrugosus*, and *T. puber*. *Teloschistes capensis*, which is endemic to south-west Africa, is basically characterized by rather erect, densely branched, terete or slightly flattened, slightly tomentose, orange-yellowish lobes with prosoplectenchymatous cortical layers composed of conglutinated periclinal hyphae covered also by small hyaline hairs (Figs 8.12, 8.14). Occasionally the lobes can be much less pigmented by anthraquinones. This species is well-known from extensive terricolous populations in the central Namib desert, but further south within the distributional range it occurs mainly on branches of shrubs. The apothecia, which are frequently developed in large numbers in corticolous populations, are usually provided with rather long fibrils (Figs 8.16, 8.17) similar to the apothecia of *T. puber*. *Teloschistes puber*, which is also endemic to south-west Africa, is characterized mainly by the more dorsiventral lobes which are partly ecorticate on the lower side (Fig. 8.18). *Teloschistes puber* and *T. capensis* otherwise share common features of typical, longish fibrils, the usually densely tomentose lobes and prosoplectenchymatous cortical layers covered with small hyaline hairs (Fig. 8.19). In contrast to *T. capensis*, however, *T. puber* only rarely occurs on the ground (Almborn 1989).

The third species in this aggregate, *Teloschistes perrugosus*, is distributed along the southern coast and in eastern Africa, and replaces *T. capensis* geographically. It differs mainly from *T. capensis* in the slightly broader lobes and, in addition, the scattered apothecia usually lack prominent fibrils (Figs 8.20, 8.21).

I see in this species complex an evolutionary pattern related to the development of ciliate apothecia, the structure and form of the lobes including the presence of tiny hairs. The limits between the various taxa is far from clear. From the evolutionary point of view an infra-specific arrangement of these entities would perhaps be much more appropriate.

This complex might appear even more complicated if we also consider a South American taxon, *T. peruensis*, which is known mainly from the coastal deserts of Peru (Thomson and Iltis 1968). This desert species does not differ much from locally sterile forms of *T. capensis* from the Namib desert. *Teloschistes peruensis* is characterized partly by the larger fissures on the lobes which can also be found on *T. capensis* (Fig. 8.13). The slightly tomentose lobes and the, occasionally, almost unpigmented, mainly sterile lobes are rather similar in both these species. The larger fissures could in fact be environmentally controlled, since

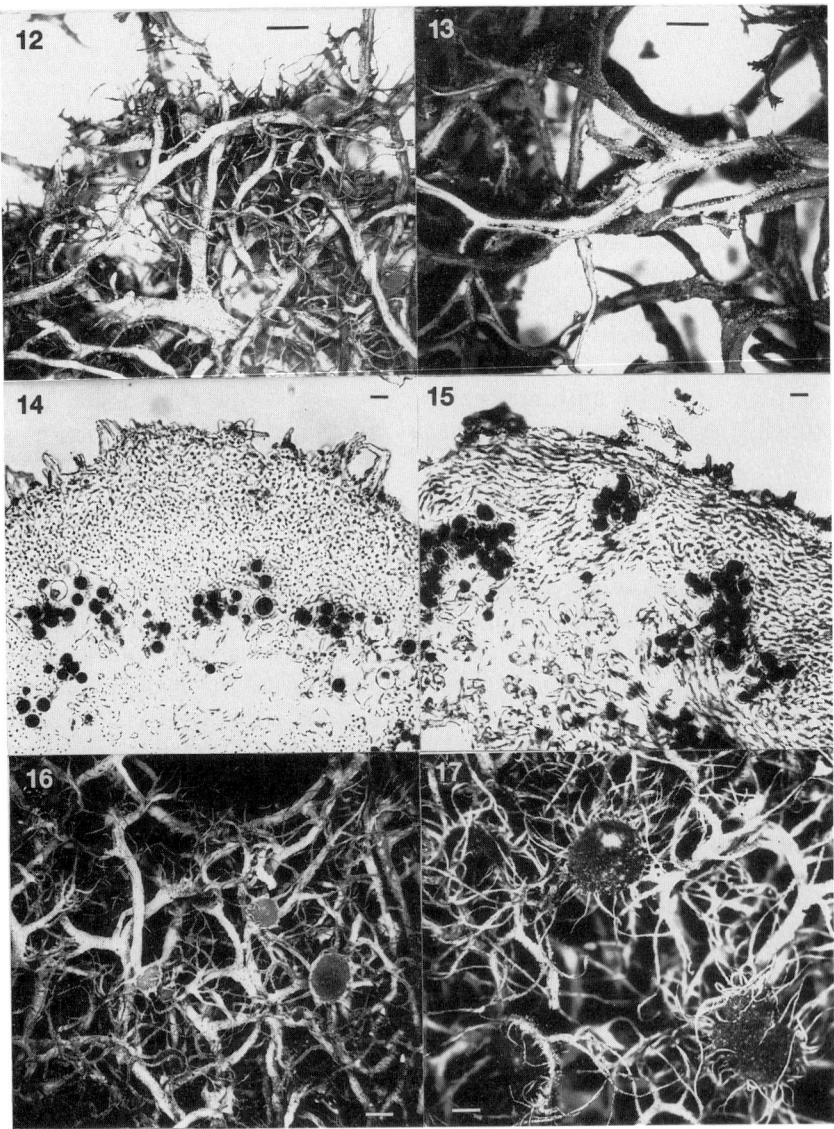

**Figs 8.12–17.** Morphology and anatomy in species aggregates in the genus *Teloschistes*. (Fig. 8.12) *T. capensis*, thallus portion showing scattered fibrils and also smaller fissures in the cortex of lobes, *Vězda*, Lich. sel. exs. 1274 (LD). (Fig. 8.13) *T. peruensis*, thallus portion showing lobes with larger fissures, *Domley* 5 (LD). (Fig. 8.14) cross-section of thallus of *T. capensis* showing hyaline hairs, *Kärnefelt* 8607-1 (LD). (Fig. 8.15) cross-section of thallus of *T. peruensis* showing hyaline hairs, *Rose* 12473 (LD). (Fig. 8.16) fertile thallus portion of *T. capensis*, *Almborn*, Lich. Afr. 72 (LD). (Fig. 8.17) richly ciliate apothecia in *T. capensis*, 1946 *Leighton* 1724 (LD). Bar in Figs 8.12, 13, 16, and 17 = 1 mm, in Figs 8.14 and 8.15 = 10μm.

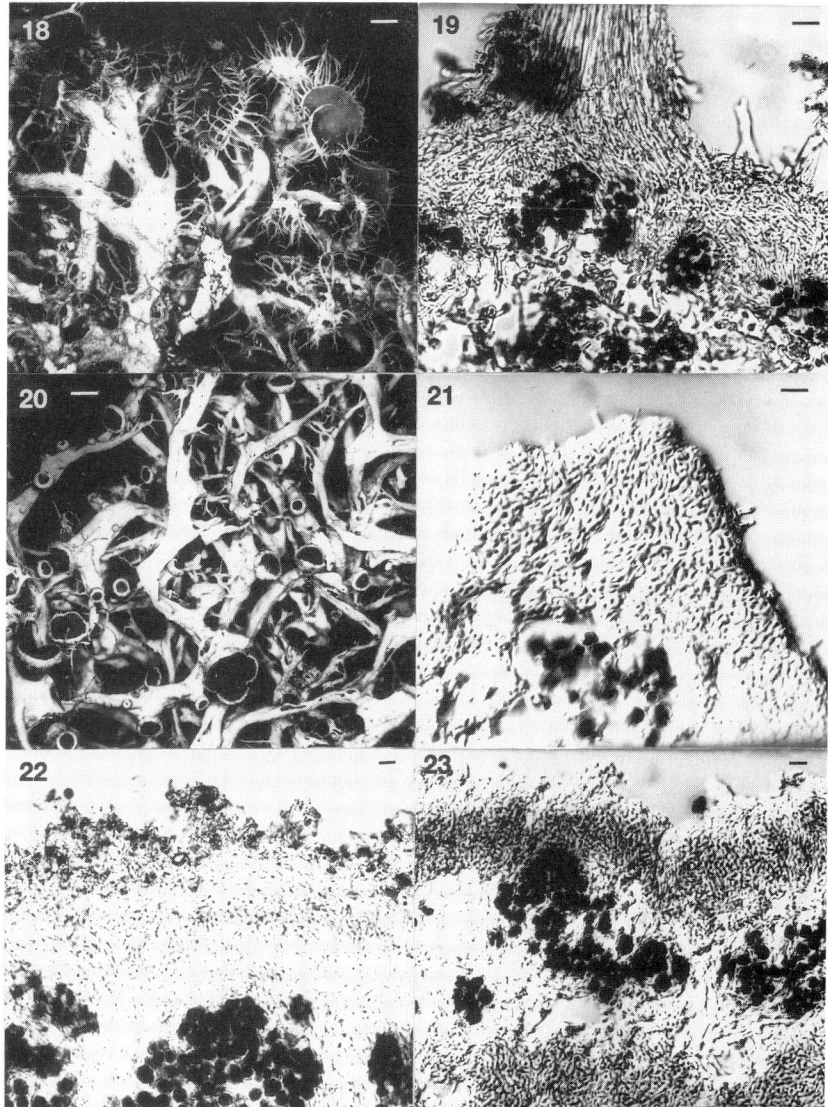

**Figs 8.18–23.** Morphology and anatomy in species aggregates in the genus *Teloschistes* and species related to the subfruticose *Caloplacae*. (Fig. 8.18) *T. puber*, thallus portion showing more dorsiventral lobes and cilate apothecia, 1941 *Garside* 5040 a (LD). (Fig. 8.19) cross-section of thallus of *T. puber* showing cortex with basal portion of a fibril and hyaline hairs. (Fig. 8.20) *T. perrugosus*, fertile thallus portions showing scattered fibrilles and mainly non-ciliate apothecia, 1962 *Kofler* 11 a (LD). (Fig. 8.21) cross-section of thallus of *T. perrugosus* showing scattered hyaline hairs, *Kofler* 11 a (LD). (Fig. 8.22) cross-section of thallus of *T. scorigenus* showing massive cortical layers, *Follmann*, Lich. exs. sel. 77 (LD). (Fig. 8.23) cross-section of thallus of *T. chrysocarpoides* showing rather massive cortical layers, *Mattick* 68-9 (B). Bar in Figs 8.18 and 8.20 = 1 mm, in Figs 8.19, 21, 22, and 23 = 10 μm.

these ecorticate portions would allow the absorption of the limited amount of water more easily. The cortical layers are characterized by the same kind of conglutinated, periclinal hyphae covered by small hyaline hairs as in *T. capensis* (Fig. 8.15).

The Peruvian deserts are similar in many ways to the coastal deserts of Namibia, where lichen vegetation is locally rather luxuriant, entirely dependent on fog from the sea. The widely geographically separated populations of Namibia and Peru are naturally difficult to explain if we claim that they actually represent the same taxon. However, there are at least 6 other lichen species known, which show this South American–African disjunction (Kärnefelt 1990).

## 3. Caloplaca

In the large genus *Caloplaca* there are many difficult species aggregates in the African flora, but in this connection I will choose only one example from the subfruticose group of species, to illustrate the problem of gradual speciation and intrageneric affinities. The subfruticose species group within *Caloplaca* was earlier investigated by Poelt and Pelleter (1984) who recognized three species: *C. bonae-spei*, *C. eudoxa*, and *C. mauritanica* from the African continent. *Caloplaca bonae-spei*, restricted to the south-west region at the Cape of Good Hope, is mainly characterized by the yellowish to orange, more or less dorsiventral subfruticose lobe portions with scattered pseudocyphellae (Fig. 8.24), and anatomical layers composed of rather massive scleroplectenchymatous hyphae with embedded crystals and algae (Fig. 8.25). *Caloplaca eudoxa*, distributed further north and known mainly from the Namib coastal desert, where it grows firmly attached to scattered rocks, mainly differs from *C. bonae-spei* in the pigmentation of the similarly dorsiventral lobes. The lobes appear to be partly more grayish and almost unpigmented, or locally also more reddish or dark orange. There are otherwise no clear structural or anatomical differences between these two species (Figs 8.26, 8.27, 8.28). A similar variation in pigmentation is observed in several other entities within the Teloschistaceae occurring in this part of Africa such as in the earlier-discussed species in the genus *Teloschistes*, where at least *T. capensis* and *T. puber* generally appear to be either much more deeply orange-pigmented or partly unpigmented in the northern part of the distributional area than southern populations which are generally characterized by more bright-orange pigmented individuals. It is difficult to establish whether this variation in pigmentation is entirely environmentally related or genetically determined, and whether the variation in these two species of *Teloschistes* is of the same nature as in the subfruticose species of *Caloplaca* discussed. Presumably we are observing a separation of allopatric

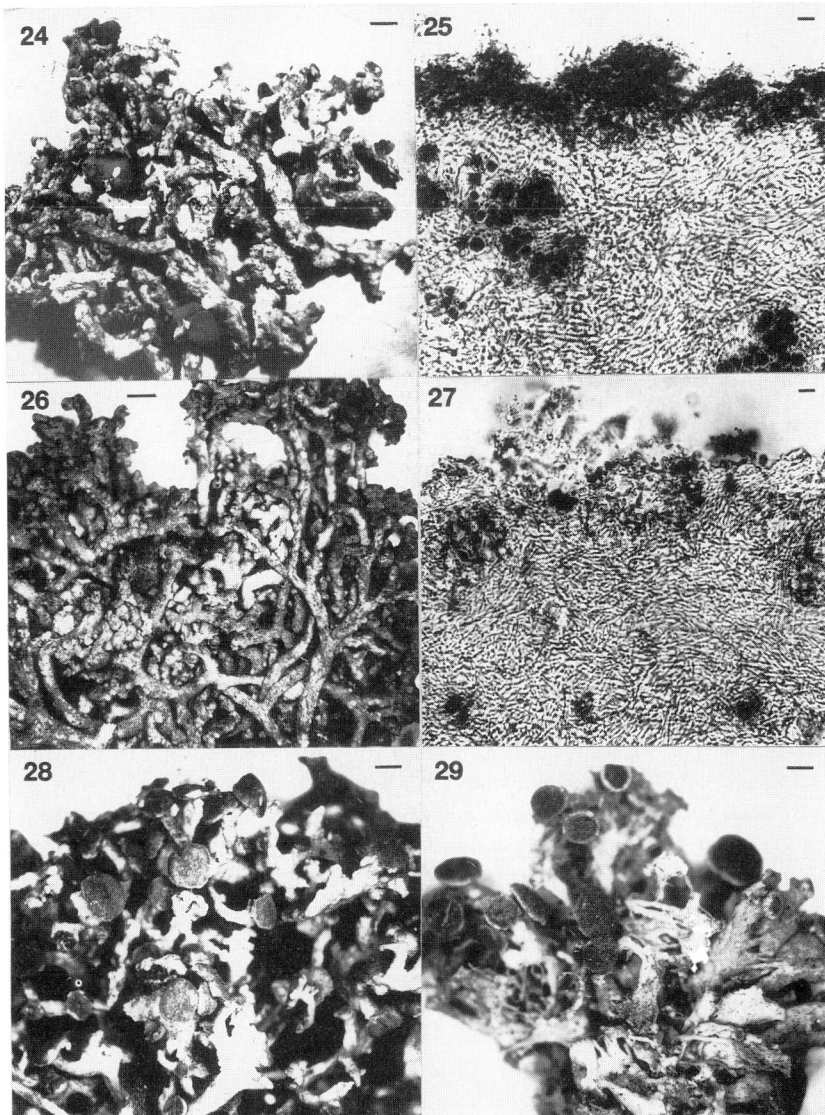

**Figs 8.24–29.** Morphology and anatomy of subfruticose species of *Caloplaca* and related species of *Teloschistes*. (Fig. 8.24) *C. bonae-spei*, thallus portion, *Almborn and Kärnefelt* 8410-07 (LD). (Fig. 8.25) cross-section of thallus of *C. bonae-spei* showing a massive scleroplectenchymatous cortical layer, *Almborn and Kärnefelt* 8414-01 (LD). (Fig. 8.26) *C. eudoxa*, thallus portion, *Almborn*, Lich. Afr. 74 (LD). (Fig. 8.27) cross-section of thallus of *C. eudoxa* showing a massive scleroplectenchymatous cortical layer, *Füncke* 6736 (W). (Fig. 8.28) *C. eudoxa*, thallus portion similar to *T. chrysocarpoides* in Fig. 8.29, *Füncke* 6736 (W) and *Mattick* 68–9 (B). Bar in Figs 8.24, 26, 28, and 29 = 1 mm, in Figs 8.25 and 8.27 = 10 μm.

populations and the evolution of species in this aggregate of subfruticose *Caloplacae*, but in my opinion it is far from clear that two species have actually evolved at the present time.

An interesting link from the same geographical region in south-west Africa towards these subfruticose species, and especially *C. eudoxa*, might exist in *Teloschistes chrysocarpoides*, characterized by brownish-red apothecia, rather narrow and usually grayish lobes and by rather massive prosoplectenchymatous cortical layers overlying the algae (Figs 8.23, 8.29). Furthermore, to the group of subfruticose *Caloplacae* the connection to another species of *Teloschistes*, *T. scorigenus* known from the Canary Islands is also extremely interesting (Kärnefelt 1989). This species is characterized by anatomical layers which, in many ways, reminds one of the subfruticose species of *Caloplaca* (Fig. 8.22).

## Evolution of higher levels in the Teloschistales

Macroevolution involves changes of greater proportions than we can observe in speciation. These changes concern characteristics which are distinctive for major groups such as genera or families and higher levels, and this knowledge can only be reached by comparative morphological studies or through the studies of fossils (Grant 1977). Among fungi, including also the lichenized groups, it is well-known that almost no fossil records exist (Thomas and Spicer 1987) and this is why we must rely entirely upon comparative morphological studies to estimate phylogenetic trends and presumed macroevolution in different groups.

The Teloschistales, comprising the families Letrouitiaceae and Teloschistaceae, has been considered as a good example of a natural group among the lichenized Ascomycotina; founded mainly on structures in the ascus, as well as structures in the ascomata and conidiomata combined with presence of certain chemical substances in the majority of species. The two families which, in addition, appear as rather natural entities, basically separated by differences in the apical ascus structures, type of ascospore discharge, and kind of ascospore septation, must have a very ancient phylogenetic history. The Letrouitiaceae, comprising only one genus and *c*. 15 species, is confined to subtropical and tropical regions, whereas the Teloschistaceae, containing the majority of the nearly 600 species included in the order, occurs in all of the major biomes (Kärnefelt 1989).

Even though these families are now, in the light of our present knowledge of the group, regarded as rather natural, ancient phylogenetic groups, it has long been the general impression that many species and, in particular, genera, in the Teloschistaceae are difficult to define and form unnatural entities or polyphyletic groups.

The phylogenetic relationships between the 10 accepted genera in the Teloschistaceae, i.e. *Apatoplaca, Caloplaca, Cephalophysis, Fulgensia, Ioplaca, Seirophora, Teloschistes, Xanthodactylon, Xanthopeltis,* and *Xanthoria* including many recognized species groups were studied according to Hennigian principles (Duncan and Stuessy 1984; Humphries and Funk 1984; Wiley 1981). These species aggregates included many recognizable groups in the supergenus *Caloplaca*, i.e. the *Caloplaca aurea*, *C. calicioides*, the *C. carphinea*, the *C. cerina*, the *C. congrediens*, the *C. eudoxa*, the *C. flavescens*, the *C. nivalis*, the *C. ochracea*, and the *C. saxicola* groups; in the genus *Fulgensia*, the *F. australis* and the *F. fulgens* groups; in the genus *Teloschistes*, the *T. capensis*, the *T. hypoglaucus*, and the *T. lacunosus* groups; and in the genus *Xanthoria*, the *X. fallax* and the *X. parietina* groups. The analyses which were carried out through the PAUP package system based on 10 respectively 24 terminal taxa and 18 character states, in many ways confirmed the view that many of the accepted or otherwise recognized genera and species aggregates, constitute unnatural groups based on homoplasious character states (Kärnefelt 1989). It is interesting to observe that the characters studied in various representative species within the most well-known genera yielded the worst resolution and that the small genera, *Apatoplaca* and *Cephalophysis* including *Caloplaca variabilis*, at least from these analyses, appear to be rather well-defined groups (Figs 8.30, 8.31).

The interpretation of unresolved cladograms is naturally very difficult and results could not directly be interpreted as representing the

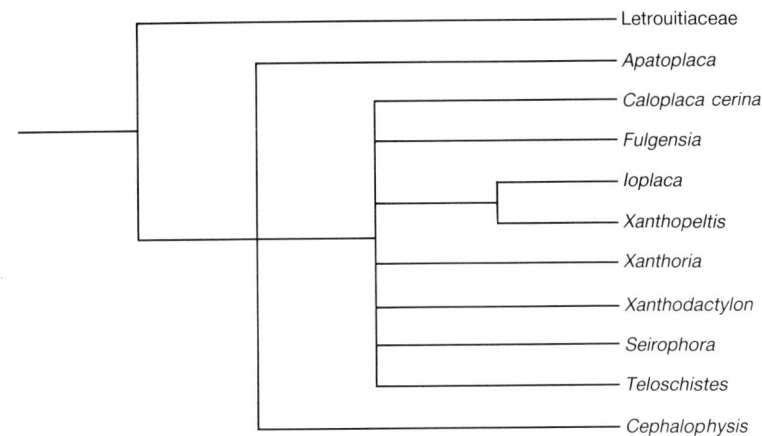

**Fig. 8.30.** Strict consensus tree constructed from 29 equally parsimonious cladograms. Terminal taxa correspond with text.

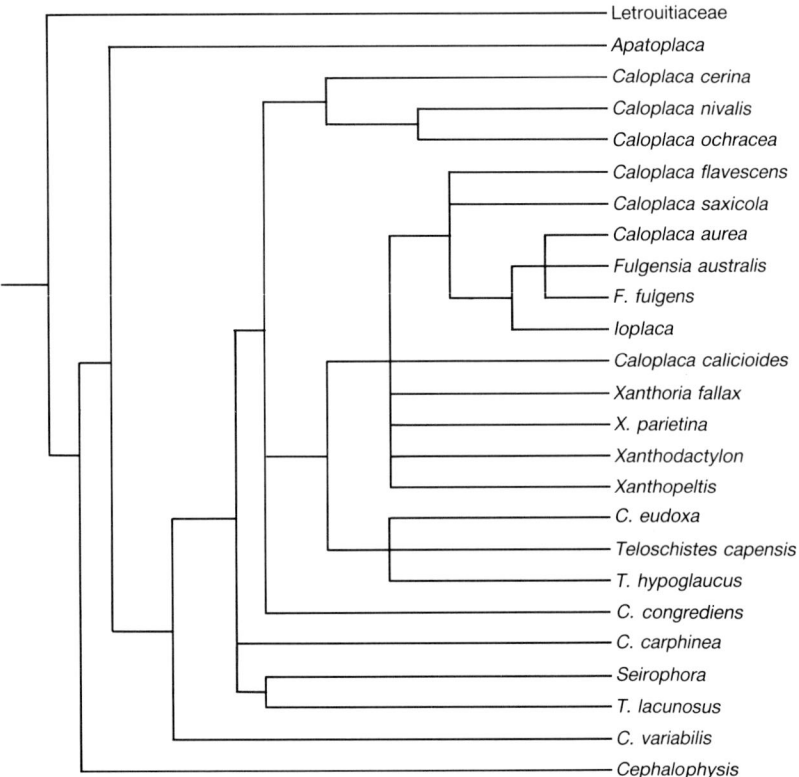

**Fig. 8.31.** Strict consensus tree constructed from 2 equally parsimonious cladograms. Terminal taxa correspond with text.

phylogenetic history of a group. The characters studied, were however, selected carefully from our knowledge of representative species within the different groups and, thus, not taken as general characters representing a group. The results, in my opinion, could have been better if the studied species, for example, those included in the genera *Teloschistes* or *Xanthoria* and presumably united with unique characters are actually closely related. One reason for all unresolved aggregates in the cladograms characterized by homoplasious characters stages might be the existence of a large number of weakly-defined species. In my opinion the existence of these badly-defined species is not because of our lack of observing the right characters but rather because of a presumably very slow evolutionary rate within this group of lichens.

## Slow evolutionary rates

The main factors affecting evolutionary rates generally involve a relationship between the population and the environment, and the characteristic of a population (Grant 1977). Concerning evolutionary rates in the Teloschistaceae, my impression is that we are observing a number of structural and chemical changes in populations which will eventually, in the course of time, give rise to reproductively isolated and stable species. Similarly, a number of different species aggregates have not become structurally sufficiently isolated to be recognized as discrete genera (Kärnefelt 1989).

Concerning the cause of major disjunctions in this group of lichens exemplified by the distribution of *X. mendozae*, and the presumed great age of certain species, I have, together with many others, become more convinced that we should look to the ancient history of the earth and plate tectonics for the answers (Galloway 1988*b*; Gradstein *et al.* 1983; Humphries 1983; Owen 1983). It is not impossible to estimate the age of a recent entity without knowledge of fossil records, if the geological history is relatively as well-known as it is in the Southern Hemisphere. However, there are only a few examples among plants in general which show that a species has existed virtually unchanged through time and space, of which the maidenhair is the most well-known example. *Ginkgo biloba* is a living fossil in the true sense of the word since it can be traced back to the early Mesozoic era (Gifford and Foster 1989). I assume that certain species of lichens also, which appear to belong to ancient Gondwanaland distributional types, could be at least as old as the *Ginkgo*.

## Conservation

The majority of species discussed in this article are known only from scattered localities in rather isolated regions and, therefore, could be regarded as maintaining a natural protection. Before this symposium I believed that none of these lichen populations were actually endangered by no agency other than, perhaps, extensive collections by lichenologists. But I am well aware of the fact that the impact of any type of environmental change would be most dramatic on these isolated geographical populations.

A dramatic decline in the earlier extensive terrestrial populations of *Teloschistes capensis* in Namibia was recently documented and the cause of the decline was simply mechanical damage from vehicles (Giess 1989). The populations which I had the pleasure of studying in 1986 obviously had been much more extensive some 20 years earlier. It is impossible to predict the nature of future environmental changes, or of their impact on the discussed lichen populations, but in the end it seems obvious now that we must accept the responsibility ourselves. If it is also true that many lichen species actually have an ancient history and have existed little changed over millions of years, we do not have the right to eradicate them from the earth at one stroke.

## Acknowledgements

I should like to thank Dr David Galloway for valuable comments on this manuscript and for inviting me to this important conference on tropical lichens. Furthermore, I wish to thank the Swedish Natural Science Research Council for financial support.

## References

Almborn, O. (1963). Studies in the lichen family Teloschistaceae. 1. *Botaniska Notiser*, **116**, 161–71.

Almborn, O. (1989). Revision of the lichen genus *Teloschistes* in central and southern Africa. *Nordic Journal of Botany*, **8**, 521–37.

Arvidsson, L. (1983). A monograph of the lichen genus *Coccocarpia*. *Opera Botanica*, **67**, 1–96 ['1982'].

Asahina, Y. (1954). Lichenologische Notizen. *The Journal of Japanese Botany*, **29**, 289–93.

Dillon, L.S. (1978). *Evolution, concepts and consequences*. Moseby Company, St. Louis.

Duncan, T. and Stuessy, T.F. (1984). *Cladistics: perspectives on the reconstruction of evolutionary history*. Columbia University Press, New York.

Elix, J.A., Johnston, J., and Armstrong, P.A. (1986). A revision of the genus *Xanthoparmelia* in Australasia. *Bulletin of the British Museum (Natural History)*, **15**, 163–362.

Futuyma, D.J. (1986). *Evolutionary biology*. Sinauer Association Inc., Sunderland, Massachusetts.

Galloway, D.J. (1985). *Flora of New Zealand lichens*. Wellington.

Galloway, D.J. (1988a). Studies in *Pseudocyphellaria* (lichens). I. The New Zealand species. *Bulletin of the British Museum (Natural History)*, **17**, 1–267.

Galloway, D.J. (1988b). Plate tectonics and the distribution of cool temperate Southern Hemisphere macrolichens. *Botanical Journal of the Linnean Society*, **96**, 45–55.

Giess, W. (1989). Einiges zu unserer Flechtenflora. *Dinteria*, **20**, 30–2.

Gifford, E.M. and Foster, A.S. (1989). *Morphology and evolution of vascular plants*. W.H. Freeman and Company, San Francisco.

Gradstein, S.R.T., Pócs, T., and Vana, J. (1983). Disjunct Hepaticae in tropical America and Africa. *Acta Botanica Hungarica*, **29**, 127–71.

Grant, V. (1977). *Organismic evolution*. W.H. Freeman and Company. San Francisco.
Hale, M.E. (1986). New species of the lichen genus *Xanthoparmelia* from southern Africa (Ascomycotina: Parmeliaceae). *Mycotaxon*, **27**, 563–610.
Hale, M.E. (1989). A monograph of the lichen genus *Karoowia* Hale (Ascomycotina: Parmeliaceae). *Mycotaxon*, **35**, 177–98.
Hawksworth, D.L. (1982). Co-evolution and the detection of ancestry in lichens. *Journal of Hattori Botanical Laboratory*, **52**, 323–9.
Humphries, C.J. (1983). Biogeographical explanation and the southern beeches. In *Evolution, time and space* (ed. R.W. Sims, J.H. Price, and P.E.S. Whalley), pp. 335–65. Systematics Association Special Volume No. **23**, Academic Press, London.
Humphries, C.J. and Funk, V.A. (1984). Cladistic methodology. In *Current concepts in plant taxonomy* (ed. V.H. Heywood and D.M. Moore), pp. 323–62. Systematics Association Special Volume No. **25**, Academic Press, London.
Kärnefelt, I. (1989). Morphology and phylogeny in the Teloschistales. *Cryptogamic Botany*, **1**, 147–203.
Kärnefelt, I. (1990). Evidence of a slow evolutionary change in the speciation of lichens. *Bibliotheca Lichenologica* (in press).
Leuckert, C. and Poelt, J. (1989). Studien über die *Lecanora rupicola*-Gruppe in Europa (Lecanoraceae). *Nova Hedwigia*, **49**, 121–67.
Lövtrup, S. (1987). What has survived of Darwin's theory? *Evolutionary Trends in Plants*, **1**, 64–9.
Owen, H.G. (1983). Some principles of physical palaeogeography. In *Evolution, time and space: The emergence of the biosphere* (ed. R.W. Sims, J.H. Price, and P.E.S. Whalley), pp. 85–114. Systematics Association Special Volume No. **23**, Academic Press, London.
Poelt, J. and Pelleter, U. (1984). Zwergstrauchige Arten der Flechtengattung *Caloplaca*. *Plant Systematics and Evolution*, **148**, 51–84.
Räsänen, V. (1939). II. Contribucion a la Flora liquenologica Sudamericana. *Anales de la Sociedad Cientifica Argentina*, **127**, 133–47.
Scharloo, W. (1987). Neo-Darwinism and trans-specific evolution. In *Systematics and evolution: a matter of diversity* (ed. P. Hovenkamp et al.), pp. 207–17. Utrecht University.
Sipman, H.J.M. (1983). A monograph of the lichen family Megalosporaceae. *Bibliotheca Lichenologica*, **18**, 1–241.
Stebbins, G.L. (1987). Is Darwinism dead? The facts say no! *Evolutionary Trends in Plants*, **1**, 69–72.
Swinscow, T.D.V. and Krog, H. (1988). *Macrolichens of East Africa*. British Museum (Natural History), London.
Tehler, A. (1983). The genera *Dirina* and *Roccellina* (Roccellaceae). *Opera Botanica*, **70**, 1–86.
Thomas, B.A. and Spicer, R.A. (1987). *The evolution and palaeobiology of land plants*. Croom Helm, London and Sydney.
Thomson, J.W. and Iltis, H.H. (1968). A fog-induced lichen community in the coastal desert of southern Peru. *The Bryologist*, **71**, 31–4.

Vainio, E. (1900). Lichenes. In *Beiträge zur Kenntnis der afrikanischen Flora, XII* (ed. H. Schinz), *Mémoires l'Herbier Boissier*, **20**, 4–5.

Wiley, E.O. (1981). *Phylogenetics: the theory and practice of phylogenetic systematics.* Wiley Interscience, New York.

Zahlbruckner, A. (1926). Afrikanische Flechten (Lichenes). *Botanische Jahrbücher für Systematik, Pflanzengeschichte und Pflanzengeographie*, **60**, 468–552.

# 9. Lichenological studies in Ecuador

L. ARVIDSSON
*Naturhistoriska Museet, PO Box 7283, S-402 35 Göteborg, Sweden*

### Abstract

The lichen flora of mainland Ecuador is still insufficiently known. For the purposes of collecting material, four expeditions were undertaken during the period 1972-1985. The herbarium at GB now comprises *c.* 10 000 lichen specimens from Ecuador, of which the majority are epiphytic macrolichens. About 160 genera have been identified so far. The collecting work was concentrated in the Andes, particularly in forests between 1000 and 4000 m altitude. The lower montane rainforest (800-2500 m), the cloud forest (2500-3400 m), and the grass páramo (3400-4000 m) are especially rich in lichens. The most important macrolichen genera of these zones are: *Sticta*, *Leptogium*, *Erioderma*, *Lobaria*, *Usnea*, *Coccocarpia*, *Pseudocyphellaria*, *Everniastrum*, *Hypotrachyna*, *Heterodermia*, and *Parmotrema*. It is premature to present a detailed survey of the phytogeography of the area. Apart from cosmopolitan taxa, the lichen flora of Ecuador includes neotropical, pantropical, American-African, circum-Pacific, and southern, and northern temperate elements. Examples of lichens found in some selected habitats are given and many lichens are recorded here for the first time from Ecuador.

### Introduction

Ecuador covers a total land area of 283 561 km² on the western side of the South American continent between 1° 20′ N and 5° 04′ S and is bounded on the north by Colombia, to the east and south by Peru, and to the west by the Pacific Ocean. The whole region, which is divided into 5 coastal, 4 eastern and 10 sierra provinces lies within the tropics, but its climatic conditions are dominated by features of the land surface. The Andean cordilleras traverse the country from north to south and divide Ecuador into three main parts: the coastal plain (La

*Tropical Lichens: Their Systematics, Conservation, and Ecology* (ed. D.J. Galloway), Systematics Association Special Volume No. 43, pp. 123-34. Clarendon Press, Oxford, 1991. © The Systematics Association, 1991.

Costa), the Andes (La Sierra), and the eastern lowland (El Oriente). The coastal lowlands, from the Pacific Ocean to the western edge of the Andes have an altitudinal range from sea level to *c.* 600 m. Large areas in the south and west here have a relatively low precipitation, with vegetation types including coastal desert and semi-desert, savannah, deciduous, and semi-deciduous forest. Lowland rainforest is found only in the north (Esmeraldas Province).

The Andes are formed essentially of two main parallel chains, joined or intersected by ridges or spurs separating intermontane basins. Several such basins occur in Ecuador at altitudes of 2500–3000 m above sea level. In both cordilleras, volcanos of more than 6000 m occur. The main vegetation types of the Andes are briefly discussed below. Finally, the eastern lowland is a plateau of *c.* 200–300 m, densely covered in rainforest. For further details of Ecuadorean vegetation see Harling (1979).

The lichen flora of mainland Ecuador is imperfectly known. There are only two minor papers dealing with this subject, namely those of Müller Argoviensis (1879), and Zahlbruckner (1905). The former deals with lichens collected by M.A. André, the latter is an account of the collection by D.H. Meyer from the Ecuadorean high Andes. Material from Ecuador is cited occasionally in monographs and revisions (Ahti 1961, 1983; Ahti and Stenroos 1986; Arvidsson 1983; Degelius 1974, 1986; Galloway and Jørgensen 1987; Hale 1975, 1976; Jørgensen 1973, 1975, 1989; Jørgensen and Galloway 1989; Kärnefelt 1980, 1986; Lamb 1977; Lamb and Ward 1974; Llano 1950; Santesson 1942*a*, 1944, 1952; Sipman 1983; Stenroos 1989; Walker 1985). An account of Ecuadorean species of *Pseudocyphellaria* has recently appeared (Galloway and Arvidsson 1990). New species, based on material from mainland Ecuador, were recently described by Esslinger (1989), Galloway (1989), Kalb (1988), Moberg (1990), Nash *et al.* (1987*a*, *b*), and Sipman (1986). An unpublished list of lichens from Ecuador adds many new records for the country (A. Aptroot, personal communication). The Galápagos, however, are better known lichenologically. For details see Weber (1986).

It was not until the early 1970s that the lichen flora of mainland Ecuador was investigated in any depth. Four expeditions were made by L. Arvidsson *et al.*, in 1972, 1979, 1983, and 1985. Their collecting work excluded the Galápagos and was concentrated on the Andes, especially the humid forests between 1000 and 4000 m. Some isolated visits to the lowlands were also made but the lichen flora here is rather impoverished compared with that of the cordilleras. A survey of the collections, together with a map of localities visited was published by Arvidsson (1986). The herbarium at GB now comprises *c.* 10 000 specimens, of which 90 per cent were collected during the four above excursions. Even though many crustaceous lichens were gathered, activity was focused primarily on foliose and fruticose species. Epiphytic lichens are more frequently collected than saxicolous and terricolous ones. In addition, Ecuadorean lichens have also been collected by various other botanists during recent years. Important material is found in, for example, AAU, ASU, BM, COLO, NY, QCA, S, U, and herb. Kalb (Neumarkt). Material from Ecuador was also distributed in Kalb's exsiccate '*Lichenes Neotropici*', fascicle X (1988).

Subsequent identification work was assisted by several specialists. Species of *Baeomyces, Cladina, Cladonia, Coccocarpia, Collema, Diploschistes, Everniastrum, Hypotrachyna, Megalospora, Parmelina, Parmotrema, Pseudocyphellaria, Pseudoparmelia, Punctelia, Stereocaulon, Teloschistes* and others are already identified. Determination still proceeds in *Anzia, Erioderma, Leptogium, Menegazzia, Nephroma, Peltigera, Physciaceae, Ramalina, Siphula, Sphaerophorus* and various pyrenolichens and foliicolous lichens.

The most frequently collected genera in the herbarium at GB are presented below, together with the number of specimens: *Leptogium*—1600, *Sticta*—1500, *Erioderma*—400, *Usnea*—400, *Hypotrachyna*—400, *Coccocarpia*—300, *Heterodermia*—300, *Lobaria*—300, *Pannaria/Parmeliella*—200, *Everniastrum*—200, *Pseudocyphellaria*—200, followed by *Parmotrema, Stereocaulon, Peltigera*, and *Cladonia*.

## Some lichen-rich habitats

As stated earlier, the Andes are the best known areas from a lichenological point of view in Ecuador. Brief comments on the lichen flora in four main vegetation types (lower montane rainforest, cloud forest, páramo and interandean desert, and semi-desert) are given below. Since many lichen groups are still unidentified, the names presented here are incomplete sample tests of the total number of species that exist in these habitats. Many lichens occur in more than one of the vegetation types discussed below. However, the various species are mentioned only once here, and that is from the habitat in which they are most frequent. For details of floristic composition of vegetation types see Harling (1979).

### 1. Lower montane rainforest

Lower montane rainforest (Fig. 9.1) gradually replaces lowland forest at an altitude of *c*. 700–800 m on both sides of the Andes. It continues to about 2500 m where it merges into cloud forest. The annual rainfall, especially on the eastern slopes of the Andes, seems to be much higher than it is in the lowlands. In the Mera region (Rio Pastaza Valley) it is

**Fig. 9.1.** Lower montane rainforest. Ecuador, prov. El Oro, near Piñas, alt. 1000 m. (Photo *M. Neuendorf*, 15 February 1977.)

said to be *c.* 5000 mm. Lower montane rainforest represents a number of habitats depending on altitude, exposure, substrate etc. For instance, road cuttings in certain stages of recolonization may be rich in lichens, such as *Cladonia subsquamosa*, *Dictyonema glabratum*, *Peltigera laciniata*, and species of *Baeomyces*, *Placopsis*, and *Stereocaulon*. Foliicolous lichens are very common in these humid forests. Phorophytes of the families Clusiaceae, Melastomataceae, and Rubiaceae being particularly rich. Identification of the various foliicolous lichens collected here is proceeding, and among species found may be mentioned: *Aulaxinea minuta*,

*Byssoloma leucobleharum, Coccocarpia domingensis, C. stellata, Croodiscus coccineus, Cryptothecia candida, Echinoplaca pellicula, Phyllophiale alba, Porina epiphylla, P. limbulata, P. phyllogena, Strigula elegans, S. subtilis, Tapellaria epiphylla, Tricharia albostrigosa,* and species of the genera *Byssolecania, Calenia, Gyalectidium, Mazosia, Sporopodium, Tricharia,* and *Trichothelium.*

Corticolous lichens are frequently collected in this zone. Solitary trees along roads or in pastures are usually covered in lichens. Common macrolichen genera are *Coccocarpia, Erioderma, Hypotrachyna, Leptogium, Parmotrema, Pseudocyphellaria, Sticta, Usnea,* and various pyrenolichens, e.g. *Astrothelium, Mycomicrothelia, Pleurotrema, Porina, Pyrenula,* and *Trypethelium,* and graphidaceous taxa. Some examples of species found are given in Table 9.1.

## 2. Cloud forest

Cloud forest, upper montane rainforest, or elfin forest (Fig. 9.2) occurs between *c.* 2500–3400 m. It is generally best developed in the eastern cordillera, and is characterized as a low, microphyllous forest with the trees overloaded with mosses, filmy ferns, and lycopods. It is exceptionally rich in lichens, especially species of *Erioderma, Everniastrum, Heterodermia, Hypotrachyna, Leptogium, Sticta,* and *Usnea.* Also, within this zone there are many different habitats. Road cuttings are often prominent in the cordilleras and many lichens are regularly found here, such as *Baeomyces fungoides, B. imbricatus, Cladia aggregata, Cladina polia, Cladonia calycantha, C. lopezii, Heterodermia circinalis, H. leucomelos, Peltigera pulverulenta, Stereocaulon novogranatense, S. ramulosum, S. strictum* var. *compressum,* and species of *Everniastrum, Oropogon* and *Placopsis.*

Many corticolous species from lower montane rainforest also occur here, together with subtropical and temperate species. The uppermost part of the cloud forest (*Ceja de la montaña* or *Ceja andina*) near the transition zone with the páramo, having low trees and bushes covered in lichens is perhaps the most lichen-rich habitat of all in Ecuador. Of special interest here is the presence of a number of undescribed species of *Erioderma* (Arvidsson and Jørgensen, in preparation). These occur mainly in the eastern cordillera. Table 9.2 gives some examples of corticolous lichens found in cloud forest.

## 3. Grass páramo

Grass páramos or pajonales (Fig. 9.3) occupy most ground between 3400 and 4000 m. Downwards they border on the low *ceja* scrub or, nowadays, often on cultivated or deforested land. The clumps of grass are generally intermingled with herbs, mosses, and lichens. In sheltered places bushes or small trees (*Polylepis*) may form rather large

**Table 9.1.** Some corticolous lichens from lower montane rainforest in Ecuador

| | |
|---|---|
| *Bulbothrix goebelii* | *Leptogium olivaceum* |
| *Catillaria endochroma* | *L. phyllocarpum* |
| *Chiodecton sanguineum* | *L. punctulatum* |
| *Coccocarpia erythrocardia* | *L. stipitatum* |
| *C. erythroxyli* | *L. vesiculosum* |
| *C. pellita* | *L.* spp. |
| *Coenogonium leprieuri* | *Lobaria peltigera* |
| *Collema leptaleum* | *L.* spp. |
| *Dictyonema sericeum* | *Megalospora tuberculosa* |
| *Dimerella lutea* | *Parmeliella pannosa* |
| *Erioderma verruculosum* | *Parmelina dissecta* |
| *E. wrightii* | *P. horrescens* |
| *Everniastrum arsenei* | *Parmotrema conformatum* |
| *E. latilobum* | *P. latissimum* |
| *Heterodermia barbifera* | *P. mellissii* |
| *H. comosa* | *P. peralbidum* |
| *H. corallophora* | *P. reticulatum* |
| *H. flabellata* | *P. subisidiosum* |
| *H. isidiophora* | *P. viridiflavum* |
| *H. lutescens* | *Physcia coronifera* |
| *H. speciosa* | *P. decorticata* |
| *H. verruculifera* | *Pseudocyphellaria arvidssonii* |
| *H. vulgaris* | *P. aurata* |
| *Hypotrachyna chlorina* | *P. clathrata* |
| *H. costaricensis* | *P. dozyana* |
| *H. degelii* | *Pseudoparmelia sphaerospora* |
| *H. dentella* | *Pseudopyrenula subgregaria* |
| *H. formosana* | *Polychidium dendriscum* |
| *H. imbricatula* | *Sticta lenormandii* |
| *Leioderma sorediatum* | *S. tomentosa* |
| *Leptogium cyanescens* | *S.* spp. |
| *L. diaphanum* | *Trypethelium aeneum* |
| *L. digitatum* | *Usnea* spp. |

stands. The corticolous lichen flora of these scrubs is often rich, including for instance *Cetrariastrum andense, C. dubitans, C. ecuadoriense, Hypotrachyna andensis, H. brevirhiza, H. caraccensis, H. ensifolia, H. gigas, H. laevigata, H. sinuosa, Pseudocyphellaria bartlettii, P. encoensis, P. intricata, Sphaerophorus melanocarpus,* and species of *Anzia, Everniastrum, Leptogium, Menegazzia, Oropogon, Sticta,* and *Usnea.*

The terricolous communities include several distinct associations

**Fig. 9.2.** Lower part of cloud forest. Ecuador, prov. Morona-Santiago, eastern slopes of Cordillera Oriental along road Gualaceo–General Leonidas Plaza, alt. 2800 m. The forest is in parts covered with bamboo (*Chusquea*) seen in the lower left hand section of the photo. (Photo *L. Arvidsson*, 10 February 1979.)

**Fig. 9.3.** Grass páramo. Ecuador, prov. Pichincha, eastern slopes of Cerro Iliniza, alt. *c.* 4200 m. Bunch grasses and low bushes (*Loricaria*) with intermingled lichens *Alectoria ochroleuca*, *Oropogon loxensis*, *Thamnolia vermicularis*, and species of *Cladonia*, *Stereocaulon* etc. (Photo *L. Arvidsson*, 7 March 1972.)

**Table 9.2.** Some corticolous lichens from cloud forest in Ecuador

| | |
|---|---|
| Brigantiaea leucoxantha | Hypotrachyna reducens |
| Catinaria versicolor | H. rockii |
| Cladonia ceratophylla | H. singularis |
| C. ochrochlora | Leioderma glabrum |
| Coccocarpia flavicans | Leptogium andinum |
| C. palmicola | L. burgessii |
| Collema glaucophthalmum | L. laceroides |
| Erioderma spp. | L. papillosum |
| Everniastrum arvidssonii | L. resupinans |
| E. catawbiense | L. spp. |
| E. cirrhatum | Lobaria spp. |
| E. columbiense | Megalospora admixta |
| E. fragile | M. sulphurata |
| E. nigrociliatum | Normandina pulchella |
| E. sorocheilum | Oropogon loxensis |
| E. vexans | Pannaria conoplea |
| Flavopunctelia flaventior | P. rubiginosa |
| Heterodermia casarettiana | P. tavaresii |
| H. circinalis | Parmotrema arnoldii |
| H. galactophylla | P. bangii |
| H. japonica | P. crinitum |
| H. leucomelos | P. gardneri |
| H. obscurata | Physcia lopezii |
| H. spp. | Pseudocyphellaria crocata |
| Hypotrachyna bogotensis | Sphaerophorus melanocarpus |
| H. densirhizinata | Sticta fuliginosa |
| H. endochlora | S. humboldtii |
| H. longiloba | S. weigelii |
| H. microblasta | S. spp. |
| H. physcioides | Teloschistes exilis |
| H. prolongata | T. flavicans |
| H. pulvinata | Usnea spp. |

from the wet páramos with species of *Cladia*, *Cladina*, *Peltigera* etc. to very exposed, almost desert-like areas with genera like *Diploschistes*, *Stereocaulon*, *Xanthoparmelia*, and *Usnea*. Table 9.3 gives some examples of terricolous and saxicolous lichens found in the grass páramo.

*4. Interandean desert and semi-desert*

These areas are confined to a number of more or less deep valleys which lie in the rain shelter of the cordillera. Two such valleys (Rio Leon Valley and the Catamayo Valley) were visited. The vegetation in

**Table 9.3.** Some terricolous and saxicolous lichens from grass páramo in Ecuador

| | |
|---|---|
| *Alectoria ochroleuca* | *Leprocaulon congestum* |
| *Arthrorhaphis alpina* | *L. gracilescens* |
| *A. citrinella* | *Omphalina foliacea* |
| *Baeomyces rufus* | *Pachyospora verrucosa* |
| *Cetraria delisei* | *Parmelinopsis swinscowii* |
| *Cladia aggregata* | *Peltigera austroamericana* |
| *C. fuliginosa* | *Punctelia stictica* |
| *Cladina arcuata* | *Psoroma hypnorum* |
| *C. boliviana* | *Siphula fastigiata* |
| *C. confusa* | *S. pteruloides* |
| *Cladonia aleuropoda* | *Solorina saccata* |
| *C. andesita* | *Stereocaulon crambidiocephalum* |
| *C. chlorophaea* | *S. glareosum* |
| *C. didyma* | *S. meyeri* |
| *C. furcata* | *S. obesum* |
| *C. macrophyllodes* | *S. pityrizans* |
| *C. mexicana* | *S. tomentosum* |
| *C. pocillum* | *S. verruciferum* var. *surreptans* |
| *C. squamosa* | *S. vesuvianum* |
| *C. verruculosa* | *Thamnolia vermicularis* |
| *Coelocaulon muricatum* | *Usnea durietzii* |
| *Diploschistes cinereocaesius* | *Xanthoparmelia distincta* |
| *D. muscorum* | *X. cotopaxiensis* |
| *Everniopsis trulla* | *X. mougeotii* |
| *Flavoparmelia ecuadoriensis* | *X. subsorediata* |
| *Hypogymnia bitteri* | *X. vagans* |
| *Leprocaulon albicans* | |

such areas (Fig. 9.4) is sparse and dominated by species of *Acacia*, the bottle-shaped trunks of *Chorisia insignis*, *Croton* spp., and various cacti. Annual rainfall here is reported to be less than 300 mm (Harling 1979). Lichens from the families Parmeliaceae and Physciaceae are prominent here, together with species of *Caloplaca* and *Collema*. Table 9.4 gives some examples of corticolous lichens from this xeromorphic habitat.

**Table 9.4.** Some macrolichens from the interandean semi-deserts

| | |
|---|---|
| *Collema conglomeratum* | *Pseudoparmelia leucoxantha* |
| *C. neglectum* | *P. texana* |
| *C. texanum* | *Punctelia subrudecta* |
| *Parmelina lindmanii* | *Pyxine berteriana* |
| *Parmotrema andinum* | *Teloschistes chrysophthalmus* |
| *P. hababianum* | *T. hypoglaucus* |
| *P. subsumtum* | *Xanthoparmelia farinosa* |
| *P. tinctorum* | *X. kurokawae* |
| *Physcia integrata* | *X. ulcerosa* |
| *P. rolfii* | *Xanthoria mendozae* |
| *P. sorediosa* | |

**Fig. 9.4.** Dry deciduous forest and semi-desert. Ecuador, prov. Loja, *c.* 10 km south of Catamayo, alt. *c.* 1500 m. (Photo *L. Arvidsson*, 20 February 1979.)

## Acknowledgements

I am grateful to the following specialists for their help in lichen identification: Professor T. Ahti, Helsinki; Dr O. Almborn, Lund; Dr A. Aptroot, Utrecht; Professor G. Degelius, Göteborg; Dr J. Elix, Canberra; Dr D. Galloway, London; Dr M. Hale, Washington; Professor

P.M. Jørgensen, Bergen; Dr I. Kärnefelt, Lund; Fil. kand. M. Lindström, Göteborg; Dr T. Lumbsch, Marburg; Professor T. Nash, Tempe; Professor R. Santesson, Uppsala; Dr H. Sipman, Berlin; Dr J. Walker, London, and Dr O. Vitikainen, Helsinki. I am also much obliged to Professor G. Harling, Göteborg, for help in various ways and to Fil. kand. Per-Olof Martinsson, Göteborg for photographic assistance.

## References

Ahti, T. (1961). Taxonomic studies on reindeer lichens (*Cladonia*, subgenus *Cladina*). *Annales Botanici Societatis Zoologicae Botanicae Fennicae Vanamo*, **32**, 1–160.

Ahti, T. (1983). Taxonomic notes on some American species of the lichen genus *Cladonia*. *Annales Botanici Fennici*, **20**, 1–7.

Ahti, T. and Stenroos, S. (1986). A revision of *Cladonia* sect. *Cocciferae* in the Venezuelan Andes. *Annales Botanici Fennici*, **23**, 229–38.

Arvidsson, L. (1983). A monograph of the lichen genus *Coccocarpia*. *Opera Botanica*, **67**, 1–96.

Arvidsson, L. (1986). The lichen flora of Ecuador. In *Current scandinavian botanical research in Ecuador* (ed. B. Øllgaard and U. Molau). *Reports from the Botanical Institute, University of Aarhus*, No. **15**, pp. 13–19.

Degelius, G. (1974). The lichen genus *Collema* with special reference to the extra-European species. *Symbolae Botanicae Upsalienses*, **20**(2), 1–215.

Degelius, G. (1986). Studies in the lichen family Collemataceae. V. Notes on some interesting *Collema* species. *Nordic Journal of Botany*, **6**, 345–9.

Esslinger, T.L. (1989). Systematics of *Oropogon* (Alectoriaceae) in the New World. *System. Bot. Monogr. vol.* **28**.

Galloway, D.J. (1989). Nomenclatural notes on *Pseudocyphellaria*. IV. Some South American taxa. *Lichenologist*, **21**, 88–9.

Galloway, D.J. and Arvidsson, L. (1990). Studies in *Pseudocyphellaria* (lichens). II. Ecuadorean species. *Lichenologist*, **22**, 103–35.

Galloway, D.J. and Jørgensen, P.M. (1987). Studies in the lichen family Pannariaceae. II. The genus *Leioderma* Nyl. *Lichenologist*, **19**, 345–400.

Hale, M. (1975). A revision of the lichen genus *Hypotrachyna* (Parmeliaceae) in tropical America. *Smithsonian Contributions to Botany*, **25**, 1–73.

Hale, M. (1976). A monograph of the lichen genus *Parmelina* Hale (Parmeliaceae). *Smithsonian Contributions to Botany*, **33**, 1–60.

Harling, G. (1979). The vegetation types of Ecuador—a brief survey. In *Tropical Botany* (ed. K. Larsen and L. Holm-Nielsen), pp. 165–74. London, Academic Press.

Jørgensen, P.M. (1973). On some *Leptogium* species with short *Mallotium* hairs. *Svensk Botanisk Tidskrift*, **67**, 53–8.

Jørgensen, P.M. (1975). Contribution to a monograph of the *Mallotium*-hairy *Leptogium* species. *Herzogia*, **3**, 433–60.

Jørgensen, P.M. (1989). *Omphalina foliacea*, a new basidiolichen from America. *Nordic Journal of Botany*, **9**, 89–95.

Jørgensen, P.M. and Galloway, D.J. (1989). Studies in the lichen family Pannariaceae. III. The genus *Fuscoderma*, with additional notes and a revised key to *Leioderma*. *Lichenologist*, **21**, 295–301.

Kalb, K. (1988). *Lichenes Neotropici, Fascikel* **X** (no. 401–50). Neumarkt.

Kärnefelt, I. (1980). *Everniastrun andense* sp. nov., a neotropical páramo lichen. *Botaniska Notiser*, **133**, 387–94.

Kärnefelt, I. (1986). The genera *Coelocaulon* and *Cornicularia* and formerly associated taxa. *Opera Botanica*, **86**, 1–90.

Lamb, I.M. (1977). A conspectus of the lichen genus *Stereocaulon* (Schreb.) Hoffm. *Journal of the Hattori Botanical Laboratory*, **43**, 191–355.

Lamb, I.M. and Ward, A. (1974). A preliminary conspectus of the species attributed to the imperfect lichen genus *Leprocaulon* Nyl. *Journal of the Hattori Botanical Laboratory*, **38**, 499–553.

Llano, A. (1950). *A monograph of the lichen family Umbilicariaceae in the Western Hemisphere*. Office of Naval Research. Washington.

Moberg, R. (1990). The lichen genus Physcia in Central and South America. *Nordic Journal of Botany*, **10** (seen in manuscript).

Müller Argoviensis, J. (1879). Les lichens Neo-Grenadins et Ecuadoriens récoltés par M. Ed. André. *Revue Mycologique*, **1**, 160–71.

Nash, T., Elix, J., and Johnston, J. (1987a). *Flavoparmelia ecuadoriensis*, a new species in the Parmeliaceae (Ascomycotina). *Mycotaxon*, **28** (2), 257–9.

Nash, T., Elix, J., and Johnston, J. (1987b). New species, new records, and a key for *Xanthoparmelia* (lichenized Ascomycotina) from South America. *Mycotaxon*, **28**, 285–96.

Santesson, R. (1942a). Two interesting new species of the lichen genus *Parmelia*. *Botaniska Notiser 1942*, 325–30.

Santesson, R. (1942b). The South American Cladinae. *Arkiv för Botanik*, **30A** (10), 1–27.

Santesson, R. (1944). Contributions to the lichen flora of South America. *Arkiv för Botanik*, **31A** (7), 1–28.

Santesson, R. (1952). Foliicolous lichens. I. *Symbolae Botanicae Upsalienses*, **12**(1), 1–590.

Sipman, H. (1983). A monograph of the lichen family Megalosporaceae. *Bibliotheca Lichenologica*, **18**. J. Cramer, Vaduz.

Sipman, H. (1986). Notes on the lichen genus *Everniastrum* (Parmeliaceae). *Mycotaxon*, **26**, 235–51.

Stenroos, S. (1989). Taxonomic revision of the *Cladonia miniata* group. *Annales Botanici Fennici*, **26**, 237–61.

Walker, J. (1985). The lichen genus *Usnea* subgenus *Neuropogon*. *Bulletin of the British Museum (Natural History) Botany*, **13**(1), 1–130.

Weber, W.A. (1986). The lichen flora of the Galapágos Islands, Ecuador. *Mycotaxon*, **27**, 451–97.

Zahlbruckner, A. (1905). Flechten, im Hochlande Ecuadors gesammelt von Prof. D. Hans Meyer im Jahre 1903. *Beihefte zum Botanischen Zentralblatt*, **19**(2), 75–84.

# 10. Notes on the lichen flora of the Guianas, a neotropical lowland area

H.J.M. SIPMAN
*Botanical Garden and Botanical Museum, Königin Luise Strasse 6–8, D-1000 Berlin 33, West Germany*

## Abstract

Evaluation of recent collections of lichenized fungi from the Guianas resulted in a new survey of the genera. This showed *Lecanorales* to be the principal order, comprising 40 per cent of all species, principally in the families Cladoniaceae, Pyxinaceae, and Ramalinaceae, and the Graphidales and Pyrenulales to be the second and third important groups with 20 per cent and 14 per cent, respectively. Fifty-five species are recorded for the first time from the Guianas, mainly foliicolous taxa. A preliminary evaluation of distribution patterns is given, confirming the principal division into a coastal zone and an inland zone. The westernmost part is also distinct and shows a strong relationship with the Guiana Highlands area of Venezuela. Concerning biogeographical relations of the Guianas lichen flora, the following elements are proposed: pantropical; neotropical + african; neotropical; amazonian; Guiana Highlands. Observations on foliicolous lichens show that well-developed forest may harbour nearly 70 species. Open or humid forests are much poorer.

## Improved survey of the lichen flora of the Guianas

Subsequent to the preparation of a checklist (Hekking and Sipman 1988) numerous new collections have become available. A botanical exploration trip to Guyana by A. Aptroot and the author in 1985 yielded over 2000 collections (mainly in B) and associated visits by

*Tropical Lichens: Their Systematics, Conservation, and Ecology* (ed. D.J. Galloway), Systematics Association Special Volume No. 43, pp. 135–50. Clarendon Press, Oxford, 1991. © The Systematics Association, 1991.

A. Aptroot to Surinam and French Guiana, about 1000 collections (in U). Some 2000 collections were made from tree crowns near Saül in French Guiana and Mabura Hill in Guyana, by D. Montfoort and R. Ek, resp. H. Cornelissen and H. ter Steege (mainly in U) who were investigating the ecology of epiphytes using climbing techniques. Part of these collections were treated by Aptroot (1988), Aubel (1991) and Sipman (1990). These increased collections allowed an improved, but not a definitive survey of the genera and higher taxonomic entities.

Arrangement is largely after the proposed classification of Eriksson and Hawksworth (1986). Modifications to this arrangement are: *Mycobilimbia* is not included in Porpidiaceae, but left in a separate family, following Hafellner (1984). The Lecideaceae is used here in the Zahlbrucknerian sense to accommodate a number of uncertain taxa. The Alectoriaceae, Ramalinaceae, and Parmeliaceae are combined under the oldest name for the group, Ramalinaceae. The Phlyctidaceae have been provisionally placed in the Graphidales, since they show a resemblance to Gomphillaceae. *Mazosia* has not been treated as a separate family, since it fits well in Opegraphaceae.

The figures indicate the estimated numbers of species, comprising the number of species recorded in the literature, of recent, unpublished, sometimes provisional, identifications and of unidentified, provisionally distinguished taxa. Question marks indicate very doubtful records, which were excluded from the additions.

## ASCOMYCOTINA

*A. Discocarpous orders, alphabetically*

**Arthoniales** 19 = **3 per cent**—Arthoniaceae 18: *Arthonia* 13; *Arthothelium* 2; *Cryptothecia* 2; *Stirtonia* 1; Chrysotrichaceae 1: *Byssocaulon* 2?; *Chrysothrix* 1.

**Caliciales** 5 = **1 per cent**—Caliciaceae 1: *Calicium* 1(2?); Coniocybaceae 1: *Chaenotheca* 1; Sphinctrinaceae 1: *Pyrgidium* 1; unkn. family 2: *Tylophoron* 2.

**Gyalectales** 16 = **3 per cent**—Gyalectaceae 16: *Belonia* 1?; *Coenogonium* 8?; *Dimerella* 6.

**Graphidales** 117 = **20 per cent**—Asterothyriaceae 1: *Asterothyrium* 1; Graphidaceae 55: *Cyclographina* 2; *Glyphis* 1(2?); *Graphina* 19; *Graphis* 19; *Gyrostomum* 1; *Medusulina* 1; *Phaeographina* 7; *Phaeographis* 5; *Sacrographa* 1(7?); Phlyctidaceae 2: *Phlyctella* 1; *Phlyctis* 1; Thelotremataceae 37: *Chroodiscus* 2; *Leptotrema* 1; *Myriotrema* 3; *Ocellularia* 20; *Polystroma* 1; *Thelotrema* 11; Gomphyllaceae 24: *Actinoplaca* 2; *Aulaxina* 5; *Calenia* 2; *Echinoplaca* 7; *Gyalectidium* 1; *Gyalideopsis* 1; *Tricharia* 6.

**Lecanorales** 236 = **40 per cent**—Acarosporaceae 2: *Biatorella* 2;

Bacidiaceae 16: *Bacidia* 6; *Biatora* 2?; *Catinaria* 3; *Eschatogonia* 1; *Phyllopsora* 2; *Physcidia* 3; *Squamacidia* 1; Brigantiaeaceae 1: *Brigantiaea* 1; Cladoniaceae 42: *Cladia* 2; *Cladina* 5; *Cladonia* 35; Coccocarpiaceae 12: *Coccocarpia* 12; Collemataceae 9: *Collema* 1(2?); *Leptogium* 7; *Physma* 1(2?); Crocyniaceae 2: *Crocynia* 2(3?); Ectolechiaceae 13: *Calopadia* 1; *Badimia* 5; *Lasioloma* 1; *Loflammea* 1; *Logilvia* 1; *Pyrenotrichum* 1; *Sporopodium* 2; *Tapellaria* 1; Haematommataceae 2: *Haematomma* 2; Hymeneliaceae 1: *Aspicilia* 1; Lecanoraceae 3: *Lecanora* 3; Lecideaceae s.l. 10: *Lecidea* s.l. 5; *Lecidea* gr. *piperis* 2; *Lecidella* 1?; *Lopadium* s.l. 2; *Toninia* s.l. 2; Megalosporaceae 2: *Megalospora* 2; Micareaceae 2: *Micarea* 2; Mycobilimbiaceae 1: *Mycobilimbia* 1?; Pannariaceae 10: *Erioderma* 4; *Leioderma* 2; *Pannaria* 3; *Psoroma* 1; Pilocarpaceae 11: *Byssolecania* 2; *Byssoloma* 7; *Fellhanera* 4; Pyxinaceae 47: *Buellia* 9; *Diplotomma* 1?; *Dirinaria* 5; *Heterodermia* 12; *Hyperphyscia* 2; *Physcia* 5; *Pyxine* 7; *Rinodina* 7; Ramalinaceae s.l. 48: *Bulbothrix* 7; *Canoparmelia* 1; *Hypotrachyna* 7; *Oropogon* 1; *Pannoparmelia* 1?; *Parmelia* 3?; *Parmotrema* 18; *Pseudoparmelia* 1; *Ramalina* 4; *Relicina* 2; *Usnea* 6; *Xanthoparmelia* 1; Stereocaulaceae 1: *Stereocaulon* 2?; Trapeliaceae 1: *Trapeliopsis* 1.

**Lichinales** 5 = **1 per cent**—Lichinaceae 3: *Jenmania* 1; *Leprocollema* 1?; *Lichina* 1?; *Psorotichia* 1; Peltulaceae 2: *Peltula* 2.

**Opegraphales** 34 = **6 per cent**—Opegraphaceae 33: *Chiodecton* 6; *Enterographa* 1; *Lecanactis* 3; *Mazosia* 11; *Melaspilea* 2; *Opegrapha* 7; *Schismatomma* 3; Roccellaceae 1: *Helminthocarpon* 1.

**Peltigerales** 12 = **2 per cent**—Lecotheciaceae 1: *Polychidium* 1; Lobariaceae 11: *Dendriscocaulon* 1; *Lobaria* 2; *Pseudocyphellaria* 2; *Sticta* 6.

**Pertusariales** 5 = **1 per cent**—Pertusariaceae 5: *Ochrolechia* 1; *Pertusaria* 4.

**Teloschistales** 8 = **1 per cent**—Letrouitiaceae 4: *Letrouitia* 4; Teloschistaceae 4: *Caloplaca* 3(4?); *Teloschistes* 1.

B. *Pyrenocarpous orders, alphabetically*

**Dothideales** 15 = **2.5 per cent**—Arthopyreniaceae s.l. 15: *Anisomeridium* 2?; *Arthopyrenia* 5; *Ditremis* 2; *Leptoraphis* 1; *Mycomicrothelia* 2; *Tomasellia* 3.

**Porinales** 25 = **4 per cent**—Trichotheliaceae 25: *Clathroporina* 1; *Porina* 21; *Trichothelium* 3.

**Pyrenulales** 80 = **14 per cent**—Pyrenulaceae 39: *Anthracothecium* 5; *Melanotheca* 6; *Parmentaria* 2; *Plagiotrema* 1; *Pseudopyrenula* 5; *Pyrenastrum* 4; *Pyrenula* 15; *Pyrgillus* 1; Trypetheliaceae 40: *Astrothelium* 14; *Campylothelium* 2; *Cryptothelium* 3; *Laurera* 4; *Pleurotrema* 1; *Trypethelium* 16; Monoblastiaceae 1: *Monoblastia* 1.

**Strigulales** 14 = **2 per cent**—Strigulaceae 12: *Phylloporis* 3; *Raciborskiella*

1; *Strigula* 8; Aspidotheliaceae 2: *Aspidothelium* 2; Phyllobatheliaceae 1: *Phyllobathelium* 1.

**Verrucariales** 3 = **0.5 per cent**—Verrucariaceae s.l. 3: *Endocarpon* 1; *Microtheliopsis* 1; *Normandina* 1; *Verrucaria* 2?.

## BASIDIOMYCOTINA

**Aphyllophorales** 5 = **1 per cent**—Dictyonemataceae 5: *Corella* 1; *Dictyonema* 4.

## LICHENES IMPERFECTI

3 = **0.5 per cent**—*Lepra* 1?; *Lepraria* 1; *Phyllophiale* 1; *Siphula* 1?.

The survey shows that close to 600 species in some 165 genera are known at present. The principal orders are Lecanorales, Graphidales and Pyrenulales, which together account for *c.* 75 per cent of all species. It is remarkable that the order Lecanorales, with 40 per cent of the total lichen flora, is the principal order, as it is in regions with a cooler climate. At the family level, the best-represented groups are the Cladoniaceae, Pyxinaceae, and Ramalinaceae in the Lecanorales, the Graphidaceae and Thelotremataceae in the Graphidales, and the Pyrenulaceae and Trypetheliaceae in the Pyrenulales. It is probable that the families Bacidiaceae and Opegraphaceae will need to be added, once crustose lichens are better known. Several orders are unusually poorly represented, e.g. Teloschistales and Verrucariales. Basidiolichens, which are thought to be common in the tropics, constitute no more than 1 per cent of the flora.

### Additions to the checklist of the Guianas

Here a selection of new records is presented. It concerns mainly vouchers for species in this paper. Many more new records are already known, for the Guianas as a whole or for some of the separate countries, which will be dealt with later. All identifications have been made by or verified by the author. Specimens are preserved in B, unless otherwise indicated. Collector abbreviations used are: SA: Sipman and Aptroot; CS: Cornelissen and ter Steege.

*Arthonia accolens* **Guyana**: Timehri, SA 18097x; Jawalla, SA 18447.
*Arthonia aciniformis* **Guyana**: Jawalla, SA 18447c; 1 km S of Kamarang, SA 18930; Mt. Latipu, SA 19010d; Trail to Mount Pui Pui N of Waramadan, SA 19221d, 19308d; Mabura Hill, CS 1.
*Arthonia trilocularis* **Guyana**: Jawalla, SA 18447d; 1 km S of Kamarang,

SA 18930b; Mt. Latipu, SA 19010f; Trail to Mount Pui Pui N of Waramadan, SA 19314; **Surinam**: Lelie Mountains, Lindeman c.s. 234a (U).

*Aulaxina submuralis* **Guyana**: Jawalla, SA 18443.

*Bacidia apiahica* **Guyana**: Trail to Mount Pui Pui N of Waramadan, SA 19228.

*Bacidia palmularis* **Guyana**: Jawalla, SA 18441; 1 km S of Kamarang, SA 18945d; Mt. Latipu, SA 19008b; Trail to Mount Pui Pui N of Waramadan, SA 19227f, 19308b; Mabura Hill, CS 7.

*Bacidia stanhopeae* **Guyana**: 1 km S of Kamarang, SA 18931d.

*Brigantiaea leucoxantha* **Guyana**: Kamarang, SA 18110.

*Byssolecania fumosonigricans* **Guyana**: Trail to Mount Pui Pui N of Waramadan, SA 19230i; Mabura Hill, CS 24; **Surinam**: Jodensavanne, Lindeman 4758f (U); Lelie Mountains, Lindeman c.s. 556g (U).

*Byssoloma aeruginascens* **Guyana**: 2 km N of Kamarang, SA 18215; Jawalla, SA 18449e; Mt. Roraima, N foothills, SA 18546d; Trail to Mount Pui Pui N of Waramadan, SA 19228e, 19314; Mabura Hill, CS 25.

*Byssoloma leucoblepharum* **Guyana**: Timehri, SA 17997a, 18097f; Jawalla, SA 18448e; Mt. Roraima, N foothills, SA 18546b; Mt. Roraima, N-foot (700 m), SA 18737; 1 km S of Kamarang, SA 18933d; Mt. Latipu, SA 19008c, 19128c; Trail to Mount Pui Pui N of Waramadan, SA 19229, 19312d; Mabura Hill, CS 26; **Surinam**: Brownsberg, LBB 13285c, 12312a (U).

*Calicium hyperelloides* **French Guiana**: Cayenne, Aptroot 15097 (B, U), teste Tibell.

*Catinaria versicolor* **Guyana**: Jawalla, SA 18308.

*Chaenotheca brunneola* **Guyana**: Mt. Latipu, SA 19092; Trail to Mount Pui Pui N of Waramadan, SA 19419, 19420, 19473, 19495; **Surinam**: Brownsberg, Aptroot 14898 (B, U), teste Tibell; **French Guiana**: Saül, Aptroot 15429 (B, U), teste Tibell.

*Chrysothrix candelaris* **Guyana**: Timehri, SA 18081; Trail to Mount Pui Pui N of Waramadan, SA 19475; Mt. Makarapan, Maas c.s. 7666 (U); **Surinam**: Brownsberg, Aptroot 14933 (U); **French Guiana**: Cayenne, Aptroot 15049 (B, U).

*Cladia aggregata* **Guyana**: Mt. Roraima, N foothills (camp 4), SA 18664; Mt. Roraima, N-slope (14–2000 m), SA 18825, 18872, 18884; Mt. Latipu, SA 19132, 19134; **Surinam**: Bakhuis Mts, Florschütz and Maas 2969 (U).

*Coccocarpia erythrocardia* **Guyana**: Mt. Latipu, N-foot, SA 18992; Mt. Latipu, SA 19070.

*Corella zahlbruckneri* **Guyana**: Trail to Mount Pui Pui N of Waramadan, SA 19413.

*Crocynia pyxinoides* **Guyana**: Timehri, SA 17905, 18025; Kamarang, SA 18193; 2 km N of Kamarang, SA 18262; **Surinam**: Zanderij, Narain 53 (B); Kabalebo, Zielman 9668 (B); **French Guiana**: Tonnegrande, Aptroot 15488 (B, U).

*Dimerella hypophylla* **Guyana**: Mt. Roraima, N foothills, SA 18541c; 1 km S of Kamarang, SA 18935b, 18935c; Trail to Mount Pui Pui N of Waramadan, SA 19227c, 19227d, 19308c.

*Echinoplaca affinis* **Guyana**: 2 km N of Kamarang, SA 18244e; Jawalla, SA 18408, 18442; Mt. Roraima, N foothills, SA 18546e; Mt. Roraima, N-foot (700 m), SA 18739; 1 km S of Kamarang, SA 18936d; Trail to Mount Pui Pui N of Waramadan, SA 19222d, 19309c.

*Echinoplaca diffluens* **Guyana**: Jawalla, SA 18442b; 1 km S of Kamarang, SA 18945g.

*Echinoplaca heterella* **Guyana**: Mabura Hill, CS 41.

*Echinoplaca pellicula* **Guyana**: Timehri, SA 18097n; Trail to Mount Pui Pui N of Waramadan, SA 19230k.

*Erioderma sorediatum* **Guyana**: Mt. Latipu, SA 19107a.

*Erioderma verruculosum* **Guyana**: Mt. Roraima, N-slope (14–2000 m), SA 18853d.

*Haematomma leprarioides* **Guyana**: Timehri, SA 18047.

*Logilvia gilva* **Guyana**: Mt. Roraima, N-foot (700 m), SA 18739b; Trail to Mount Pui Pui N of Waramadan, SA 19308.

*Mazosia pseudobambusae* **Guyana**: Mt. Roraima, N foothills, SA 18542g; 1 km S of Kamarang, SA 18938c; Trail to Mount Pui Pui N of Waramadan, SA 19230p; Mabura Hill, CS 50.

*Microtheliopsis uleana* **Guyana**: Trail to Mount Pui Pui N of Warmadan, SA 19225c; Mabura Hill, CS 81.

*Normandina pulchella* **Guyana**: Kamarang, SA 18108; Jawalla, SA 18307; **French Guiana**: Saül, Montfoort and Ek 1045, 1046 (U), Aptroot 15176 (B, U); Tonnegrande, Aptroot 15490 (B, U).

*Oropogon loxensis* **Guyana**: Mt. Roraima, N-slope (14–2000 m), SA 18907.

*Phylloporis phyllogena* **Guyana**: Timehri, SA 18097d; 1 km S of Kamarang, SA 18940b; Trail to Mount Pui Pui N of Waramadan, SA 19223; Mabura Hill, CS 88.

*Phylloporis platypoda* **Guyana**: Jawalla, SA 18444d, 18445; 1 km S of Kamarang, SA 18940; Mt. Latipu, SA 19009; Trail to Mount Pui Pui N of Waramadan, SA 19222f, 19317b; Mabura Hill, CS 91.

*Physcidia squamulosa* **Guyana**: 2 km N of Kamarang, SA 18274.

*Physcidia wrightii* **Guyana**: 2 km N of Kamarang, SA 18221; Mt. Roraima, N foothills, SA 18587; Mt. Roraima, N-foot (700 m), SA 18713, 18732, 18795; Mt. Latipu, SA 19114; Trail to Mount Pui Pui N of Waramadan, SA 19255, 19408, 19423, 19468, 19479; Mabura Hill, CS C-0077 (U); **Surinam**: Brownsberg, Aptroot 14938 (B, U); **French Guiana**: Saül, Montfoort and Ek 2192 (U), Aptroot 15422 (B, U).

*Polychidium dendriscum* **Guyana**: Mt. Roraima, N-slope (14–2000 m), SA 18841.

*Porina leptosperma* **Guyana**: Jawalla, SA 18445b, 18449b.

*Porina rubentior* **Guyana**: Timehri, SA 18097b; Jawalla, SA 18449c; Mt. Roraima, N foothills, SA 18542c, 18542d; 1 km S of Kamarang,

SA 18945h; Mt. Latipu, SA 19009b; Trail to Mount Pui Pui N of Waramadan, SA 19223b; Mabura Hill, CS 99.
*Porina rufula* **Guyana**: Jawalla, SA 18449; Mt. Roraima, N foothills, SA 18542, 18542b; Mt. Roraima, N-foot (700 m), SA 18738b, 18738c; 1 km S of Kamarang, SA 18941d; Mt. Latipu, N-foot, SA 19009e; Trail to Mount Pui Pui N of Waramadan, SA 19223d, 19315c; Mabura Hill, CS 103.
*Pseudocyphellaria aurata* **Guyana**: Trail to Mount Pui Pui N of Waramadan, SA 19410.
*Pyrgidium monticellum* **Guyana**: Mt. Latipu, SA 18977.
*Pyrgillus americanus* **Guyana**: Trail to Mount Pui Pui N of Waramadan, SA 19429.
*Raciborskiella janeirensis* **Guyana**: Jawalla, SA 18411b; Mt. Latipu, SA 19009g; Trail to Mount Pui Pui N of Waramadan, SA 19227b.
*Sticta fuliginosa* **Guyana**: Mt. Latipu, SA 19112; Trail to Mount Pui Pui N of Waramadan, SA 19377.
*Stirtonia sprucei* **Guyana**: Jawalla, SA 18443c; 1 km S of Kamarang, SA 18934b; Trail to Mount Pui Pui N of Waramadan, SA 19222c, 19316; Mabura Hill, CS 112.
*Strigula maculata* **Guyana**: 1 km S of Kamarang, SA 18943b.
*Strigula melanobapha* **Guyana**: Timehri, SA 18097r; Jawalla, SA 18446.
*Tapellaria epiphylla* **Guyana**: 1 km S of Kamarang, SA 18944; Trail to Mount Pui Pui N of Waramadan, SA 19220e.
*Teloschistes flavicans* **Surinam**: Brownsberg, Aptroot 14924 (U).
*Tricharia santessoniana* **Guyana**: Mt. Roraima, N-foot (700 m), SA 18739c.
*Trichothelium annulatum* **Guyana**: Jawalla, SA 18449f, 18449g; Mt. Roraima, N foothills, SA 18545f, 18945; Trail to Mount Pui Pui N of Waramadan, SA 19220d, 19228c, 19314d; Mabura Hill, CS 121.
*Tylophoron crassiusculum* **Guyana**: Jawalla, SA 18300.
*Tylophoron protrudens* **Guyana**: Timehri, SA 18009; Kamarang, SA 18184.
*Usnea baileyi* **Guyana**: Trail to Mount Pui Pui N of Waramadan, SA 19480.

## Distribution patterns

Six main geographical zones are distinguished in the Guianas, which are important in any consideration of lichen distribution: (**1**) A narrow band of low-lying, fertile alluvial soils, is found along the coast. Much of it is cultivated and human influence is strongest here. (**2**) A zone of flat, low-lying white sand extends along the coast from the Venezuelan border to the westernmost part of French Guiana. This carries extensive savannahs and light forests. Because the soil is poor, there is little human influence, except for fire. (**3**) Most of the remaining area consists of hilly country below *c.* 500 m, covered with native forest. It is virtually uninhabited and at present little influenced by man. Towards the east the bedrock changes from predominantly sandstone to predominantly granite, and accordingly forests become richer. (**4**) In the extreme west the hills rise to over 1000 m. Here the landscape is dominated by table mountains and passes into the Guiana Highlands area in Venezuela. Soils are usually very poor and considerable savannah areas exist among the forests, often on bare rock flats on top of the table lands. (**5**) One of the highest table mountains of the Guiana Highlands, Mount Roraima, is partly in Guyana. It rises to 2800 m and shows an altitudinal zone with mossy forest and scrub. (**6**) In southern Guyana and Surinam extensive savannahs exist, the result of a drier climate.

Since the lichen flora in the Guianas has been explored in only a few places with any intensity it is, consequently, difficult to distinguish between discontinuous distributions due to under-recording and real distribution patterns, coincident with the above geographical areas which have been used to identify distribution patterns.

*A. Coastal zone*

As recorded already by Aptroot (1988), taxa of the Pyxinaceae often show a strong preference for either the coastal zone or the interior, and are little influenced by elevation. Some appear to prefer the eastern or the western parts of the coastal zone, such as *Dirinaria aegialita* which is absent in the west, *D. purpurascens* which is absent from the east, *D. picta* which is more common in the east and *Pyxine cocoes* which is more common in the west. Also in the Ramalinaceae similar patterns are seen, for example, *Parmotrema praesorediosum* is known only from the coastal zone, whereas most other parmelioid genera seem to avoid the coast (Aubel 1991).

*B. Roraima*

A few taxa seem to be restricted to the higher parts of Mount Roraima, above *c.* 1500 m; such as *Oropogon loxensis* and *Polychidium dendriscum*.

*C. Western table lands*

Restricted to the table mountain area of western Guyana appear to be species such as *Cladina confusa*, *C. sprucei*, *Cladonia signata*; most Lobariaceae, such as *Pseudocyphellaria aurata*, *Sticta fuliginosa*, but not *Sticta weigelii*; *Biatorella conspersa* and some Pannariaceae like *Erioderma* and *Leioderma*; among Ramalinaceae most species of *Hypotrachyna* and *Parmotrema peralbidum*; among Pyxinaceae *Heterodermia barbifera* and *H. leucomelos*; also, *Buellia tabacina* and *Catinaria versicolor* may belong here. In general, these species are not confined to the high tables and occur also in the valleys. This group contains species with a mainly montane distribution in the neotropics, such as the Pannariaceae,

Lobariaceae, and species of *Hypotrachyna* or *Heterodermia*, as well as species which are found elsewhere mainly at lower elevations, such as *Cladonia signata*. Several of the species found in the western table lands seem to extend to the highest points elsewhere in Guyana and Surinam, including *Cladia aggregata*, *Cladonia ceratophylla*, and *Heterodermia flavosquamosa* (Sipman, 1990).

## D. Inland zone

The majority of lichens are found in scattered localities throughout the inland areas of the Guianas, without any particular pattern, except perhaps an avoiding of the coast.

No lichens have been found so far, which are restricted to the white savannah zone (2) or the interior savannah zone (6), and no differentiated distribution patterns have been found among foliicolous lichens. This is in agreement with results from other countries (Nowak and Winkler 1970), which indicate that foliicolous lichens have wide geographical and altitudinal amplitudes. Few foliicolous lichens are known from the coastal zone, however, probably because of the scarcity of well-developed forest.

A very distinct distribution pattern is shown by the family Cladoniaceae (Fig. 10.1). Taxa prefer open, savannah-type vegetation. The figures on the map concern the following geographical units: the table lands of western Guyana (zone 4); the rest of Guyana, mainly the white sand savannahs near the coast; the high mountains of interior Surinam; the white sand savannahs (zone 2) and surroundings in coastal Surinam; all of French Guiana. They represent the total numbers of species of the genera *Cladia*, *Cladina*, and *Cladonia*. Included are only reliably recorded, recently published or revised species, 22 altogether. The revision of the neotropical Cladoniaceae presently undertaken by Ahti will no doubt contain considerable additions and the true number of Cladoniaceae occurring in the Guianas is most probably well above 30.

As Fig. 10.1 shows, most species are restricted to the extreme west (zone 4), where they prefer the rocky savannahs on the table mountains. A few extend into high mountains in Surinam or the sandy savannahs, and very few extend to French Guiana. The latter are synanthropic species of decaying wood in clearings.

Concerning the geographical relationships of the lichen flora of the Guianas as a whole, the following observations may be made. Most of the species of Pyxinaceae investigated by Aptroot (1988) appear to have a pantropical distribution. Among Parmeliaceae (Aubel 1991) there appear to be a few species restricted to the neotropics, but again most are pantropical. Ahti (1984, 1986) reports on some Cladoniaceae

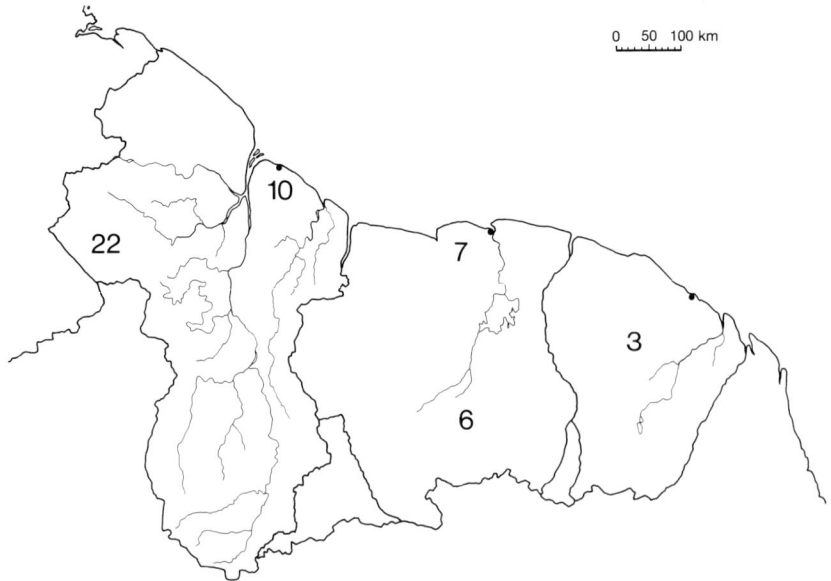

**Fig. 10.1.** Numbers of species of Cladoniaceae in different parts of the Guianas.

species restricted to the Guiana Highlands, a few of which also occur in the western part of the Guianas: *Cladina densissima* and *C. spinea*. Some further Cladoniaceae appear to be restricted to the Neotropics, such as *Cladina dendroides*, *Cladonia ceratophylla*, *C. secundana* (Stenroos 1989). Among foliicolous taxa, most species are pantropical. Some appear to be restricted to, or concentrated in the Amazon basin, such as *Mazosia praemorsa*, *M. rubropunctata*, *M. tumidula*, and *Stirtonia sprucei*. The combined neotropical + african distribution type is represented by *Byssoloma aeruginascens* and *Opegrapha filicina*. The only lichen species which seems restricted to the Guianas (assuming that an unlikely record from Spain is erroneous) and conspicuous enough to make that observation significant, seems to be *Polystroma fernandezii*.

This leads to the following conclusions: endemism is apparently very rare in the lichen flora of the Guianas; an element restricted to the Guiana Highlands is present, but not very prominent; the same counts for an element centred in the Amazon basin; a considerable proportion of species are widespread in the neotropics; also the combined neotropical + african distribution type is represented; probably the majority of the lichen species are pantropical.

## Observations on foliicolous lichens

On a collecting trip to Guyana in 1985 extensive collections of foliicolous lichens were made in nine different primary forest sites. Since such lichens are often very small and hard to recognize in the field, collecting was done by picking large numbers of leaves with recognizable fertile lichens, often over a hundred, from different habitats and of different textures. In this way some random sampling was effected in these collections and the results permit a comparison of the foliicolous lichen flora of different forest types.

Some restrictions have to be made, however. First, the collecting time varied from about two to six hours per locality. This does not seem to have influenced the conclusions, since the locality where most time was spent (nr. 5) remained the poorest site. Second, the investigated areas were not sharply delimited. Collecting was done on *c*. 100 m along and near a trail. Consequently, some variation in habitat was included. Again this does not seem to have influenced the conclusions, since the most varied locality (nr. 2) was one of the poorest sites. Another restriction was that collections were made only within reach of the ground. No information is available on the foliicolous lichen flora at higher levels in the tree-crowns. Observations in Columbia (Sipman, unpublished) indicate that there are fewer species higher up in the trees and that the flora there has a composition comparable to that in open forest types. The genera *Calopadia*, *Echinoplaca*, *Strigula*, and *Tapellaria* seem to be better represented there. The optimal habitat for foliicolous lichens in the forest appears to be the foliage of small, slow-growing woody plants, probably often dormant young trees.

Information on the foliicolous lichen flora of tropical forests is scarce. Nowak and Winkler (1970) recorded 45 species from five forest sites around 50 and 2000 m on the Sierra Nevada de Santa Marta, Columbia. The number of species per site ranged from 5 to 19, without clear altitudinal preferences. Later, Nowak and Winkler (1975) reported 57 foliicolous lichens from forest in Choco, Columbia, without an indication of the locality size.

The taxonomy of foliicolous lichens, although greatly clarified in recent years by Santesson (1952), Sérusiaux and Vězda, is still far from completely known. Only two-thirds of the collected taxa could be identified to species level. Therefore, both figures of completely identified species and those including unidentified taxa are given in the table below. The latter have of course a somewhat hypothetical status, but give a more complete impression.

*Description of localities*

**1.** EAST DEMERARA district. Timehri, Dakara Creek, Thompsons farm. Alt. *c.* 10 m. 6°29′N, 58°15′W. 2 February 1985. Moderately dry forest (Walaba forest) on poor, white sand.
**2.** UPPER MAZARUNI district. Pakaraima mountains, *c.* 2 km. N of Kamarang. Alt. *c.* 500 m. 5°53′N, 60°38′W. 4 February 1985. *c.* 25 m tall, virgin forest along rivulet, grading into light forest on rocky plateau with little soil and *c.* 10 m tall, well-lit savannah–forest on ridge.
**3.** UPPER MAZARUNI district. Jawalla village, at confluence of Kukui river and Mazaruni river. Alt. *c.* 500 m. 5°40′N, 60°29′W. 6 February 1985. *c.* 25 m tall, virgin forest on sandy soil near river.
**4.** UPPER MAZARUNI district. E-bank of Waruma river, *c.* 4 km S of confluence with Kako river (camp 1). Alt. *c.* 550 m. 5°28′N, 60°47′W. 9 February 1985. *c.* 25 m tall, virgin forest on river bank.
**5.** UPPER MAZARUNI district. N-slope of Mt. Roraima, above E-bank of Waruma river (camp 5). Alt. *c.* 700 m. 5°17′N, 60°46′W. 12–19 February 1985. *c.* 25 m tall, virgin, wet, mossy forest.
**6.** UPPER MAZARUNI district. Kamarang, along trail to Waramadan. Alt. *c.* 500 m. 5°52′N, 60°37′W. 23 February 1985. *c.* 25 m tall, disturbed forest near village.
**7.** UPPER MAZARUNI district. N-foot of M. Latipu, *c.* 8 km N of Kamarang. Alt. *c.* 600 m. 5°57′N, 60°38′W. 24 February 1985. *c.* 15 m tall forest on poor sandy soil.
**8.** UPPER MAZARUNI district. Trail from Kamarang river to Pui Pui mountain, at landing site *c.* 5 km NW of Waramadan. Alt. *c.* 600 m. 5°48′N, 60°47′W. 1 March 1985. *c.* 25 m tall forest near river.
**9.** UPPER MAZARUNI district. Trail from Kamarang river to Pui Pui mountain, *c.* 10 km N of Waramadan. Alt. *c.* 800 m. 5°57′N, 60°45′W. 28 February 1985. Light forest on rocky savannah with poor soil.

*List of foliicolous lichens collected on nine forest plots in Guyana*

| Lichen | Locality | | | | | | | | |
|---|---|---|---|---|---|---|---|---|---|
| | 1 | 2 | 3 | 4 | 5 | 6 | 7 | 8 | 9 |
| *Actinoplaca strigulacea* | | | | | | 6 | | | |
| *Anisomeridium foliicolum* | | | 3 | | | 6 | | 8 | |
| *Arthonia accolens* | 1 | | 3 | | | | | | |
| *Arthonia aciniformis* | | | 3 | | | 6 | 7 | 8 | 9 |
| *Arthonia cyanea* | | 2 | 3 | 4 | | 6 | | | 9 |
| *Arthonia trilocularis* | | | 3 | | | 6 | 7 | | 9 |
| *Aspidothelium fugiens* | | | | | | | | 8 | |
| *Aulaxina minuta* | | | 3? | | | 6 | | | |
| *Aulaxina quadrangula* | 1 | | | | | 6 | | | 9 |

*List of foliicolous lichens collected on nine forest plots in Guyana* (cont.)

|  | Locality | | | | | | | | |
|---|---|---|---|---|---|---|---|---|---|
| Lichen | 1 | 2 | 3 | 4 | 5 | 6 | 7 | 8 | 9 |
| *Aulaxina submuralis* |  |  | 3 |  |  |  |  |  |  |
| *Bacidia apiahica* |  |  |  |  |  |  |  | 8 |  |
| *Bacidia brasiliensis* |  |  | 3 |  |  | 6 |  | 8 | 9 |
| *Bacidia palmularis* |  |  | 3 |  |  | 6 | 7 | 8 | 9 |
| *Bacidia stanhopeae* |  |  |  |  |  | 6 |  |  |  |
| *Badimia dimidiata* |  |  | 3 | 4 |  |  | 7 | 8 | 9 |
| *Byssolecania deplanata* | 1 |  |  |  |  | 6 | 7 | 8 | 9 |
| *Byssolecania fumosonigricans* |  |  |  |  |  |  |  | 8 |  |
| *Byssoloma aeruginascens* |  | 2 | 3 | 4 |  |  |  | 8 | 9 |
| *Byssoloma leucoblepharum* | 1 |  | 3 | 4 | 5 | 6 | 7 | 8 | 9 |
| *Byssoloma subdiscordans* |  | 2 | 3 | 4 | 5 | 6 |  |  | 9 |
| *Byssoloma tricholomum* | 1 |  |  |  |  |  |  |  | 9 |
| *Calenia conspersa* |  |  | 3 |  |  |  |  | 8 |  |
| *Calenia submaculans* |  | 2 | 3 | 4 |  | 6 |  | 8 |  |
| *Chroodiscus coccineus* | 1 |  | 3 | 4 |  | 6 |  | 8 | 9 |
| *Coccocarpia epiphylla* |  |  |  | 4 |  |  |  |  |  |
| *Coccocarpia tenuissima* |  |  | 3 | 4 |  |  |  | 8 | 9 |
| *Cryptothecia candida* |  |  |  |  |  |  |  | 8 |  |
| *Dimerella hypophylla* |  |  |  | 4 |  | 6 |  | 8 | 9 |
| *Echinoplaca affinis* |  | 2 | 3 | 4 | 5 | 6 |  | 8 | 9 |
| *Echinoplaca diffluens* |  |  | 3 |  |  | 6 |  |  |  |
| *Echinoplaca pellicula* | 1 |  |  |  |  |  |  | 8 |  |
| *Gyalectidium filicinum* |  |  | 3 |  |  | 6 |  | 8 | 9 |
| *Lasioloma arachnoideum* |  |  | 3 | 4 |  | 6 |  | 8 | 9 |
| *Loflammea flammea* |  |  |  |  |  | 6 |  |  |  |
| *Logilvia gilva* |  |  |  |  | 5 |  |  |  | 9 |
| *Mazosia melanophthalma* | 1 |  | 3 | 4 | 5 | 6 | 7 | 8 | 9 |
| *Mazosia phyllosema* | 1 |  | 3 | 4 | 5 | 6 |  | 8 |  |
| *Mazosia pilosa* | 1 |  | 3 |  |  | 6 |  | 8 | 9 |
| *Mazosia praemorsa* | 1 | 2 | 3 | 4 | 5 | 6 | 7 | 8 | 9 |
| *Mazosia pseudobambusae* |  |  |  | 4 |  | 6 |  | 8 |  |
| *Mazosia rotula* | 1 | 2 | 3 | 4 |  | 6 | 7 | 8 | 9 |
| *Mazosia rubropunctata* | 1 |  |  |  |  |  |  |  | 9 |
| *Mazosia tumidula* |  | 2 | 3 | 4 |  |  |  | 8 |  |
| *Microtheliopsis uleana* |  |  |  |  |  |  |  | 8 |  |
| *Opegrapha filicina* |  |  | 3 |  |  | 6 |  |  | 9 |
| *Phyllobathelium epiphyllum* | 1 |  | 3 |  |  |  |  | 8 |  |
| *Phyllophiale alba* |  |  |  |  |  | 6 |  | 8 | 9 |
| *Phylloporis phyllogena* | 1 |  |  |  |  | 6 |  | 8 |  |

List of foliicolous lichens collected on nine forest plots in Guyana (cont.)

| Lichen | Locality | | | | | | | | |
|---|---|---|---|---|---|---|---|---|---|
| | 1 | 2 | 3 | 4 | 5 | 6 | 7 | 8 | 9 |
| Phylloporis platypoda | | | 3 | | | 6 | 7 | 8 | 9 |
| Porina epiphylla | 1 | | 3 | | | 6 | 7 | 8 | 9 |
| Porina fulvella | | 2 | | | | 6 | | 8 | |
| Porina imitatrix | | | | | | | | 8 | |
| Porina leptosperma | | | 3 | | | | | | |
| Porina rubentior | 1 | | 3 | 4 | | 6 | 7 | 8 | |
| Porina rufula | | | 3 | 4 | 5 | 6 | 7 | 8 | 9 |
| Raciborskiella janeirensis | | | 3 | | | | 7 | 8 | |
| Stirtonia sprucei | | | 3 | | | 6 | | 8 | 9 |
| Strigula concreta | | | | | | | 7 | | |
| Strigula elegans | 1 | | 3 | 4 | 5 | 6 | 7 | 8 | 9 |
| Strigula maculata | | | | | | 6 | | | |
| Strigula melanobapha | 1 | | 3 | | | | | | |
| Strigula nemathora | | | 3 | | | 6 | | 8 | 9 |
| Strigula subtilissima | | 2 | 3 | 4 | 5 | | 7 | 8 | 9 |
| Tapellaria epiphylla | | | | | | 6 | | 8 | |
| Tricharia carnea | 1 | | | | | | | | |
| Tricharia dilatata | | | 3 | | | | | | |
| Tricharia santessoniana | | | | | 5 | | | | |
| Tricharia urceolata | 1 | | | | | | | 8 | |
| Trichothelium annulatum | | | 3 | 4 | | 6 | | 8 | 9 |
| Completely identified taxa, total = 69 | 21 | 10 | 43 | 23 | 11 | 41 | 17 | 46 | 34 |
| Including unidentified taxa, total = 108 | 22 | 14 | 54 | 35 | 12 | 57 | 7 | 69 | 45 |

These results show that over 100 species of foliicolous lichens, nearly 70 of which were completely identified, occur in the forest undergrowth in a fairly small area of Upper Mazaruni, Guyana. The total number of foliicolous lichens known from the Guianas has been raised from 49, as mentioned in the checklist (Hekking and Sipman 1988), to 85.

On a single plot, nearly 70 species have been found. This figure does not seem to be exceptional, since comparable investigations in forests near Araracuara in the Amazon basin in Columbia (Sipman, unpublished), revealed a plot with 89 species. Particularly well represented are the genera *Byssoloma*, *Mazosia*, and *Porina*. Among the commonest species are *Arthonia aciniformis*, *Bacidia palmularis*, *Badimia dimidiata*, *Byssolecania deplanata*, *Byssoloma aeruginascens*, *B. leucoblepharum*, *B. subdiscordans*, *Calenia submaculans*, *Chroodiscus coccineus*, *Echinoplaca affinis*, most species of *Mazosia*, *Phyllobathelium epiphyllum*, *Phylloporis* spp., *Porina epiphylla*, and *P. rufula*-group, *Strigula elegans*, *S. subtilissima*, and *Trichothelium annulatum*.

Field observations indicate that some species of *Echinoplaca* and *Tricharia* are also frequent, but under-represented, because they are often without apothecia and thus difficult to recognize. The genera *Arthonia*, *Byssoloma*, *Byssolecania*, *Dimerella*, and *Opegrapha*, and *Porina epiphylla* seem to prefer shaded conditions, whereas *Badimia dimidiata*, *Calenia*, *Chroodiscus*, and the species of *Mazosia* prefer more light. Species of *Coccocarpia* seem to prefer forest near streams. Certain taxa seem to be pioneers on young leaves such as *Calenia*, *Echinoplaca*, and *Tricharia*. Very coriaceous leaves are by no means only those being colonized. Evidently only the age of the leaves plays a role, and thin, old leaves may have ample lichen growth.

A considerable variation exists between plots: species numbers range from 12 to 68. The richest plots (3, 6 and 8) are from well-developed rain forests on rich soils near a river with (sometimes man-made) gaps in the canopy. Among the poorest sites, two (2 and 7) are from low and light forest on very poor substrates, and one (5) a very humid, mossy forest. Locality 1, equally with a fairly low number of species, is from a dry type of forest. Locality 4 represents a transition to the very humid forest of plot 5.

Although no detailed information was collected on numbers of foliicolous lichen species in strongly disturbed, secondary or planted forests, it seems likely that in such habitats, numbers will be lower. Unpublished observations in a selectively logged forest in French Guiana (Piste de St.-Elie) showed that the increased amount of light penetrating into the lowest levels of the forest had a strong effect on all lichens near soil level, which were all discoloured and moribund. Such conditions evidently result in the restriction of the lichen flora to elements which resist high light intensities. Secondary forests tend to be composed of fast-growing species, which generally are unattractive to foliicolous lichens. Further, the uniform conditions in planted forests are likely to lead to an impoverished foliicolous lichen flora there as well.

## References

Ahti, T. (1984). The status of *Cladina* as a genus segregated from *Cladonia*. *Beiheft zur Nova Hedwigia*, **79**, 25–61.

Ahti, T. (1986). New species and nomenclatural combinations in the lichen genus *Cladonia*. *Annales Botanici Fennici*, **23**, 205-20.

Aptroot, A. (1988). *Pyxinaceae* (Lichenes). In *Flora of the Guianas*, Ser. E, Fasc. 1 (ed. A.R.A. Görts-van Rijn), pp. 1-60. Koenigstein; Koeltz. ["1987"].

Aubel, R. van (1991). *Parmeliaceae* (Lichenes). In *Flora of the Guianas*, Ser. E, Fasc. 2 (ed. A.R.A. Görts-van Rijn). (in preparation)

Eriksson, O. and Hawksworth, D.L. (1986). Outline of the Ascomytes—1986. *Systema Ascomycetum*, **5**, 185-324.

Hafellner, J. (1984). Studien in Richtung einer natürlicheren Gliederung der Sammelfamilien Lecanoraceae und Lecideaceae. *Beiheft zur Nova Hedwigia*, **79**, 241-371.

Hekking, W.H.A. and Sipman, H.J.M. (1988). The lichens reported from the Guianas before 1987. *Willdenowia*, **17**, 193-228.

Nowak, R. and Winkler, S. (1970). Foliicole Flechten der Sierra Nevada de Santa Marta (Kolumbien) und ihre gegenseitigen Beziehungen. *Österreichische Botanische Zeitschrift*, **118**, 456-85.

Nowak, R. and Winkler, S. (1975). Foliicolous lichens of Chocó, Colombia, and their substrate abundances. *Lichenologist*, **7**, 53-8.

Santesson, R. (1952). Foliicolous lichens. I. A revision of the taxonomy of the obligately foliicolous, lichenized fungi. *Symbolae Botanicae Upsalienses*, **12**, 1-590.

Sipman, H.J.M. (1990). Lichenotheca Latinoamericana a museo botanico berolinensi edita, fasciculum primum. *Willdenowia*, **19**, 543-51.

Stenroos, S. (1989). Taxonomic revision of the *Cladonia miniata* group. *Annales Botanici Fennici*, **26**, 237-61.

# 11. Aspects of the foliose lichen flora of the southern-central coast of São Paulo State, Brazil[1]

M.P. MARCELLI

*Instituto de Botânica, C.P. 4005, CEP 01051, São Paulo – SP, Brazil*

### Abstract

The southern-central littoral of São Paulo State at 23°25′S has an equatorial climate with 2000 ±500 mm annual rainfall. Principal habitats for lichens include mangroves and restinga formations, rocky shores, and urban environments and substrates. The present work compares the foliose lichen floras of these habitats. At present 161 species in 28 genera are recognized, most of them (84 per cent) foliose. The family Parmeliaceae contributes some 56 per cent of the foliose lichen flora, followed by Physciaceae (17 per cent), Pannariaceae (7 per cent), Stictaceae, Coccocarpiaceae, Collemataceae, and others. Fruticose species (*Usnea* and *Ramalina*) account for only about 14 per cent of the flora. Parmeliaceae are much more abundant than any other family and *Parmotrema* alone represents 66 per cent of that family with many species of the *P. gardneri* complex. All species with cyanobacterial symbionts are found in mangroves. Many species with green algae are specific to certain habitats, substrates or hosts. Higher air humidity and wind protection are necessary for most of the species to become fertile.

### Introduction

São Paulo in south-eastern Brazil is the most heavily populated and economically developed state of the country. It is also a region where

[1] This work was supported by Grant No. 303282/87-8 from CNPq (Conselho Nacional de Desenvolvimento Científico e Tecnológico).

*Tropical Lichens: Their Systematics, Conservation, and Ecology* (ed. D.J. Galloway), Systematics Association Special Volume No. 43, pp. 151-70. Clarendon Press, Oxford, 1991. © The Systematics Association, 1991.

extensive and intensive alterations of the natural flora have occurred. Deforestation, atmospheric pollution, and urbanization have drastically increased during this century, especially since the 1950s.

The São Paulo coastal cities are isolated from the more developed interior of the state by a chain of mountains about 700–800 m high (Serra do Mar) which are covered in tropical rainforest. These cities are located on a sand strip ('restinga') ranging from 2 to 5 km between the sea and the mountains.

Not long ago, due to its isolation by the mountains the flora along the coast was well preserved since access to the beaches and farms was very difficult. Starting in the 1970s, an intensive process of urbanization took place after the construction of better roads. Property speculation increased considerably and the urban zone around the cities extended several times. The flat, sandy soil facilitated deforestation and enormous tracts of forests, shrub and dune vegetation have disappeared from the area under study. This process of urbanization, the situation of the roads and the presence of an old and important harbour city (Santos) at the central littoral of the state caused a deterioration gradient of the natural flora from Santos to Cananéia (a small fishing-town near the state south frontier) and the Ilha do Cardoso (Cardoso Island; a protected natural reserve).

Close to the Tropic of Capricorn, this part of the coast, between 23° and 25°, has an 'equatorial ever humid' climate, and the annual temperature mean is a little over 20 °C. Frost never occurs and monthly mean temperatures show very little variation. However, absolute records of 3.2 °C and 40.2 °C have already been obtained at Itanhaém (Lamberti 1969). Annual precipitation is 2000±500 mm (Blanco and Godoy 1966), and usually it rains more than 60 mm during the drier months (Godoy and Ortolani 1962). The mean monthly relative humidity ranges between 76 per cent and 90 per cent (Lamberti 1969).

Extensive beaches are the principal components of the land–sea interface and behind them the sandy soil covered by 'restinga' vegetation extends into the foothills of Serra do Mar. Restinga vegetation has a characteristic flora that changes in structure according to soil humidity and distance from the sea. In higher elevations and near the ocean it is shrubby, but in other areas large trees more than 15 m high occur. Bogs are common and the inner microclimate varies from sunny to shady and from dry to inundated providing a variety of different habitats, each one with its characteristic lichen flora. Foliose and fruticose species occur only in situations in which sunlight reaches trunks and branches.

Mangroves are found at river margins and mouths and on protected areas of the coast. Their structure varies, with shrubby to arboreal (c. 17 m high) vegetation found, but the soil is always soaked and their deeper areas are never too shady for lichens (as happens in the restinga). However, foliose and fruticose species appear only in the more illuminated regions (Marcelli 1987).

Rocky shores are rare. Places where the mountains reach the coast are very few in the southern–central part of the state. They occur at Itanhaém, Peruibe, and Ilha do Cardoso, and granite, quartz, basalt, and sedimentary rocks are the main components in these places. These rocks generally occur at the foot of rounded hills which are covered by grassy fields or pluvial forests.

Trees in the urban zone are almost exclusively *Terminalia cattapa* L. (90 per cent), an African Combretaceae introduced as shade trees. In addition, palms are always present.

The present work discusses the foliose lichen flora of the coastal region between Santos and Cananéia. The composition of species, genera, families, and morphobiological groups, in mangroves, restingas, rocky shores, and central urban zones are compared.

## The lichen flora

Specimens collected so far (more than 5000) belong to 161 non-crustose species (28 genera) and 135 foliose species (22 genera). Table 11.1 shows the number of species found by genus in the habitats studied, together with abbreviations used in the figures for generic names. The morphobiological groups used follow Marcelli (1987): FOLVE = foliose with green algae, FOLAZ = foliose with cyanobacteria, FRUTI = fruticose, FILAM = filamentose. A complete list of the foliose and filamentose taxa presently known is given below. To facilitate cross-reference to Marcelli (1987) where quantitative mangrove data can be found, the code used for mangrove species is provided. Species marked (*) are uncertain synonyms but well-established morphological and/or ecological varieties of some other identified species. (1) species found in mangroves and restingas. (2) species found only in mangroves. (3) species found only in restingas and/or beaches. (4) species found only on rock shores. (5) species found only on coastal hills. Orders and Families follow Poelt (1973).

*Ascolichens*

    1. Graphidales

        Gyalectaceae
            *Coenogonium leprieurii*    3
            *Coenogonium moniliforme*    3

**Table 11.1** Numbers of species from 28 non-crustose/squamulose lichen genera found on southern-central São Paulo State coasts in the habitats investigated. They are arranged in alphabetical order by abbreviations (Abbr.) used in the generic figures

| Abbr. | Genus | Total | Mangrove | Restinga | Rocks |
|---|---|---|---|---|---|
| Bb | *Bulbothrix* | 5 | 2 | 5 | 0 |
| Cc | *Coccocarpia* | 10 | 10 | 5 | 1 |
| Cl | *Corella* | 1 | 1 | 0 | 0 |
| Cn | *Coenogonium* | 2 | 0 | 2 | 0 |
| Cr | *Cora* | 1 | 1 | 1 | 0 |
| Dr | *Dirinaria* | 5 | 1 | 4 | 2 |
| Dt | *Dictyonema* | 1 | 1 | 1 | 0 |
| Er | *Erioderma* | 3 | 3 | 1 | 0 |
| Ht | *Heterodermia* | 8 | 5 | 8 | 1 |
| Hy | *Hypotrachyna* | 3 | 3 | 2 | 0 |
| Lb | *Lobaria* | 2 | 2 | 1 | 0 |
| Lp | *Leptogium* | 6 | 6 | 6 | 1 |
| Pc | *Polychidium* | 1 | 1 | 1 | 0 |
| Pe | *Parmeliella* | 1 | 1 | 0 | 0 |
| Ph | *Physcia* | 7 | 4 | 3 | 2 |
| Pl | *Parmelina* | 7 | 2 | 6 | 1 |
| Pm | *Physma* | 1 | 1 | 1 | 0 |
| Pn | *Pannaria* | 5 | 5 | 1 | 0 |
| Pp | *Pseudoparmelia* | 9 | 1 | 9 | 1 |
| Ps | *Pseudocyphellaria* | 2 | 0 | 2 | 0 |
| Pt | *Parmotrema* | 51 | 23 | 35 | 10 |
| Px | *Pyxine* | 3 | 1 | 1 | 1 |
| Rl | *Relicina* | 1 | 1 | 1 | 0 |
| Rm | *Ramalina* | 11 | 7 | 2 | 2 |
| St | *Sticta* | 3 | 1 | 3 | 1 |
| Tl | *Teloschistes* | 2 | 2 | 2 | 1 |
| Un | *Usnea* | 9 | 9 | 3 | 0 |
| Xt | *Xanthoparmelia* | 1 | 0 | 0 | 1 |
| Total | | 161 | 94 | 106 | 25 |

2. Lecanorales
    Lichinaceae
        *Polychidium dendriscum*     Polden    1
    Stictaceae
        *Lobaria crenulata*     Lobcre    1
        *Lobaria peltigera*     Lobpel    1
        *Pseudocyphellaria aurata*     Psecyp    1
        *Pseudocyphellaria clathrata*        3
        *Sticta weigelii*     Stiwei    1
        *Sticta sinuosa*        3
        *Sticta variabilis*        3
    Collemataceae
        *Leptogium austroamericanum*     Lepaus    1
        *Leptogium isidiosellum*     Lepisi    1,4
        *Leptogium marginellum*     Lepmar    1
        *Leptogium moluccanum*     Lepmol    1
        *Leptogium phyllocarpum*     Lepphy    1
        *Leptogium ulvaceum*     Lepulv    1
        *Physma byrsaenum*     Phybyr    1
    Coccocarpiaceae
        *Coccocarpia asterella*     Cocast    1
        *Coccocarpia domingensis*        3,5
        *Coccocarpia erythroxili*     Cocery    2
        *Coccocarpia palmicola*     Coccro    1,4
        *Coccocarpia pellita*     Cocpel    1
        *Coccocarpia smaragdina*     Cocsma    1
        *Coccocarpia marcellii*     CoccoB    2
        *Coccocarpia* sp. 2 (*)—*domingensis*     CoccoA    1
        *Coccocarpia* sp. 3 (*)—*pellita*        1
        *Coccocarpia* sp. 4 (*)—*pellita*        1
    Pannariaceae
        *Erioderma unguigerum*     Eriphy    1
        *Erioderma wrightii*     Eriwri    2
        *Erioderma* sp. 1        1
        *Pannaria isidioidea*     Panisi    2
        *Pannaria lurida*     Panlur    1
        *Pannaria mosenii*     Panmos    1
        *Pannaria stylophora*     Pansty    2
        *Parmeliella nigrocincta*     Pannig    1
    Parmeliaceae
        *Bulbothrix goebelii*     Pargoe    1
        *Bulbothrix isidiza*        3

| | | |
|---|---|---|
| *Bulbothrix laevigatula* | Parlae | 1 |
| *Bulbothrix oliveirae* | | 3 |
| *Bulbothrix tabacina* | | 3 |
| *Hypotrachyna bahiana* | Parbah | 1 |
| *Hypotrachyna dentella* | Parden | 1 |
| *Hypotrachyna formosana* | Parfor | 2 |
| *Parmelina antillensis* | Parant | 2 |
| *Parmelina spumosa* | Parspu | 1 |
| *Parmelina* cf. *versiformis* | | 3 |
| *Parmelina* sp. 1 | | |
| *Parmelina* sp. 2 | | 3 |
| *Parmelina* sp. 3 | | 3 |
| *Parmelina* sp. 4 | | 3 |
| *Parmotrema argentinum* | Pararg | 1 |
| *Parmotrema austrosinense* | | 5 |
| *Parmotrema cetratum* | Parcet | 1,5 |
| *Parmotrema chinense* | | 3 |
| *Parmotrema* cf. *argentinum* | Parcfa | 1 |
| *Parmotrema* cf. *expansum* | | 4 |
| *Parmotrema* cf. *gardneri* | | 1,5 |
| *Parmotrema* cf. *permutatum* | | 5 |
| *Parmotrema* cf. *reticulatum* | Parcfr | 1 |
| *Parmotrema* cf. *tinctorum* (\*) *sorediate* | | 3 |
| *Parmotrema consors* | | 3 |
| *Parmotrema crinitum* | Parcrn | 2 |
| *Parmotrema cristiferum* | Parcrt | 1 |
| *Parmotrema dilatatum* | Pardil | 1 |
| *Parmotrema fasciculatum* | | 3 |
| *Parmotrema fumarprotocetraricum* | | 2 |
| *Parmotrema gardneri* | Pargar | 1 |
| *Parmotrema internexum* | Parint | 2 |
| *Parmotrema lobulatum* | ParmoC | 1 |
| *Parmotrema macrocarpum* | Parmac | 1 |
| *Parmotrema madilyneae* | Parmad | 1 |
| *Parmotrema melanothrix* | Parmel | 1 |
| *Parmotrema michauxianum* | Parmix | 1 |
| *Parmotrema mordeni* (\*) (*praesorediosum* ?) | | 4 |
| *Parmotrema neotropicum* | Parneo | 2 |
| *Parmotrema praesorediosum* | Parpre | 1,5 |
| *Parmotrema reticulatum* | Parret | 1 |
| *Parmotrema sancti-angeli* | | 3,5 |
| *Parmotrema subarnoldii* | | 3 |
| *Parmotrema subisidiosum* | Parsbi | 2 |
| *Parmotrema sulphuratum* | Parsul | 1 |
| *Parmotrema tinctorum* | | 3,5 |
| *Parmotrema tinctorum* (\*) *isidiate/sorediate* | | 3 |
| *Parmotrema ultraluscens* | Parult | 1,4 |

| | | |
|---|---|---|
| *Parmotrema* sp. 1 | | 4 |
| *Parmotrema* sp. 2 | | 4 |
| *Parmotrema* sp. 5 | | 3 |
| *Parmotrema* sp. 6 | | 4 |
| *Parmotrema* sp. 7 | | 4 |
| *Parmotrema* sp. 9 | | 4 |
| *Parmotrema* sp. 10 | | 5 |
| *Parmotrema* sp. 11 | | 5 |
| *Parmotrema* sp. 13 | | 4 |
| *Parmotrema* sp. 14 | | 5 |
| *Parmotrema* sp. 16 | | 3 |
| *Parmotrema* sp. 17 | | 5 |
| *Parmotrema* sp. 18 | | 5 |
| *Parmotrema* sp. 19 | | 3 |
| *Parmotrema* sp. 20 | | 3 |
| *Parmotrema* sp. 21 | | 3 |
| *Parmotrema* sp. 22 | | 1,5 |
| *Parmotrema* sp. 23 | | 4 |
| *Pseudoparmelia* cf. *amazonica* | | 3 |
| *Pseudoparmelia* cf. *caroliniana* | | 3 |
| *Pseudoparmelia crozalsiana* | | 3 |
| *Pseudoparmelia salacinifera* | | 3 |
| *Pseudoparmelia sphaerospora* | Parsph | 1 |
| *Pseudoparmelia texana* | | 3 |
| *Pseudoparmelia* sp. 1 | | 5 |
| *Pseudoparmelia* sp. 2 | | 3 |
| *Pseudoparmelia* sp. 3 | | 3 |
| *Relicina abstrusa* | Parabs | 2 |
| *Xanthoparmelia* sp. 1 | | 4 |

Physciaceae

| | | |
|---|---|---|
| *Dirinaria aegialita* | | 1,4 |
| *Dirinaria* cf. *applanata* | | 4 |
| *Dirinaria* cf. *flava* | | 3 |
| *Dirinaria picta* | | 3 |
| *Dirinaria* sp. 1 | | 1 |
| *Heterodermia comosa* | Hetcom | 3 |
| *Heterodermia corallophora* | Hetcor | 1 |
| *Heterodermia dendritica* var. *propagulifera* | Hetpro | 1 |
| *Heterodermia obscurata* | Hetobs | 1 |
| *Heterodermia speciosa* | Hetspe | 1 |
| *Heterodermia squamulosa* | | 3 |
| *Heterodermia tremulans* | Hettre | 1 |
| *Heterodermia vulgaris* | | 3 |
| *Physcia alba* var. *linearis* | Phylin | 1 |
| *Physcia alba* var. *obcessa* | Phyobc | 1 |
| *Physcia albicans* var. *hypomela* | Phycri | 1 |

| | | |
|---|---|---|
| *Physcia convexa* | | 4 |
| *Physcia sorediosa* | Physor | 2 |
| *Physcia* sp. 2 | | 3 |
| *Physcia* sp. 3 | | 3 |
| *Pyxine rhizophorae* | Pyxriz | 2 |
| *Pyxine meissneri connectens* | | 3 |
| *Pyxine* sp. 2 | | 4 |

*Basidiolichens*

  1. Aphyllophorales

    Dictyonemataceae

| | | |
|---|---|---|
| *Cora pavonia* | Corpav | 1 |
| *Corella brasiliensis* | Corela | 1 |
| *Dictyonema sericeum* | Dicser | 1 |

Fruticose and filamentose species together constitute about 16 per cent of the non-crustose/squamulose flora and will not be treated here. The major part of the non-crustose/squamulose species are lichens containing green algae (FOLVE) (Figs 11.1, 11.2 and 11.7). Among foliose lichens, the Parmeliaceae alone account for more than half (76) of the species (Fig. 11.3).

Only six families of foliose lichens are represented in the southern-central São Paulo State coastal vegetation, and among these, *Parmotrema*

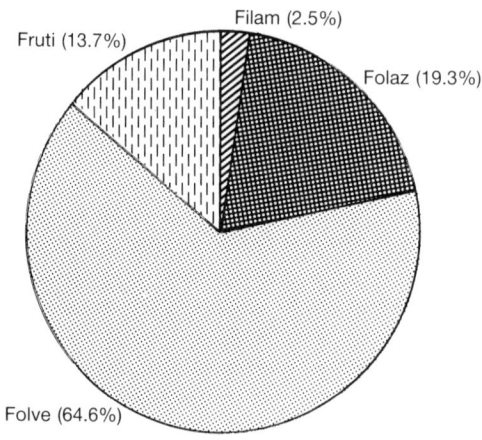

**Fig. 11.1.** Non-crustose lichens of the São Paulo coast; morphobiological groups composition.

# The foliose lichen flora of the southern-central coast of São Paulo State 159

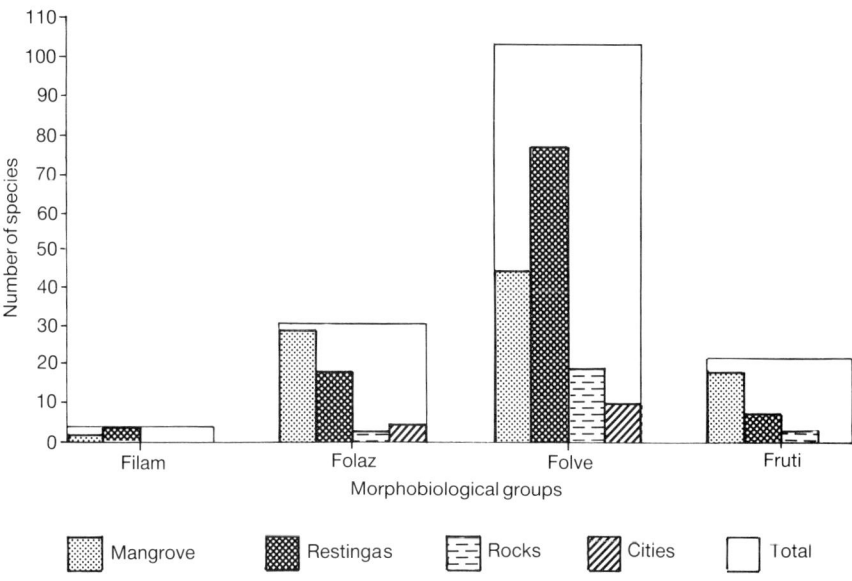

**Fig. 11.2.** Habitat preference of the morphobiological groups.

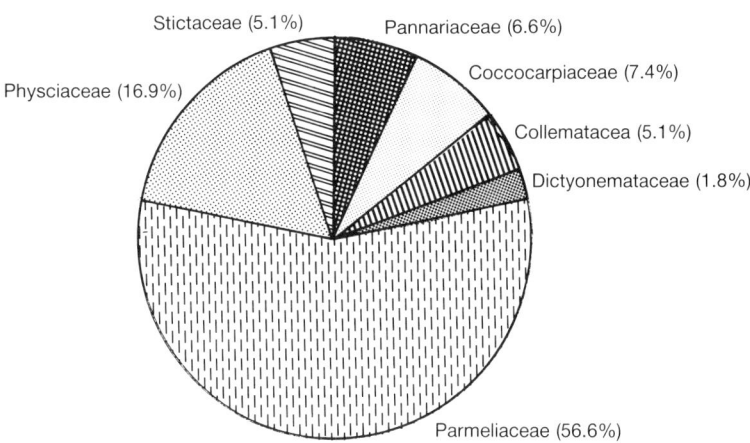

**Fig. 11.3.** Foliose lichen families of the southern-central São Paulo State coast.

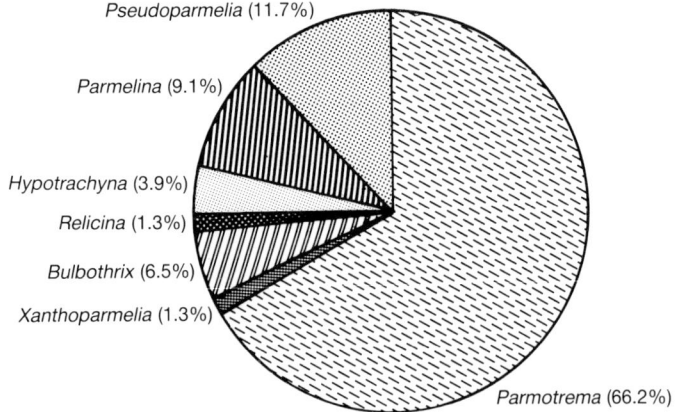

**Fig. 11.4.** Parmeliaceae of the southern-central São Paulo State coast.

represents *c.* 66 per cent (51 species) of the family (Fig. 11.4). Thus, *Parmotrema* is the floristically most important genus of the São Paulo coast (*c.* 38 per cent of the foliose; 32 per cent of the non-crustose/squamulose). It is also the most abundant.

The second most important foliose family is the Physciaceae (Fig. 11.3). However, since its species have generally small thalli and are not abundant they are easily overlooked except for saxicolous species (chiefly *Dirinaria*). *Heterodermia* is floristically the most important genus, but *Dirinaria* is fairly abundant (Fig. 11.5).

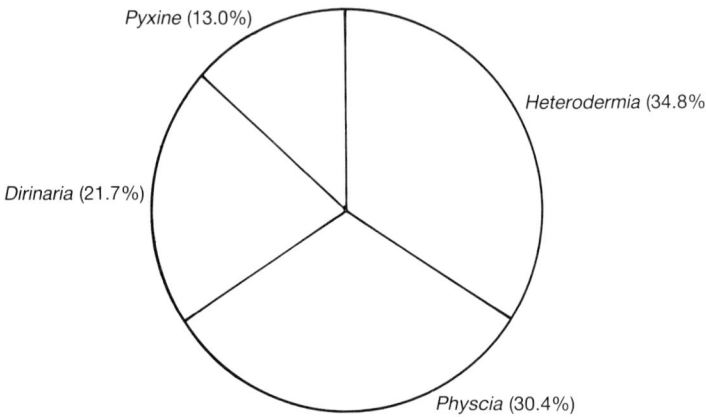

**Fig. 11.5.** Physciaceae of the southern-central São Paulo State coast.

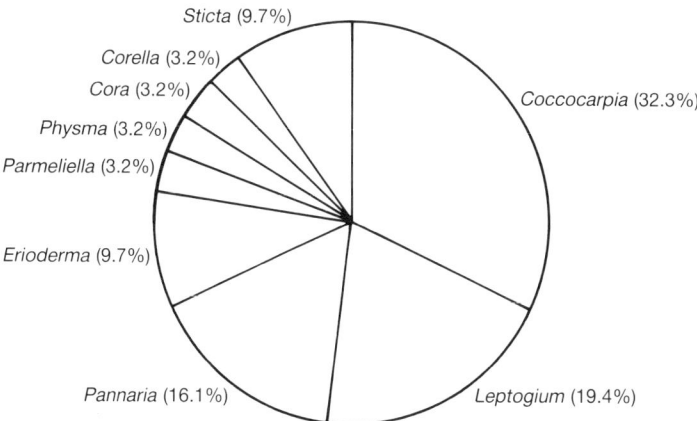

**Fig. 11.6.** Cyanobacteria-containing foliose genera of the southern-central São Paulo State coast.

Other families are chiefly those containing cyanobacteria with the more important genera being *Coccocarpia*, *Leptogium*, and *Pannaria* (Fig. 11.6). Only *Leptogium* is relatively abundant outside very humid habitats.

## Habitat distribution

*Urban areas*

The poverty of the urban lichen floras is seen by comparing Fig. 11.7 and Fig. 11.8. The few species found are not abundant and have little cover, except for *Dirinaria picta*, a characteristic weedy species of urban regions. The most common tree in the cities (*Terminalia cattapa* L.) is not a good substrate for native lichen species, perhaps because of the strong shadow provided by the crown, in spite of having an exposed trunk. In some humid places it can be covered by *Leptogium* species (chiefly *L. isidiosellum*), generally mixed with specimens of *Physcia albicans* and adnate species of *Heterodermia*, chiefly *H. dendritica* var. *propagulifera*.

On wind-exposed trees growing near the sea it is possible to find *Parmotrema tinctorum*, *P. consors*, *P.* cf. *reticulatum*, *P. praesorediosum*, and species of *Heterodermia* and *Dirinaria*, sometimes associated with species of *Coccocarpia* and *Leptogium* that grow inside bark crevices or near the soil.

Old, tall palms have a little diversified flora when growing near

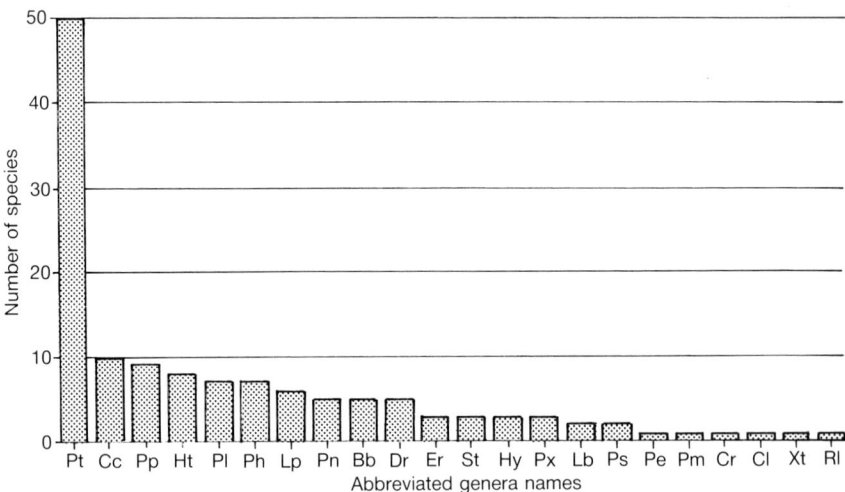

**Fig. 11.7.** Foliose lichens of the southern-central São Paulo coast; the numbers of species by genus.

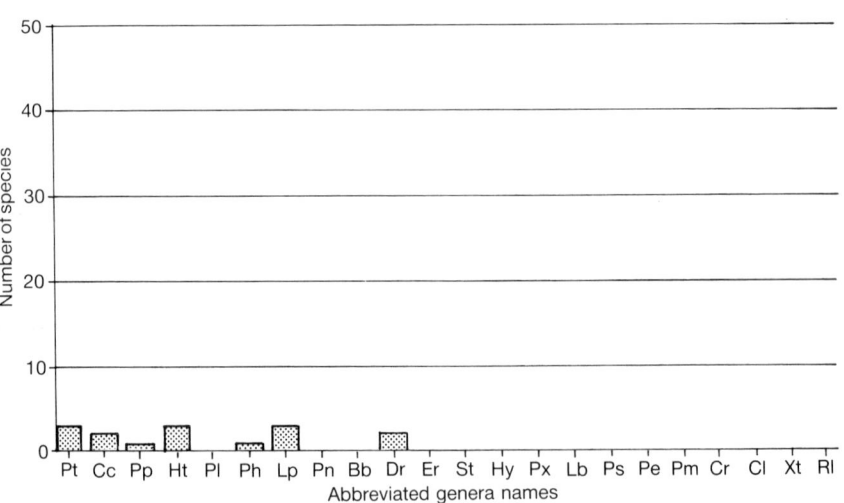

**Fig. 11.8.** Foliose lichens of the southern-central São Paulo coast; the number of species by genus in cities.

beaches; principal lichens include *Parmotrema praesorediosum*, *P. reticulatum*, and other species of *Parmotrema*, *Dirinaria*, *Bulbothrix*, and *Hypotrachyna*.

## Rocky shores

The non-crustose lichen flora of the few extant rocky shores is abundant and presents a large biomass but does not show great diversity (Fig. 11.9). Cyanobacterial species are rare and appear only inside crevices or under the protection of grasses and shrubs, or in places where underground freshwater bathes the rock surfaces. Fertile species and individuals are also very rare and are found only in crevices and on areas of rocks sheltered from wind. It is possible that the high temperature of the substrate and its associated desiccation are responsible for this scarcity. Black rocks are devoid of lichens because of the temperatures they reach. *Parmotrema* is by far the most abundant and important non-crustose genus on rocks.

## Mangroves

The lichen flora of the S/SE mangroves of Brazil has *c.* 300 species of which 90 are non-crustose/squamulose (Marcelli 1987) and found all over the São Paulo State.

*Parmotrema* is ecologically the most important foliose genus (23 species) followed by those containing cyanobacteria such as *Coccocarpia*,

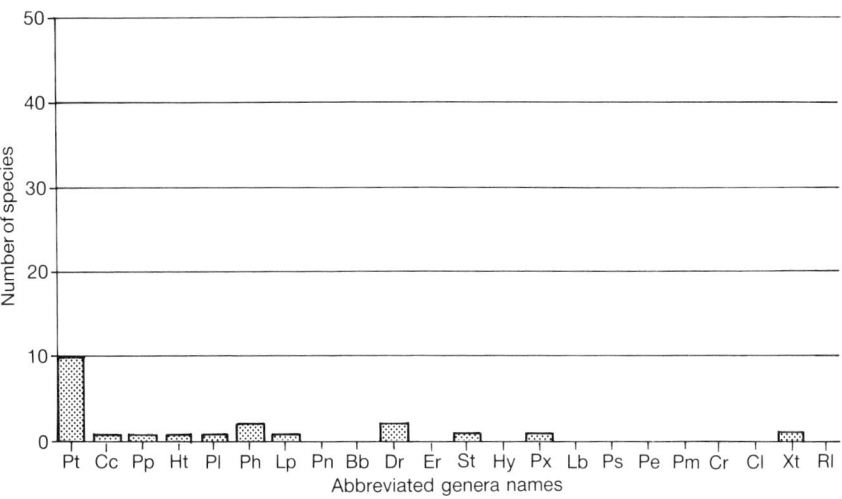

**Fig. 11.9.** Foliose lichens of the southern-central São Paulo coast; the number of species by genus on rocky shores.

*Leptogium, Pannaria, Erioderma, Sticta,* and the green algae-containing Physciaceae (Fig. 11.10) and *Lobaria* species, which have high diversity and great abundance, mainly in sunny microclimates.

Due to the association of high illumination and humidity all the coastal species containing cyanobacteria grow inside mangroves and are usually bigger, healthier, and more abundant than in any other habitat, except for *Leptogium isidiosellum* which is more frequent outside mangroves.

*Parmotrema internexum, Pannaria,* and species of *Lobaria* grow almost exclusively on *Rhizophora mangle. Parmotrema crinitum* and *Leptogium austroamericanum* show a preference for *R. mangle* but also grow on *Laguncularia racemosa. Parmotrema madilynae, P. dilatatum, P.* cf. *reticulatum,* and fruticose lichens prefer *L. racemosa* as a substrate (Marcelli 1987). *P. internexum* and *P. crinitum* are among the 10 ecologically most important *R. mangle* lichens in the sunny mangrove microclimates, whereas *L. austroamericanum* and *L. moluccanum* are important in shaded areas. *Parmotrema madilynae, P.* cf. *reticulatum, P. dilatatum, Erioderma unguigerum,* and *Usnea* species are among the 10 ecologically most important *L. racemosa* lichens in the sunny microclimatic mangrove regions, whereas *Sticta weigelii* and *Coccocarpia* sp. 2 have this same importance in shaded situations.

*Restinga and beach vegetation*

Inside the higher restinga forests the environment is too dark for non-

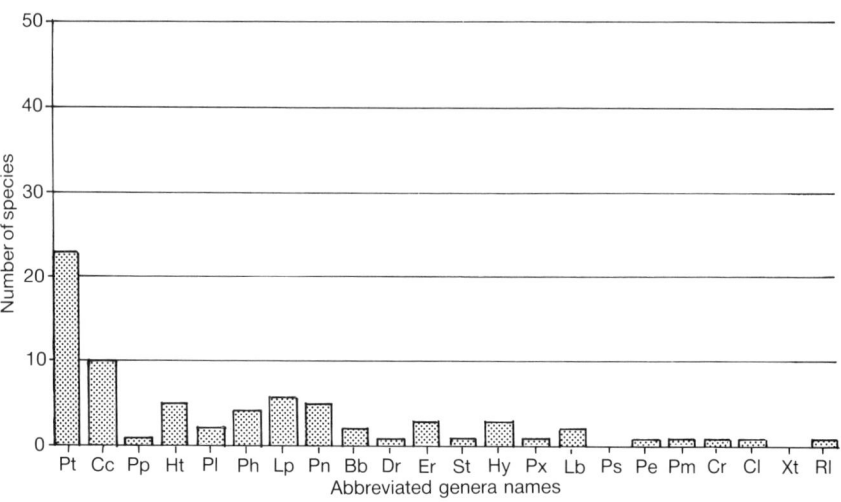

**Fig. 11.10.** Foliose lichens of the southern-central São Paulo coast; the number of species by genus in mangroves.

crustose species which grow only along the edges of glades, trails, streams, and ponds, where the sunlight reaches trunks and branches. The most striking characteristic of the restinga lichen flora is not the foliose flora but the very abundant and colourful crusts of Trypetheliaceae (absent from mangroves) and Graphidaceae, and soil Cladoniaceae.

The restinga soil has a great abundance of *Cladonia* species whenever the illumination is like a 'clear shadow' as under glades, shrubs or open places in the forest. However, they can grow in direct sunlight where the soil is commonly damp.

Restinga vegetation grows over ancient dunes and although the soil surface is undulate, the canopy is almost horizontal. Often the interdune space is almost at sea level or even below the water table. Because of this, restinga swamps are frequent; here the trees produce few leaves and the microclimate is similar to that of the mangroves and has a very rich associated hygrophylic foliose/fruticose lichen flora where *Parmotrema dilatatum*, *P. cristiferum*, *P. sulphuratum*, and *P. reticulatum* are common. *Leptogium isidiosellum*, *L. marginellum*, *L. moluccanum*, and *Sticta weigelii* are the most common hygrophilic foliose lichens in shade and are not particularly frequent. However, they cover large areas of the trunks when they do occur and usually they are mixed with specimens of other species of *Leptogium*, *Sticta*, *Heterodermia*, *Lobaria*, and *Coccocarpia*. Trees in places which are at the same time shady and dry have commonly *Parmotrema praesorediosum* and *P. gardneri* as epiphytes.

Shrubby restinga vegetation grows near beaches and along dune ridges and grows in glades of the tall restinga. It supports a heliophilic green algae containing lichen flora dominated by *Parmotrema madilynae*, *P. dilatatum*, and *Dirinaria picta*, with a noteworthy presence of *Heterodermia* and *Physcia* species. Where shrubs and small trees are wind-protected and the microclimate becomes wetter and clear but without direct sunlight, a large biomass of lichens covers branches and twigs. Laciniate species of *Parmotrema* such as *P.* cf. *argentinum*, *P. michauxianum*, and *P. lobulatum*, mixed with *Bulbothrix*, *Parmelina*, *Leptogium*, *Erioderma*, and ascendant *Heterodermia* species are the most common foliose lichens. Cyanobacterial species are not abundant even in the most humid situations and the Parmeliaceae is the dominant group among the non-crustose restinga lichens (Fig. 11.11). Fruticose species are rare except for *Teloschistes flavicans* and some species of *Ramalina* which are very frequent on twigs of beach shrubs and on branches of wind-exposed hill trees.

*Coastal hills*

Coastal hills are very rare in the study area; the principal site sampled being *c.* 150 km north of Santos (São Sebastião city), quite distant from

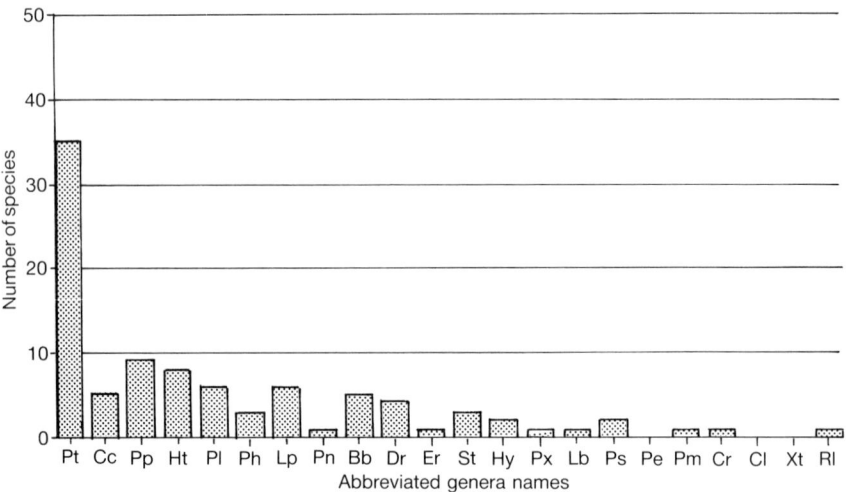

**Fig. 11.11.** Foliose lichens of the southern-central São Paulo coast; the number of species by genus in restingas.

the main area studied. These hills are rounded, generally 50–150 m high and covered with natural fields having some isolated trees and shrubs. These have several lichen species that are not common in or are absent from mangroves or restinga vegetation, such as *Parmotrema austrosinense*, *P. sancti-angeli*, *P.* cf. *permutatum*, and *P. tinctorum*.

## Comparison of habitats

In Fig. 11.12 the composition of the non-crustose/squamulose lichen floras of the principal habitats investigated is compared. In the cities there is only a small number of non-abundant species which include no fruticose or filamentose lichens. Rocky shores have a few more species than urban areas, but they are abundant and the composition of the flora is very characteristic and specific, sharing almost no species with any other habitat.

Mangrove and restinga have a significantly higher number of species but the composition of the flora and the relative abundance within these two habitats is rather different. Mangroves have many more FOLAZ and FRUTI species than the restinga vegetation which, in turn, has almost twice as many FOLVE taxa as the mangroves. With respect to Parmeliaceae, mangroves are almost exclusively populated by species of *Parmotrema* whereas several other species of *Pseudoparmelia*, *Parmelina*, and *Bulbothrix* which are present in the restinga vegetation

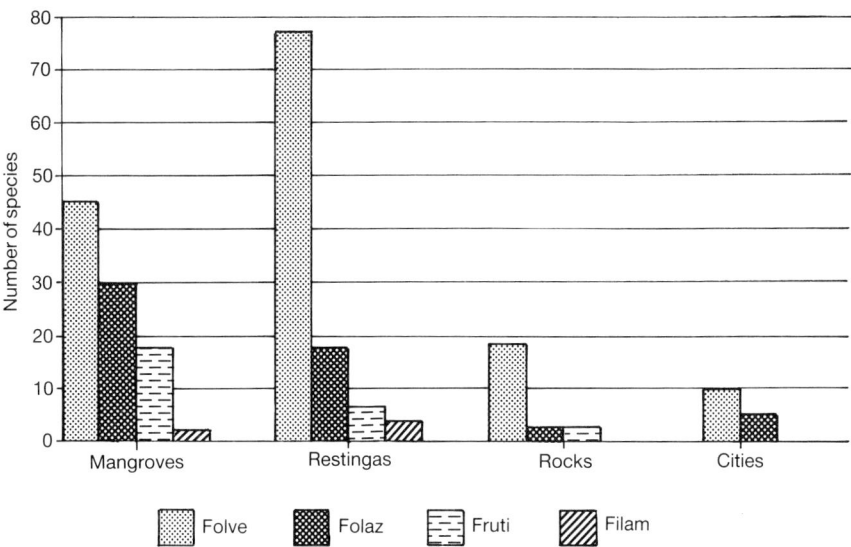

**Fig. 11.12.** Non-crustose lichens of the southern-central São Paulo coast; morphobiological composition of the habitats.

are not represented in mangroves (compare Fig. 11.10 and Fig. 11.11). Even the *Parmotrema* species found in both places are not the same, except for *P. cristiferum*, *P. gardneri*, *P. dilatatum*, *P. praesorediosum*, and *Parmelina subfatiscens*. Isidiate species of *Parmotrema* are much more abundant in mangroves than in the restinga, where they are uncommon and grow only near the ground. *Dirinaria*, represented in mangroves only by one specimen, *D. aegialita*, is fairly common in restingas and on rocky shores. *Sticta weigelii* is the only *Sticta* species found in mangroves. *Pannaria* and *Erioderma*, abundant in mangroves, are rare in restinga vegetation.

In summary, mangroves have a more hygrophilic lichen flora while that of restingas is a more heliophilic one. However, since swampy areas with microclimates identical to those of mangroves are also formed in the restinga vegetation it is possible that the difference between these habitats is one of host specificity. In this respect it is interesting that FOLAZ species depend much more on microclimate and less on the substrate than do FOLVE or FRUTI taxa. This is because FOLAZ species almost always grow in humid places where the substrate humidity is not so important (Marcelli 1987). Fertile specimens are rare for almost all species, except in some protected and humid or shady places.

## Species identity

Many of the coastal species are common pantropical lichens also present in Africa and/or Australia and easily identified. The major problem is in the specific identity of taxa in the Parmeliaceae because, in addition to the few common species there are many undescribed taxa and others which are either very poorly studied and/or ill-defined. The lack of a comprehensive modern bibliography and the almost total absence of modern lichen herbaria for comparative study makes the work of Brazilian lichenologists an exceedingly difficult task. Sometimes it takes months for someone to verify if a particular lichen has already been described.

The recorded lichen flora of Brazil is basically that collected more than 50 years ago by botanists primarily interested in knowing the vegetation of various areas of South America. Because the coastal flora is virtually unknown and even many *Parmotrema* species remain to be described much basic taxonomic work is needed before we can entertain studies on biogeography and phylogeny. Table 11.2 shows the problems of identification for the São Paulo's coastal Parmeliaceae. About one-third to a half of the collected species remain unidentified. Most of the saxicolous species are commonly sterile and very closely attached and cannot easily be collected. This explains the absence in older literature of data on such abundant taxa. The same problem exists for the adnate and/or infrequent species of *Pseudoparmelia* and *Parmelina*.

**Table 11.2.** Number of identified and unidentified species of Parmeliaceae in southern-central São Paulo State coasts

| Genus | Total number of species | Species of doubtful identity | Species unidentified |
|---|---|---|---|
| *Parmotrema* | 51 | 5 | 19 |
| *Pseudoparmelia* | 8 | 2 | 2 |
| *Parmelina* | 7 | 1 | 4 |
| *Xanthoparmelia* | 1 | — | 1 |
| *Relicina* | 1 | — | — |
| *Bulbothrix* | 5 | — | — |
| *Hypotrachyna* | 3 | — | — |
| Total | 76 | 8 | 26 |

For saxicolous species in the Parmeliaceae it was possible to get good specific names only for *Parmotrema ultraluscens* and *P. mordeni*, neither being the most common species. Among the remaining foliose lichens the most common are species of *Dirinaria* found in direct sunlight and *Leptogium isidiosellum* in shady sites. Marginally sorediate *Parmotrema* species normally fall within the *P. gardneri* complex (including also P negative species) and 13 morphological groups were defined including *P.* cf. *gardneri*, *P. praesorediosum* and variations, *P. mordeni*, *P. chinense*, *P. subarnoldii*, the true *P. dilatatum* (in full agreement with Vainio's description) and some of its variations.

## Conclusions

The natural habitats of the São Paulo coast are little known environments and not only for lichens. They have a large number of species, many of which are undescribed. Some important species complexes are well-represented and certainly a number of taxonomic problems can be solved by careful collections and habitat investigations, chiefly in relation to the Parmeliaceae.

Unfortunately, these natural habitats are disappearing at an alarming rate through active speculation concerning land and property causing deforestation of enormous areas of virgin restingas and mangroves. These habitats remained intact and undisturbed for many years before new roads made access easy for people looking for 'beach houses'.

Today, except for the Ilha do Cardoso reserve, shrubby and subarboreal restingas are very rare and are found only in small patches around some beaches. One very good and representative subarboreal patch of restinga which was collected for this work in Praia Grande city, no longer exists. The major part of the material studied from Itanhaém city was collected by the author about 10 years ago. Since then subarboreal restingas are gone forever.

## References

Blanco, H.G. and Godoy, H. (1966). *Carta anual das chuvas do estado de São Paulo*. Instituto Agronômico da Secretaria da Agricultura do Estado de São Paulo.

Godoy, H. and Ortolani, A.A. (1962). *Carta climática do estado de São Paulo, 1945–1962*. Instituto Agronômico. Climatologia Agricola. Secretaria da Agricultura do Estado de São Paulo.

Lamberti, A. (1969). Contribuição ao conhecimento da ecologia das plantas do manguezal de Itanhaém. *Boletim da Faculdade de Filosofia, Ciências e Letras da Universidade de São Paulo*, 317. *Botânica*, **23**, 1–217.

Marcelli, M.P. (1987). Ecologia dos líquens dos manguezais da região sul-sudeste do Brasil com especial atenção ao de Itanhaém (SP). Unpublished D. Phil. thesis. Universidade de São Paulo.

Poelt, J. (1973). Classification. In *The Lichens* (ed. V. Ahmadjian and M.E. Hale Jr.), pp. 599–632. Academic Press, New York.

# 12. Ultrastructure of subtropical crustose lichens

S.C. TUCKER, S.W. MATTHEWS, and R.L. CHAPMAN
*Department of Botany, Louisiana State University, Baton Rouge, Louisiana, 70803-1705, USA*

### Abstract

Ultrastructure in thirteen species of subtropical lichens from Louisiana, USA was studied. All belong to genera prevalent in the tropics and all have trentepohliaceous photobionts. Homogeneous cross-wall ultrastructure identified *Trentepohlia* as photobiont in species studied of *Anisomeridium*, *Arthonia*, *Pyrenula*, and *Trypethelium*. *Cephaleuros*, as photobiont in the lichen *Strigula complanata*, has a similar homogeneous cross-wall structure, but is easily distinguished on other characteristics. A pronounced circular wall thickening around the central cluster of plasmodesmata identified *Phycopeltis* as photobiont in species of *Anisomeridium* (different to the above species), *Graphis*, *Mazosia*, *Opegrapha*, *Porina*, and *Schismatomma*. Previously, *Graphis scripta*, *Schismatomma*, and *Anisomeridium* had been reported to have *Trentepohlia* instead. Haustoria were present in all thirteen crustose lichen species and all had the intraparietal type (as defined by Honegger) in which the algal cell wall invaginates and surrounds the invading fungal haustorium. In some haustoria of several of the taxa, however, the algal wall around the haustorial tip undergoes thinning or degeneration. If the algal wall is completely degraded around the haustorium, this results in a condition approaching Honegger's 'intracellular' type of haustorium in which there is a shift toward a parasitic role for the mycobiont.

### Introduction

As the systematics of tropical lichens become better known, it will become possible to see how these taxa differ from the better known lichens of temperate and arctic zones. Most ultrastructural research has been conducted on temperate species, as surveyed by Ahmadjian (1964), Peveling (1973), Galun (1988), and Tschermak-Woess (1988). Of the two most abundant groups of green algal photobionts, *Trebouxia* has been studied extensively (see Peveling 1973; Galun 1988 for references), but *Trentepohlia* has received relatively little attention. Light microscopy of species of *Arthonia*, *Graphis*, and *Opegrapha* species containing *Trentepohlia* revealed the presence of haustoria which were said to lack an algal wall sheath (Tschermak 1941). Some information on the ultrastructure of lichenized trentepohliaceous taxa was reported for the lichens *Chiodecton sanguineum* (Ellis 1975; Withrow and Ahmadjian 1983), *Coenogonium implexum* (Meier and Chapman 1983). *Strigula elegans* (Chapman 1976a), and *Trypethelium eleuteriae* (Lambright and Tucker 1980). These lichens are all tropical species which also extend into the subtropics (Tucker 1981).

We need more information on tropical taxa regarding the taxa of algae present as photobionts: which ones are prevalent and which are rare. Are photobionts more variable in a species in the tropics than in related species elsewhere? Are cellular organelles in tropical lichens similar to those in temperate taxa? Are haustoria more or less prevalent in the tropics compared to temperate taxa? What type of haustorium is most common in the tropics? Most significantly, we would like to know whether tropical lichens have a tendency toward a more parasitic role than in temperate lichens. Growth appears more rapid in tropical lichens, as for example in foliicolous species attaining the fruiting condition in the life span of an evergreen leaf as its substrate. Corticolous colonies are often inordinately large, indicating an ability of the lichen to pace its growth to that of the substrate. Structural details of this apparently rapid growth for tropical species of lichens have not been investigated.

Our aims in this paper are: (1) to determine the genus of photobiont, on the basis of diagnostic wall characteristics among trentepohliaceous algae, and (2) to determine the kind of haustorium, if present. A more detailed presentation of our results from this study is given in Matthews *et al.* 1989.

The ultrastructure of haustoria of tropical lichens has received almost no attention. Bracker and Littlefield (1973) dealt mostly with haustoria of pathogenic fungi and Honegger (1985) studied haustoria of trebouxioid photobionts. The haustoria of the pyrenocarpous lichens, another common tropical assemblage, have received little attention from ultrastructuralists. Galun *et al.* (1973) reported on the ultrastructure of haustoria of two desert species of *Dermatocarpon* from

Israel. The current study focuses on several pyrenocarp species as well as several Discomycetes, which are tropical and subtropical crusts. The pyrenocarps together with graphids, Arthoniaceae and Thelotremaceae are major components of the tropical lichen flora.

## Materials and methods

Colonies of thirteen species of lichen crusts were collected in bottom-land hardwood forests in Baton Rouge, Louisiana, USA. Twelve were corticolous and one was foliicolous on leaves of *Magnolia grandiflora*. Sources and substrates are indicated in Table 12.1. Pieces of each sample were set aside as vouchers and deposited in the lichen herbarium at Louisiana State University. Pieces of the remaining samples were subdivided into 1 mm pieces and fixed in the field in 3.5 per cent glutaraldehyde, 1 mM $CaCl_2$ and 0.1 per cent Tween 80 in 0.02 M potassium phosphate buffer at pH 6.84 for 3 h at 22 °C, followed by 3 h at 4 °C. Later, in the laboratory, the samples were rinsed in potassium phosphate buffer and post-fixed overnight in 1 per cent buffered osmium tetroxide at 4 °C. They were dehydrated to 100 per cent ethanol with 2 per cent uranyl acetate included in 80 per cent ethanol during the dehydration. The material was embedded in L.R. White resin (hard grade), polymerized for 24 h at 60 °C, and sectioned with a diamond knife on a Sorval Porter-Blum MT-2 ultramicrotome. Sections 75–90 nm thick were collected on copper grids and post-stained with uranyl acetate and lead citrate. They were studied and micrographs taken at 60 or 80 kV on a JEOL 100-CX transmission electron microscope.

## Results

Twelve of the lichens studied form crusts on bark, while the last, *Strigula complanata*, grows on the upper surface of leaves. The substrate tree species are indicated in Table 12.1. Most have superficial crusts, although three pyrenocarpous species, *Pyrenula anomala*, *Trypethelium ochroleucum*, and *T. tropicum* are endoperidermal. In nine of the taxa, mycobiont hyphae were observed to invade cork cells of the substrate, while in three (*Arthonia tumidula*, *Opegrapha viridis*, *Pyrenula anomala*), hyphae appear strictly intercellular in the host. *Strigula complanata* is subcuticular on leaves. The lichen thallus has either a primitive, non-cellular, amorphous type of cortex (*Pyrenula anomala*, *Trypethelium ochroleucum*, *T. tropicum*) or a cellular, mycelial cortex overlying the algal layer.

**Table 12.1.** Photobionts in genera studied, provenance and references

| Lichen | Photobiont | Substrate tree |
|---|---|---|
| Arthoniaceae | | |
| *Arthonia* (some spp. foliicolous) | | |
| *A. rubella* (Fée) Nyl. | *Trentepohlia* (m) | *Liquidambar styraciflua* L. (b) |
| *A. tumidula* (Ach.) Ach. | *Trentepohlia* (i) | *Celtis laevigata* Willd. (b) |
| *A. radiata* (Pers.) Ach. | *Trentepohlia* (i) | |
| *Arthonia* (1 sp.) | *Trentepohlia* (p) | |
| *Arthonia* (non-foliicolous) | *Phycopeltis* (k, m) | |
| | Chlorococcaceae (m) | |
| Graphidaceae | | |
| *Graphis scripta* (L.) Ach. *s. lat.* | *Trentepohlia* (g, j, q) | *Carpinus caroliniana* Walt. (b) |
| | *Phycopeltis* (i) | |
| Lecanactidaceae | | |
| *Schismatomma rappii* | *Trentepohlia* (l) | *Platanus occidentalis* L. (b) |
| Opegraphaceae | | |
| *Mazosia* | *Phycopeltis* (m) | |
| *Mazosia sp.* | *Phycopeltis* (i) | |
| *Opegrapha mougeotii* Massal. | *Trentepohlia* (p) | *Platanus occidentalis* L. (a) |
| *O. rufescens* Pers. (*O. herpetica*) | *Trentepohlia* (p) | |
| *O. filicina* Mont. (foliicolous) | *Phycopeltis* (m) | |
| *O. viridis* Pers. *s. lat.* | *Phycopeltis* (i) | *Celtis laevigata* Willd. (b) |
| *O. dibbenii* Sérus., *O. lambinonii* Sérus. | | |
| *O. santessonii* Sérus. | *Phycopeltis* (o) | |

| | | |
|---|---|---|
| Pyrenulaceae | | |
| *Pyrenula* | | |
| *P. anomala* (Ach.) Vainio | *Trentepohlia* (f) | *Acer rubrum* var. *drummondii* |
| | *Trentepohlia* (i) | (H. and A.) Sarg. (b) |
| *P. nitida* Ach. | *Trentepohlia* (d) | |
| Strigulaceae | | |
| *Anisomeridium* | | |
| *A. subprostans* (Nyl.) R.C. Harris | *Trentepohlia* (f) | *Platanus occidentalis* L. (b) |
| *A. tuckerae* R.C. Harris | *Phycopeltis* or *Trentepohlia* (i) | *Liquidambar styraciflua* L. (b) |
| *Strigula* (foliicolous) | *Trentepohlia* (i) | |
| *S. complanata* (Fée) Mont. | *Cephaleuros* (e, f, m) | *Magnolia grandiflora* L. (a) |
| | *Cephaleuros* (i) | |
| Trichotheliaceae | | |
| *Porina* | | |
| *P. pseudofulvella* Sérus. | *Phycopeltis* (f, m) | |
| | *Phycopeltis* (n) | |
| | *Phycopeltis* (i) | *Salix nigra* Marsh. (c) |
| *P. pulla* (Ach.) Müll. Arg. *s. str.* | *Phycopeltis* (i) | *Liquidambar styraciflua* L. (c) |
| | *Phycopeltis* (i) | *Lonicera sempervirens* L. (c) |
| Trypetheliaceae | | |
| *Trypethelium* | | |
| *T. ochroleucum* (Eschw.) Nyl. | *Trentepohlia* (f, h) | *Acer rubrum* var. *drummondii* |
| | *Trentepohlia* (i) | (H. and A.) Sarg. (b) |
| *T. tropicum* (Ach.) Müll. Arg. | *Trentepohlia* (i) | *Acer rubrum* var. *drummondii* |
| | | (H. and A.) Sarg. (b) and (c) |

Location: (a) = LSU campus; (b) = Burden Research Plantation; (c) = Ben Hur Farm.

Authorities: (d) = Ahmadjian 1964; (e) = Chapman 1976a; (f) = Harris 1975; (g) = Hérisset 1946; (h) = Lambright and Tucker 1980; (i) = Matthews et al. 1989; (j) = Nakano 1988; k = Peveling 1973; (l) = Poelt and Vězda 1977; (m) = Santesson 1952; (n) = Sérusiaux 1979; (o) = Sérusiaux 1985; (p) = Tschermak 1941; (q) = Verseghy 1961.

*Photobiont*

The cell structure conforms to that of previous reports for vegetative cells of lichenized *Trentepohlia* (Tschermak 1941; Ellis 1975; Lambright and Tucker 1980; Withrow and Ahmadjian 1983), *Phycopeltis* (Good and Chapman 1978*a*), and *Cephaleuros* (Chapman 1976*a*). In *Phycopeltis* (Figs 12.1, 12.2) and *Trentepohlia* (Figs 12.3–5), the chloroplast is parietal; in *Cephaleuros* the chloroplasts are either parietal or central. Plastoglobuli (Figs 12.4, 12.5, 12.7, 12.14, 12.18) in chloroplasts were seen in all photobionts except those of *Strigula complanata* and *Trypethelium tropicum* and at least some starch grains (Figs 12.4, 12.9, 12.17, 12.23, 12.24) were seen in all taxa. Haematochrome droplets (Figs 12.1–5, 12.7, 12.12) in the cytoplasm and mitochondria are present in all.

Walls are bi- or tri-layered, and often appear thickest close to the point of haustorial penetration (Figs 12.19, 12.23, 12.25). Number of cell wall layers and thickness are given elsewhere (Matthews *et al.* 1989). Plasmodesmata are clustered at the centre of each cross-wall (Figs 12.6, 12.7–12, 12.13, 12.18) in all three algal genera. The algal family Trentepohliaceae includes aerial or terrestrial algae. Three genera which are common as lichen photobionts are *Trentepohlia*, *Cephaleuros*, and *Phycopeltis*. All are common in the free-living state in both tropics and subtropics, including the collecting sites for this study in Baton Rouge, Louisiana. One of us (R.L. Chapman) with his students and colleagues, has studied the ultrastructure of free-living taxa of two of these genera: *Cephaleuros* (Chapman 1976*b*, 1980, 1984; Chapman and Henk 1985, 1986), and *Phycopeltis* (Good and Chapman

**Fig. 12.1–12.6.** Algal cell structure. **Fig. 12.1.** Cells of *Phycopeltis sp.* in *Opegrapha viridis*. Haustorial entry was beginning (at arrow). Note large chloroplasts and haematochrome droplets in the algal cells. **Fig. 12.2.** Cells of *Phycopeltis sp.* in *Porina pulla*. Large chloroplasts, plastoglobuli, and haematochrome droplets are present. **Fig. 12.3.** *Trentepohlia sp.* in *Pyrenula anomala*. The chloroplasts contain abundant grana and haematochrome droplets. **Fig. 12.4.** Parts of two cells of *Trentepohlia sp.* in *Anisomeridium tuckerae*. Fungal hyphae are also visible. Algal cells contain peripheral chloroplasts, nuclei, haematochrome droplets, plastoglobuli, and starch. **Fig. 12.5.** Cell of *Trentepohlia sp.* in *Arthonia rubella*, showing haematochrome droplets, plastoglobuli, and a fungal haustorium in cross-section. Both algal and fungal walls ensheath the haustorium. **Fig. 12.6.** Cross-wall between two cells of *Cephaleuros virescens* in *Strigula complanata*. The plasmodesmata (arrowheads) are clustered, and the wall is homogeneous. A, algal cell; F, fungal cells; a, algal cell wall; f, fungal cell wall; h, haematochrome droplet; n, nucleus; p. plastoglobuli; s, starch grains.

Ultrastructure of subtropical crustose lichens 177

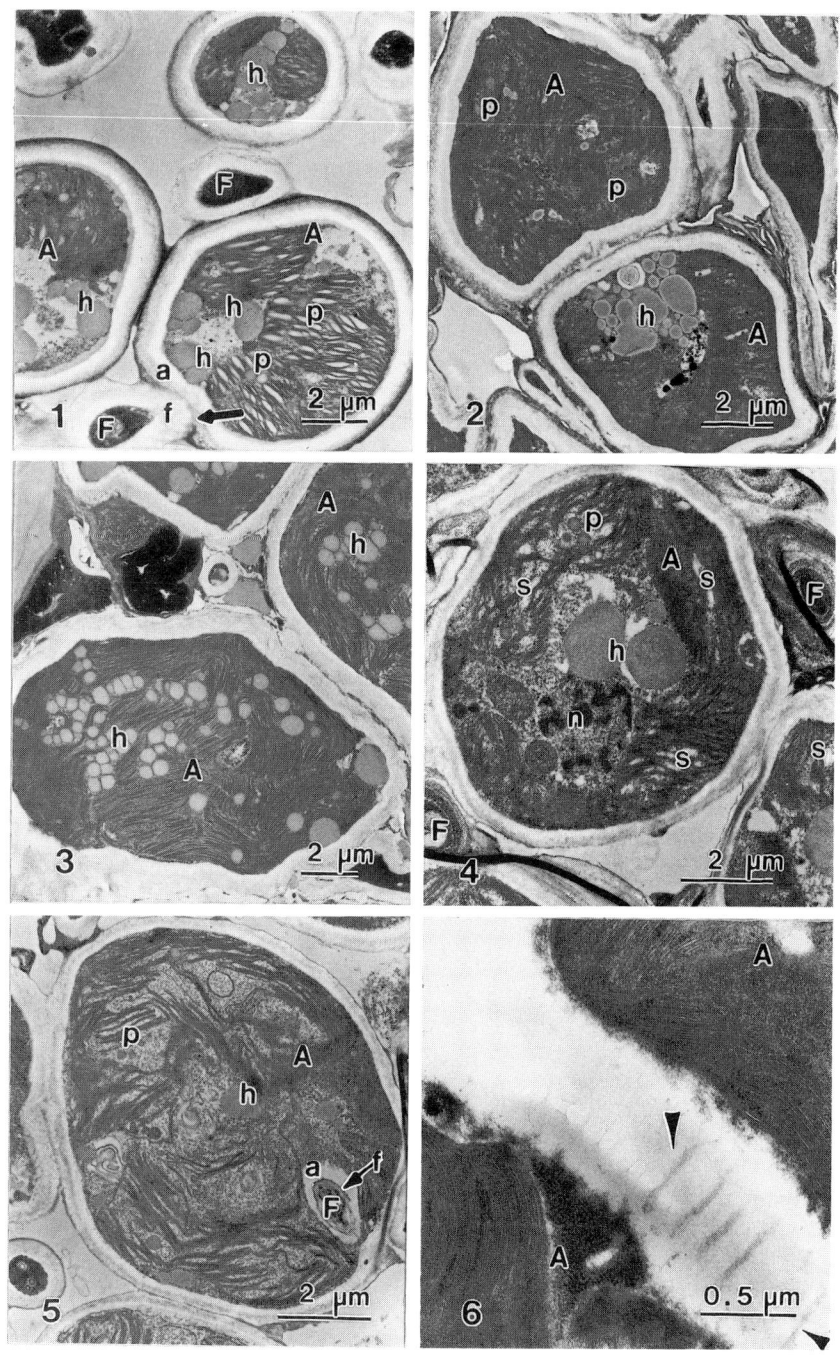

1978a, b). *Trentepohlia* was studied in the lichenized state in *Coenogonium interplexum* (Meier and Chapman 1983). *Trentepohlia*, although the most common of the three, both in the free-living and the lichenized state, is a difficult taxon systematically (Chapman 1984). The cells of *Trentepohlia* change during lichenization so that filaments sometimes cannot be detected, cell shape is atypical, and the usual pigments may be missing or scant.

The configuration of the cross-wall is a useful ultrastructural distinction. *Phycopeltis* has a circular wall thickening around the central cluster of plasmodesmata in the cross-wall, while *Trentepohlia* and *Cephaleuros* have the plasmodesmatal cluster but lack the circular wall thickening. We did not attempt to identify to species; for *Trentepohlia* at least it is doubtful that the current level of knowledge allows reasonably accurate identification to species (R. Thompson, personal communication). Homogeneous cross-wall structure, typical of *Trentepohlia*, was found in *Arthonia rubella* (Fig. 12.7), *A. tumidula* (Fig. 12.8), *Anisomeridium tuckerae* (Fig. 12.9), *Pyrenula anomala* (Fig. 12.10), *Trypethelium ochroleucum* (Fig. 12.11), and *T. tropicum* (Fig. 12.12). Homogeneous cross-walls also occur in *Cephaleuros* and in *Strigula complanata* (Fig. 12.6), but can be easily distinguished at the macroscopic level on the basis of the subcuticular position of thalli (Chapman 1976a). Cross-walls with a halonate wall thickening around the cluster of plasmodesmata, typical of free-living *Phycopeltis*, were found in *Graphis scripta* (Fig. 12.14), *Mazosia* sp. (Fig. 12.15), *Opegrapha viridis* (Fig. 12.16), *Porina pulla* (Fig. 12.17), and *Schismatomma rappii* (Fig. 12.18). Determination of wall type was questionable for *Anisomeridium subprostans* (Fig. 12.13) although it most closely resembles the *Phycopeltis* type.

*Mycobiont*

Cytological details of mycelial hyphal cells conform to published accounts (Peveling 1973; Ellis 1975; Withrow and Ahmadjian 1983); concentric bodies (Fig. 12.27, 12.30) were seen in all. The plasmalemma appeared sinuous in some: *Arthonia rubella, A. tumidula, Mazosia*

---

**Fig. 12.7–12.12.** Photobiont cross-wall in various lichen taxa. Homogeneous cross-walls with centrally clustered plasmodesmata (arrowheads), typical of *Trentepohlia*, are present. **Fig. 12.7.** *Arthonia rubella.* **Fig. 12.8.** *Arthonia tumidula.* **Fig. 12.9.** *Anisomeridium tuckerae.* **Fig. 12.10.** *Pyrenula anomala.* Cluster of plasmodesmata is seen in cross-section. **Fig. 12.11.** *Trypethelium ochroleucum.* **Fig. 12.12.** *Trypethelium tropicum.* The plasmodesmata are seen in cross-section. A, algal cell; h, haematochrome droplets; p, plastoglobuli; s, starch grains.

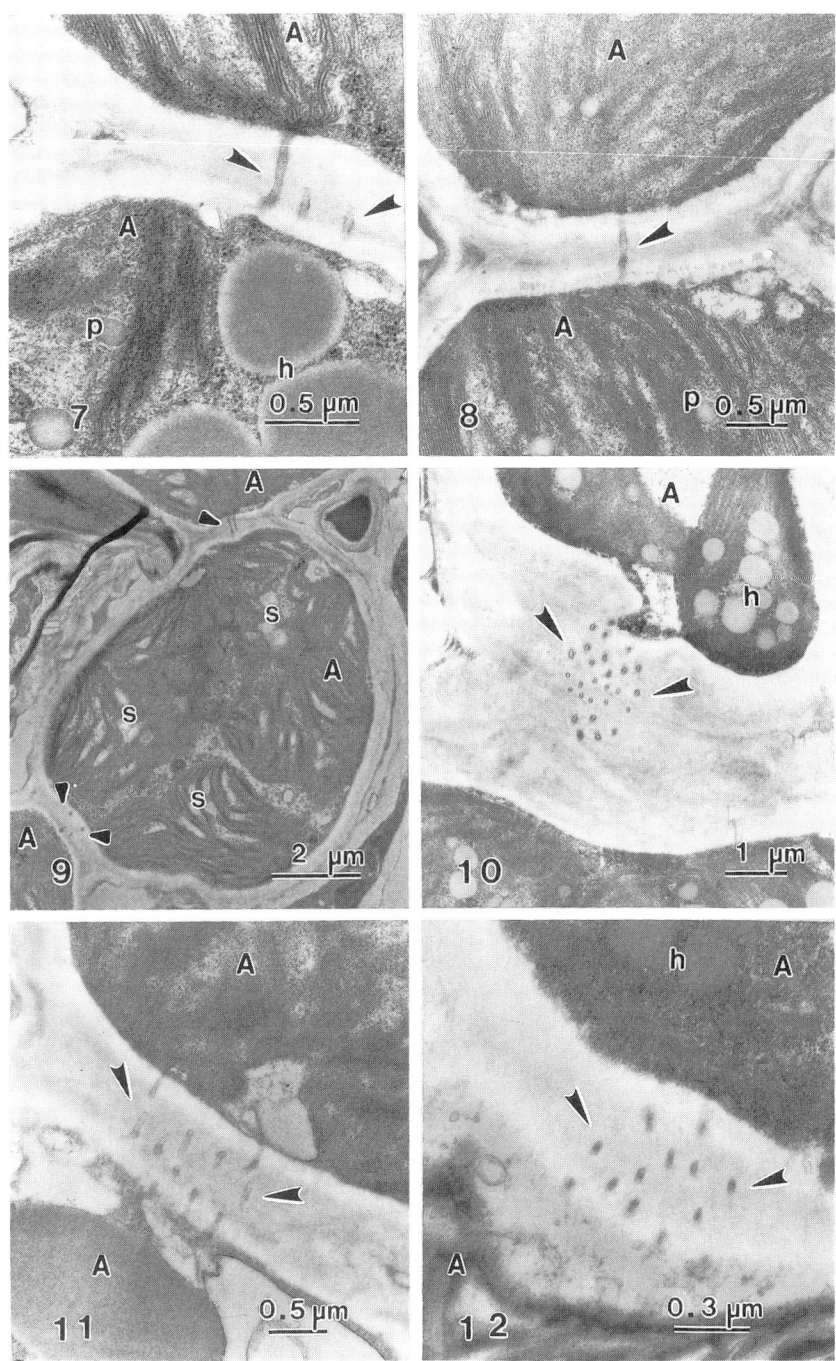

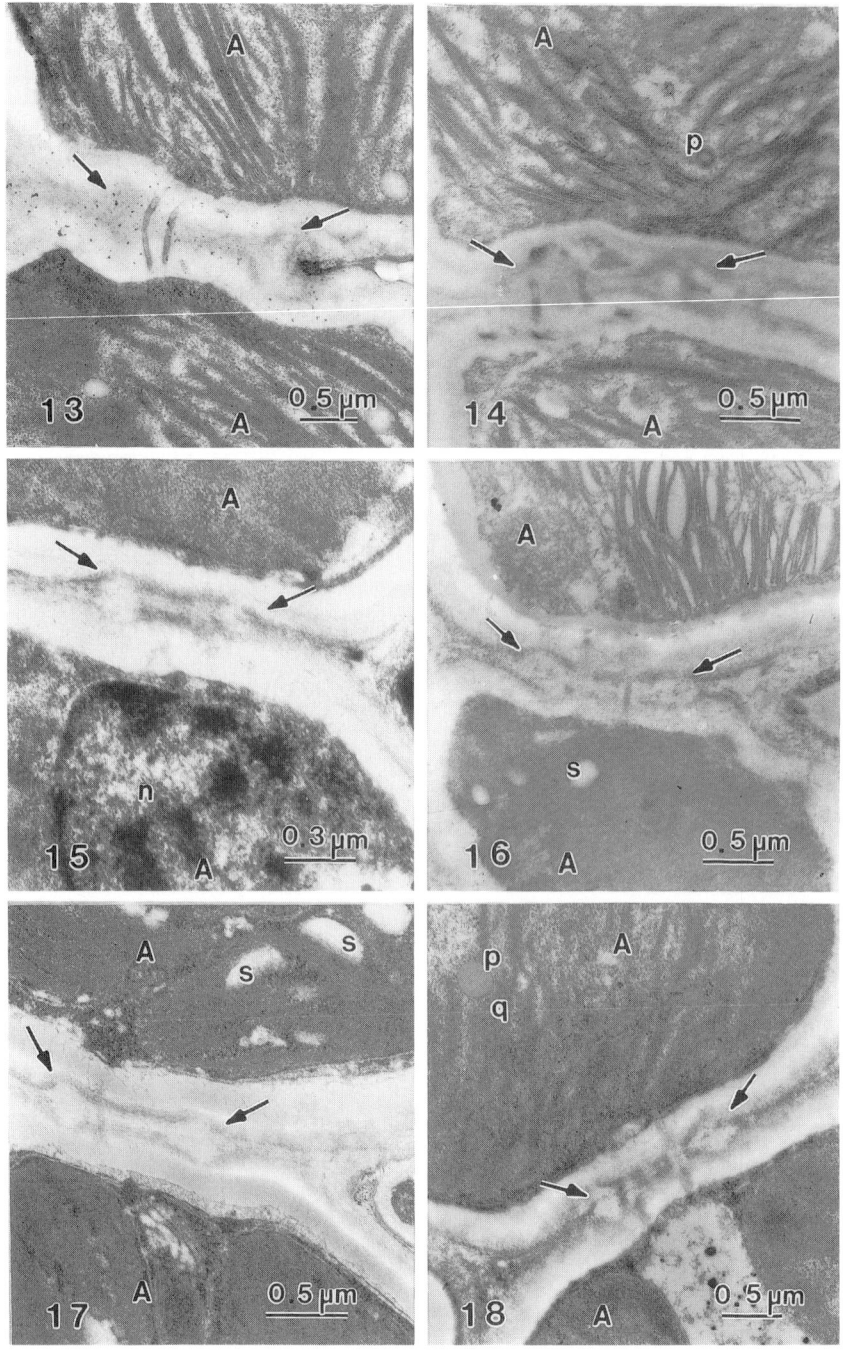

sp. (Fig. 12.23), *Opegrapha viridis*, *Pyrenula anomala* (Fig. 12.30), *Anisomeridium tuckerae*.

Well-developed haustoria were found in all thirteen species studied. The haustoria all are intraparietal Type 2 according to Honegger (1986), since the algal wall invaginates ahead of the invading haustorium. The algal cell wall invaginates at the point of contact and appears thicker around the point of entry. We observed occasional branching of the haustorial tip in *Trypethelium ochroleucum* (Fig. 12.29) and *Anisomeridium tuckerae*, and the presence of two haustoria in one algal cell in *Porina pulla* (Fig. 12.28), *Arthonia tumidula*, *Graphis scripta*, *Pyrenula anomala*, *Strigula complanata*, *Trypethelium tropicum*, and *T. ochroleucum* (Fig. 12.29). The haustorium is surrounded by the algal cell wall in close contact with the fungal cell wall (Figs 12.19, 12.21, 12.25, 12.29). The algal wall persists intact over the haustorial tip in *Anisomeridium subprostans* (Fig. 12.22), *A. tuckerae*, *Arthonia rubella*, and *A. tumidula* (Fig. 12.21), *Porina pulla* (Figs 12.27, 12.28), *Trypethelium ochroleucum* (Fig. 12.29), and *T. tropicum*. In several other taxa [*Graphis scripta*, *Mazosia* sp. (Fig. 12.24), *Opegrapha viridis*, *Schismatomma rappii* (Fig. 12.20), *Pyrenula anomala*, and *Strigula complanata* (Fig. 12.26)] the algal wall appears thinner around the haustorial tip and shows some deterioration or loss of wall.

Invaded algal cells tend to show some degeneration, although several taxa [*Arthonia rubella*, *A. tumidula* (Fig. 12.21), *Anisomeridium tuckerae*, *Opegrapha viridis*, *Pyrenula anomala*, *Schismatomma rappii* (Figs 12.19, 12.20), *Trypethelium ochroleucum* (Fig. 12.29)] have both healthy and somewhat degraded algal cells among those containing haustoria. Cells of *S. rappii* at the early stage of haustorial penetration (Fig. 12.19) look healthy, but cells with well-developed haustoria show a much higher degree of degeneration (Fig. 12.20). All algal cells containing haustoria appear degenerate in *Mazosia* sp. (Figs 12.23, 12.24), *Strigula complanata* (Figs 12.25, 12.26), and *Trypethelium tropicum* (Fig. 12.12).

In *Porina pulla* (Figs 12.27, 12.28), most of the penetrated algal cells show no evident damage.

---

**Fig. 12.13–12.18.** Photobiont cross-walls in various lichen taxa in which a cluster of plasmodesmata is surrounded by a circular wall thickening (at arrows), typical of *Phycopeltis*. **Fig. 12.13.** *Anisomeridium subprostans.* **Fig. 12.14.** *Graphis scripta.* **Fig. 12.15.** *Mazosia sp.* **Fig. 12.16.** *Opegrapha viridis.* **Fig. 12.17.** *Porina pulla.* **Fig. 12.18.** *Schismatomma rappii.* n, nucleus; p, plastoglobuli; s, starch.

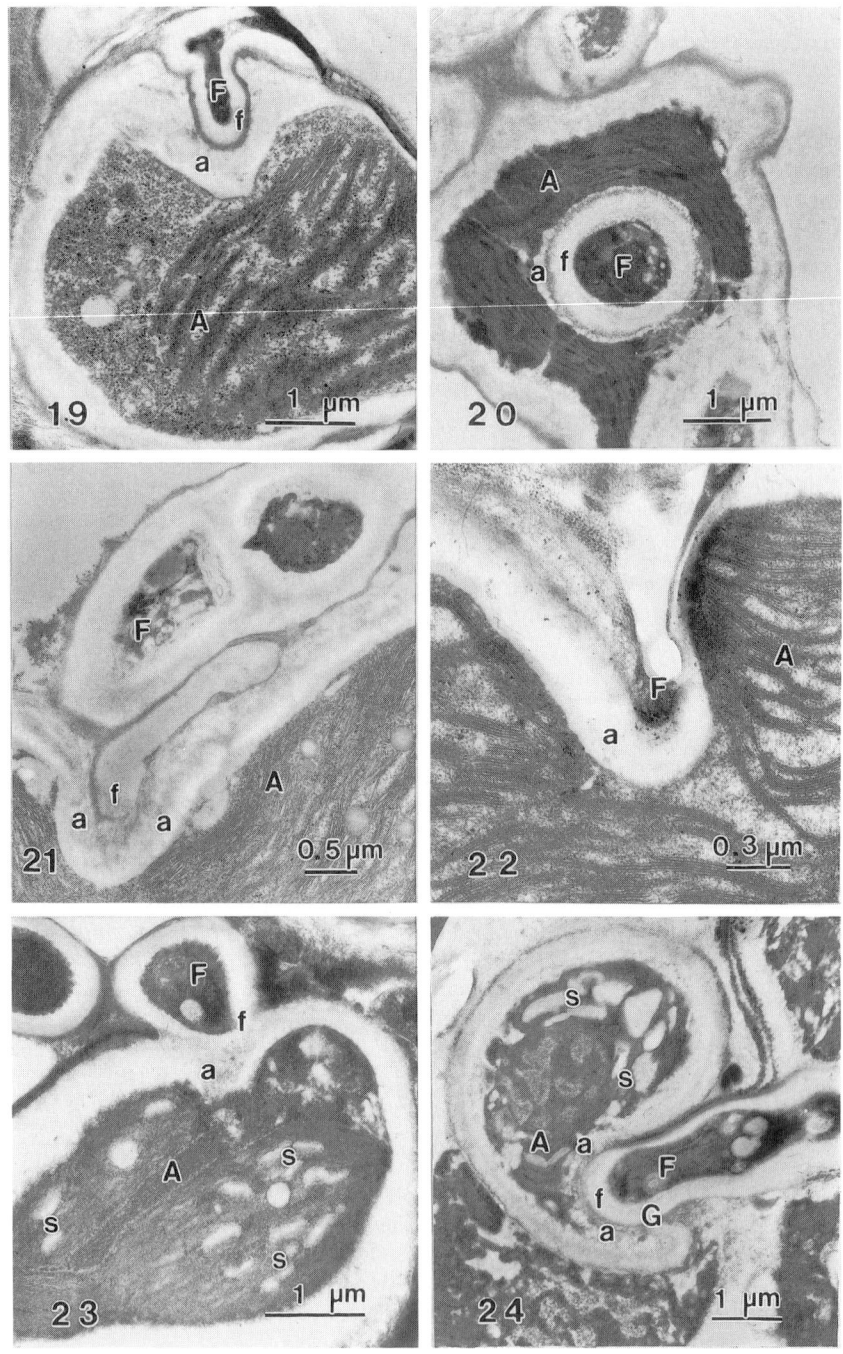

## Discussion

*Depth of lichen penetration in host tree*

The corticolous lichen species studied are strictly endoperidermal; they penetrate into the dead cork but not into the underlying living tissues of the bark. All are found on several species of substrate trees; for example, *Anisomeridium tuckerae* has been collected on six tree species and *Pyrenula anomala* on 11 tree species (Tucker and Harris 1980). This lack of specificity plus the endoperidermal limitation of the thallus argue against a parasitic role for these crusts. Lambright and Tucker (1980) found hyphae of the mycobiont (*Trypethelium eleuteriae*) between and inside dead cork cells; this type we call endoperidermal, since it penetrates only the outer periderm. Some of the cork cell walls containing hyphae had undergone thinning and deterioration, compared to walls of more interior cork cells. Cork cells with thickened, suberized walls are dead and may remain unchanged or may collapse. The lichen in *T. eleuteriae* may produce lysing agents which break down the cellulose, suberin, lignin, and other materials in the cork cell walls. Depending on the tree species, cork tissue may be periodically worn away or shed as flakes of dead tissue. Most lichens penetrate only within the outer part of the bark, into dead cork. In barks with successive cork cambial layers, parts of the non-conducting phloem including sclereids and fibres may be cut off outside a new cork cambium. These lenses of secondary phloem can either be retained as dead tissue in 'tight' barks, or sloughed off together with cork as bark scales in scaling barks.

Reports of lichen hyphal penetration into living tissues of host trees (Estevez *et al.* 1980; Ascaso *et al.* 1980), are significant if true, since this association would be parasitic. These workers studied *Evernia prunastri* on branches of *Quercus suber* (cork oak). They show sections of woody branches with massive mycelial accumulations between successive

---

**Fig. 12.19–12.24.** Fungal haustoria of various lichen taxa. **Figs 12.19, 12.20.** *Schismatomma rappii*. **Fig. 12.19.** Invagination of photobiont wall around the haustorium. Algal cell contents are in good condition. **Fig. 12.20.** Cross-section of an haustorium with the surrounding algal wall thin and irregular. Algal cell contents are degenerate. **Fig. 12.21.** *Arthonia tumidula*. Haustorium entering algal cell, while algal cell wall is being invaginated around it. **Fig. 12.22.** *Anisomeridium subprostans*. Invaginating algal cell wall around haustorial tip. **Fig. 12.23, 12.24.** *Mazosia sp*. **Fig. 12.23.** Early stage of algal wall invagination ahead of haustorium at the point of entry. **Fig. 12.24.** Later stage with algal wall over haustorial tip appearing thin and degenerate. A, algal cell; F, fungal cell or haustorium; a, algal wall; f, fungal wall; s, starch.

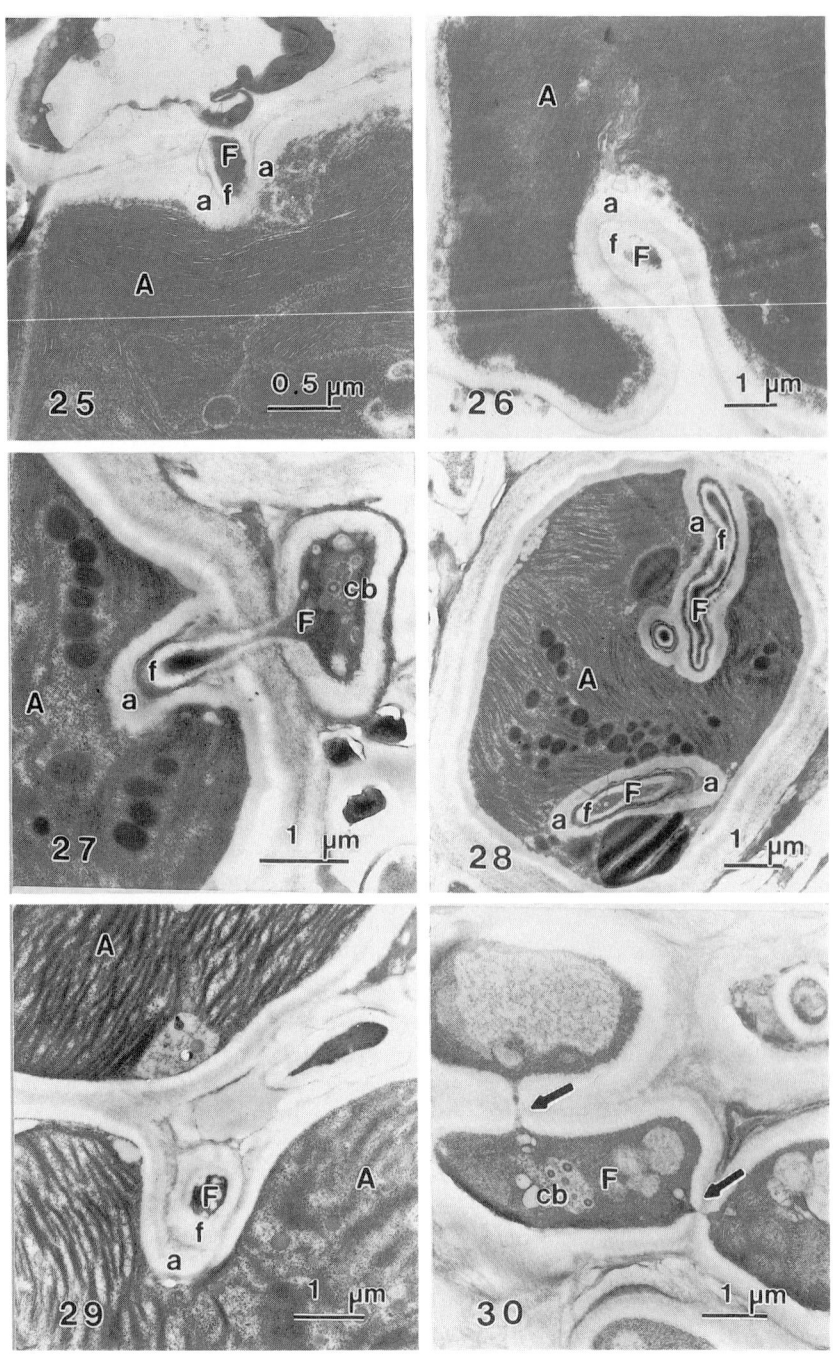

periderms, within living phloem, and in the secondary xylem inside the vascular cambium. No algal cells appear in the figures. Since *Evernia* is a fruticose lichen and is not known to have large mycelial masses in the substrate, some doubt exists that the mycelium illustrated belongs to the lichen. The pattern within the wood, in particular, resembles the infection of a wood-rotting fungus. Another argument against a parasitic role for *Evernia prunastri* is that it occurs on a wide variety of tree species, whereas parasitic fungi tend to have narrow host ranges.

*Haustorial type*

We found haustoria to be present in all thirteen tropical crusts studied. Of other reports on tropical crusts, *Strigula elegans* has haustoria (Chapman 1976a), *Chiodecton sanguineum* lacks haustoria (Ellis 1975), while *Trypethelium elueteriae* (Lambright and Tucker 1980) has appressoria. There is almost no information on what proportion of tropical lichens with trentepohliaceous photobionts have haustoria. By way of a contrast, ample information is available on trebouxioid-photobiont-bearing lichens of temperate regions, so that Tschermak (1941), Peveling (1973), and Honegger (1985) are able to correlate the type of alga–fungus contact with the level of phylogenetic specialization of the lichen. Honegger (1986) classifies types of alga–fungus interface; well-developed haustoria may be intracellular (in which the fungal wall contacts the host cell plasmalemma), or intraparietal (in which the contact is between fungal wall and algal host cell wall). Each type of haustorium has several subtypes. Honegger (1986) reported that most lichens have a mixture of types of interfaces, but with one type predominating. She generalized that unspecialized non-stratified crusts tend to have intracellular haustoria, while the more specialized stratified crusts and more complex growth forms have intraparietal haustoria. Discomycetous lichens tend to have intracellular types or none (Galun 1988). None of the lichens we studied had the anomalous structures (wall

---

**Fig. 12.25–12.30.** Fungal haustoria of various lichen species and hyphal structure. **Fig. 12.25, 12.26.** *Strigula complanata*. **Fig. 12.25.** Entry of haustorium with algal wall invaginating around it. **Fig. 12.26.** Haustorium present in algal cell lumen with ensheathing algal wall thin and degenerate or absent over haustorial tip. **Fig. 12.27, 12.28.** *Porina pulla*. **Fig. 12.27.** Inner layer of algal cell wall is invaginated around the haustorium at the point of entry. **Fig. 12.28.** Transverse and oblique sections of haustoria, surrounded by algal cell wall. **Fig. 12.29.** *Trypethelium ochroleucum*. Bi-layered algal cell wall ensheathes the entering haustoria. **Fig. 12.30.** *Pyrenula anomala*. Hyphal cells, containing concentric bodies and having septal pores (at arrows), which are not occluded by Woronin bodies. A, algal cell; a, algal cell wall; cb, concentric bodies; F, fungal cell or haustorium; f, fungal cell wall; p, plastoglobuli.

apposition, collar, or sheath around haustorium, anomalous wall material at the point of haustorium entry) which were reported in various lichenized and/or non-lichenized fungi by Honegger (1986) and by Bracker and Littlefield (1973). The walls of photobiont cells in our material (including some Discomycetes) showed the ability to invaginate and/or expand around the invading haustorium, unlike walls of host cells invaded by pathogenic fungal haustoria (Bracker and Littlefield 1973). The symbiotic mutualistic relationship of the lichens may be based at least in part on the wall growth response of the alga. In several of the taxa studied among cells containing haustoria, we observed both cells in good condition and deteriorated cells in the same preparation. The relationship may be initially mutualistic when haustoria first invade healthy algal cells and then progress with time to a more parasitic state in which the algal cell deteriorates and dies. Peveling (1973) reported on similar progressive degradation in cells invaded by haustoria in *Lecanora muralis*. Galun et al. (1970), however, suggested that algal cells in an *Aspicilia* studied normally die after a certain time, regardless of whether haustoria are present.

In several of the species we studied, occasional haustoria have a thin or degenerated algal wall ensheathing the tip, which could conceivably be classified as intracellular in Honegger's classification. Chapman (1976a) called the type of haustorium in *Strigula elegans* intracellular, since it usually had only fungal wall around the tip; however, fibrous remnants of the algal cell wall were erratically present around the haustorium, with most algal wall remaining around the base of the haustorium. Peveling (1973) also reported intracellular haustoria with remnants of the algal wall persisting. It is evident that there are transitional conditions between intracellular and intraparietal haustoria and that the intraparietal type may change over time to the intracellular type, at least in some species.

The haustoria in pyrenocarpous lichens have been inadequately investigated. The type (intraparietal) found in the seven pyrenocarpous taxa in four families in the present study differs markedly from the type (intracellular) in *Dermatocarpon* and *Verrucaria* reported by Galun et al. (1970, 1971); and Kushnir and Galun (1977). In these taxa, both algal and fungal walls are said to disintegrate at the point of contact. New wall material of anomalous origin is deposited and the haustorium penetrates without an algal wall sheath. In contrast, in our material we found invagination of the algal wall at the point of contact and maintenance of the algal wall over the tip of the haustorium. The algal wall persists over the haustorial tip in the five pyrenocarps we studied: *Anisomeridium subprostans*, *A. tuckerae*, *Porina pulla*, *Trypethelium ochroleucum*, and *T. tropicum*. There was some thinning and dissolution of the algal wall over the haustorial tip in two pyrenocarps (*Pyrenula anomala* and *Strigula complanata*). Four pyrenocarp families are represented by these taxa: Pyrenulaceae, Strigulaceae, Trichotheliaceae, and Trypetheliaceae. Clearly, additional pyrenocarps need to be studied to determine their type of haustorium. Possibly, there is a correlation of haustorial type with ecological habitat rather than with systematic affinity, since the seven tropical crusts in our study represented a diverse group in four families. The desert species studied by Galun *et al.* (1970, 1971) undoubtedly have different physiological requirements, which might be expressed in part by a different pattern of haustorial development.

Haustoria in the six non-pyrenocarp crustose lichens also are intraparietal, with a tendency in most to lose some of the algal wall distally over the haustorial tip over time. These six species belong to four different families (Arthoniaceae, Graphidaceae, Lecanactidaceae, Opegraphaceae). In four species (*Graphis scripta*, *Mazosia* sp., *Opegrapha viridis*, and *Schismatomma rappii*) at least some haustoria were observed in which the algal wall over the haustorial tip was thin or deteriorated. Two *Arthonia* species were the only exceptions in which all preparations showed that the algal wall remains intact over the haustorium. A common tendency among a diversity of tropical crusts, regardless of systematic affinity, is to show an apparent increase in intimacy of contact over time within individual algal cells. The loss of algal wall around the haustorial tip as the haustorium penetrates the algal cell lumen has also been reported in *Parmelia sulcata* (Webber and Webber 1970).

*Ecological distribution of trentepohliaceous photobionts*

*Trentepohlia*, *Phycopeltis*, and *Cephaleuros* are all common in the freeliving state in the tropics and subtropics; *Trentepohlia* usually occurs on bark, *Phycopeltis* on evergreen leaves (Good and Chapman 1978a, b), and *Cephaleuros* with different species on leaves (Chapman 1976b) or on bark (Chapman and Henk 1985). *Cephaleuros virescens* is a pathogen on leaves of tea, coffee, and other tropical evergreens, and on fruits (Marlatt and Alfieri 1981; Holcomb 1986; Hawksworth 1988). It was reported to occur on 218 vascular plant taxa in Louisiana and adjacent states (Holcomb 1985) and on over 600 species worldwide (Hawksworth 1988). *Strigula* species, the lichenized form of *C. virescens* on leaves, has been reported on numerous substrates by Holcomb; Santesson (1952) reported it on 99 host species of vascular plants. The mode of action of *C. virescens* is unusual in that algal cells do not enter tissues of the leaf. They invade below the upper cuticle and form a thin plate-like thallus over the epidermis. In some host species, the palisade mesophyll leaf tissue below the alga forms a cup-shaped wound periderm, and the mesophyll cells cut off below the alga die (Chapman and Good 1976, 1983). Hence the lesions are unsightly, but local and their large numbers may represent a sizeable diversion of metabolic energy by the host.

Trentepohliaceous photobionts are abundant in tropical and subtropical floras; for example, in 38 per cent of the 543 lichen species reported for subtropical Louisiana (Tucker 1981), trentepohliaceous photobionts can be assumed to be present. Ahmadjian (1967) estimates that about 9 per cent of the lichen taxa in temperate flora have trentepohliaceous photobionts. The lichen groups which most commonly have trentepohliaceous photobionts are Graphidaceae, Arthoniaceae, Thelotremaceae, and several pyrenocarp families.

The relative dearth of trentepohliaceous taxa in the north and their prevalence in tropics and subtropics, may be based on their vulnerability to cold, or their inability to compete under conditions of cold stress. Nash *et al.* (1987) demonstrated that *Trentepohlia* is more vulnerable to cold damage than any of the other common lichen photobionts that he tested.

*Specificity of photobiont*

Although most lichens are assumed to have one specific photobiont, several examples are known where there is variability (Table 12.1). *Arthonia* can have photobionts in several genera of algae: *Trentepohlia* (present work), *Phycopeltis* (Peveling 1973, Santesson 1952), or *Chlorococcus* (Santesson 1952). *Opegrapha* species have been reported to have either *Trentepohlia* (Tschermak 1941) or *Phycopeltis* (Santesson 1952; Sérusiaux 1985); our work revealed the presence of *Phycopeltis* in *O. viridis*. We found two other examples of lichen genera having genera of photobionts different from those previously reported, in *Schismatomma rappii* and *Graphis scripta*. Previous work on unspecified species of *Schismatomma* (Poelt and Vezda 1977) reported *Trentepohlia*, while we found *Phycopeltis*. For *Graphis scripta*, *Phycopeltis* was found in Louisiana material, while previous work had reported various species of *Trentepohlia*. *Trentepohlia umbrina* (Hérisset 1946) and *T. annulata* (Verseghy 1961) had been reported for European material and *T. lagenifera* for widespread collections from Japan (Nakano 1988). This variability of photobiont may be correlated with the highly variable morphology of *Graphis scripta* worldwide. Other lichen genera reported to have more than one genus of photobiont include *Chaenotheca* (Tibell 1984, pp. 639–40), *Endocarpon* (Tschermak-Woess 1988), *Lobaria* and other Stictaceae (Yoshimura 1971), and *Verrucaria* (Tschermak-Woess 1988). These examples, in which diverse algae are interchangeable as the photobiont in one species of lichen or in species of one genus are instructive and indicate that the fungus largely dictates morphology and that the fungus-alga interaction can tolerate such intergeneric or interspecific variability as may exist between the alternative photobiont taxa.

## Acknowledgements

The help of Richard Harris, who identified or verified the species studied, is greatly appreciated. The work was supported in part by the National Science Foundation under grants BSR-8418922 and BSR-8722514 to S.C. Tucker, and BSR-8722739 to R.L. Chapman.

## References

Ahmadjian, V. (1964). Further studies in lichenized fungi. *The Bryologist*, **67**, 87–98.

Ahmadjian, V. (1967). *The Lichen symbiosis*. Blaisdell Publishing Company, Waltham, Mass.

Ascaso, C., Gonzales, C., and Vicente, C. (1980). Epiphytic *Evernia prunastri*: Ultrastructural facts. *Cryptogamie: Bryologie, Lichenologie*, **1**, 43–51.

Bracker, C.E. and Littlefield, L.J. (1973). Structural concepts of host–pathogen interface. In *Fungal pathogenicity and the plants response* (ed. R.W.J. Byrde and C.V. Cutting), pp. 159–318. Academic Press, London.

Chapman, R.L. (1976a). Ultrastructural investigation on the foliicolous pyrenocarpous lichen *Strigula elegans* (Fée) Müll. Arg. *Phycologia*, **15**, 191–6.

Chapman, R.L. (1976b). Ultrastructure of *Cephaleuros virescens* (Chroolepidaceae; Chlorophyta). I. Scanning electron microscopy of zoosporangia. *American Journal of Botany*, **63**, 1060–70.

Chapman, R.L. (1980). Ultrastructure of *Cephaleuros virescens* (Chroolepidaceae; Chlorophyta). II. Gametes. *American Journal of Botany*, **67**, 10–17.

Chapman, R.L. (1984). An assessment of the current state of our knowledge of the Trentepohliaceae. Systematics Association. In *Proceedings of the symposium on systematics of the green algae* (ed. D.E.G. Irvine and D.M. John), pp. 233–50. Academic Press, London.

Chapman, R.L. and Good, B.H. (1976). Observations on the morphology and taxonomy of *Phycopeltis hawaiiensis* King (Chroolepidaceae). *Pacific Science*, **30**, 187–95.

Chapman, R.L. and Good, B.H. (1983). Subaerial symbiotic green algae: interactions with vascular plant hosts. In *Algal symbiosis: a continuum of interaction strategies* (ed. L.J. Goff), pp. 173–204. Cambridge University Press, New York.

Chapman, R.L. and Henk, M.C. (1985). Observations on the habit, morphology, and ultrastructure of *Cephaleuros parasiticus* (Chlorophyta) and a comparison with *C. virescens*. *Journal of Phycology*, **21**, 513–22.

Chapman, R.L. and Henk, M.C. (1986). Phragmoplasts in cytokinesis of *Cephaleuros parasiticus* (Chlorophyta) vegetative cells. *Journal of Phycology*, **22**, 83–8.

Ellis, E.A. (1975). Observations on the ultrastructure of the lichen *Chiodecton sanguineum*. *The Bryologist*, **78**, 471–6.

Estevez, P., Orus, M.I., and Vicente, C. (1980). Estudios morfologicos sobre *Evernia prunastri* de vida saprofitica. *Cryptogamie: Bryologie, Lichenologie*, **1**, 33–41.

Galun, M. (1988). The fungus–alga relation. In *CRC handbook of lichenology*, Vol. 1. (ed. M. Galun). pp. 147–58. CRC Press, Boca Raton, Florida.

Galun, M., Kushnir, E., Behr, L., and Ben-Shaul, Y. (1973). Ultrastructural investigation on the alga–fungus relation in pyrenocarpous lichen species. *Protoplasma*, **78**, 187–93.

Galun, M., Paran, N., and Ben-Shaul, Y. (1970). Structural modifications of the phycobiont in the lichen thallus. *Protoplasma*, **69**, 85–96.

Galun, M., Paran, N., and Ben-Shaul, Y. (1971). Electron microscopic study on the lichen *Dermatocarpon hepaticum*. *Protoplasma*, **73**, 457–68.

Good, B.H. and Chapman, R.L. (1978a). The ultrastructure of *Phycopeltis* (Chroolepidaceae: Chlorophyta). I. Sporopollenin of the cell walls. *American Journal of Botany*, **65**, 27–33.

Good, B.H. and Chapman, R.L. (1978b). Scanning electron microscope observations on zoosporangial abscission in *Phycopeltis epiphyton* (Chlorophyta). *Journal of Phycology*, **14**, 374–6.

Harris, R.C. (1975). A taxonomic revision of the genus *Arthopyrenia* Massal. s. lat. (Ascomycetes) in North America. Ph.D. Dissertation, Michigan State University, East Lansing, Michigan.

Hawksworth, D.L. (1988). Effects of algae and lichen-forming fungi on tropical crops. In *Perspectives of mycopathology* (ed. V.P. Agnihotri, A.K. Sarbhoy, and D. Kumar), pp. 76–83. Malhotra Publ. House, New Delhi.

Hérisset, A. (1946). Demonstration expérimentale de rôle du *Trentepohlia umbrina* dans la synthèse des Graphidées corticoles. *Compte rendu hebdomadaire des Séances de l'Académie des Sciences, Paris*, **222**, 100–2.

Holcomb, G.E. (1985). New hosts of the parasitic lichen *Strigula*. *Plant Disease*, **69**, 100.

Holcomb, G.E. (1986). Hosts of the parasitic alga *Cephaleuros virescens* in Louisiana, and new host records for the continental United States. *Plant Disease*, **70**, 1080–3.

Honegger, R. (1985). Fine structure of different types of symbiotic relationships in lichens. In *Lichen physiology and cell biology* (ed. D.H. Brown), pp. 287–302. Plenum Press, New York.

Honegger, R. (1986). Ultrastructural studies in lichens. I. Haustorial types and their frequencies in a range of lichens with trebouxioid photobionts. *New Phytologist*, **103**, 785–95.

Kushnir, I. and Galun, M. (1977). The fungus–alga association in endolithic lichens. *Lichenologist*, **9**, 123–30.

Lambright, D.D. and Tucker, S.C. (1980). Observations on the ultrastructure of *Trypethelium eleuteriae*. *The Bryologist*, **83**, 170–8.

Marlatt, R.B. and Alfieri, S.A., Jr. (1981). Host of the parasitic alga *Cephaleuros* Kunze in Florida. *Plant Disease*, **65**, 520–2.
Matthews, S.W., Tucker, S.C., and Chapman, R.L. (1989). Ultrastructural features of mycobionts and trentepohliaceous phycobionts in selected subtropical crustose lichens. *Botanical Gazette*, **150**, 417–38.
Meier, J.L. and Chapman, R.L. (1983). Ultrastructure of the lichen *Coenogonium interplexum*. *American Journal of Botany*, **79**, 400–7.
Nakano, T. (1988). Phycobionts of some Japanese species of the Graphidaceae. *Lichenologist*, **20**, 353–60.
Nash, T.H., Kappen, L., Lösch, R., Larson, D.W., and Matthes-Sears, U. (1987). Cold resistance of lichens with *Trentepohlia*- or *Trebouxia*-photobionts from the North American west coast. *Flora*, **179**, 241–51.
Peveling, E. (1973). Fine structure. In *The lichens* (ed. V. Ahmadjian and M.E. Hale, Jr.), pp. 147–82. Academic Press, New York.
Poelt, J. and Vězda, A. (1977). *Bestimmungsschlussel europaischer Flechten*. Pt. I. J. Cramer, Vaduz.
Santesson, R. (1952). Foliicolous lichens. I. *Symbolae Botanicae Upsalienses*, **12**, 1–590.
Sérusiaux, E. (1979). Two new foliicolous lichens from tropical Africa. *Lichenologist*, **11**, 181–5.
Sérusiaux, E. (1985). Goniocysts, goniocystangia, and *Opegrapha lambinonii* and related species. *Lichenologist*, **17**, 1–25.
Tibell, L. (1984). A reappraisal of the taxonomy of Caliciales. *Beiheft Nova Hedwigia*, **79**, 597–713.
Tschermak, E. (1941). Untersuchungen über die Beziehungen von Pilz und Alga in Flechtenthallus. *Oesterreichisches Botanisches Zeitschrift*, **90**, 233–307.
Tschermak-Woess, E. (1988). The algal partner. In *CRC handbook of lichenology*, Vol. 1. (ed. M. Galun), pp. 39–92. CRC Press, Boca Raton, Florida.
Tucker, S.C. (1981). Checklist of Louisiana lichens. *Louisiana Academy of Sciences, Proceedings*, **44**, 58–70.
Tucker, S.C. and Harris, R.C. (1980). New and noteworthy pyrenocarpous lichens from Louisiana and Florida. *The Bryologist*, **83**, 1–20.
Verseghy, K. (1961). A *Graphis scripta* Ach. (Lichenes gonidiumára vonatkozó vizgálatok. *Botanikae Közlemények*, **49**, 95–9.
Webber, M.M. and Webber, P.J. (1970). Ultrastructure of lichen haustoria: symbiosis in *Parmelia sulcata*. *Canadian Journal of Botany*, **48**, 1521–4.
Withrow, K. and Ahmadjian, V. (1983). The ultrastructure of lichens. VII. *Chiodecton sanguineum*. *Mycologia*, **75**, 337–9.
Yoshimura, I. (1971). The genus *Lobaria* of Eastern Asia. *Journal of the Hattori Botanical Laboratory*, **34**, 231–364.

# 13. On the importance of botanical gardens for lichens in the Asian tropics

L. ARVIDSSON

*Naturhistoriska Museet, PO Box 7283, S - 402 35 Göteborg, Sweden*

### Abstract

The importance of botanical gardens for lichens is discussed. Several factors favour a diverse lichen flora in gardens, such as (1) the presence of forests or groves; (2) a long ecological continuity; (3) the presence of many different habitats and substrates; (4) the presence of surrounding vegetation and bark that to some extent absorbs and buffers air pollution; (5) the presence of rare phorophytes which might house strongly specialized lichens. In areas radically re-moulded by man, gardens represent important refuges and can function as genetic banks for lichens. Cultivation of lichens in gardens is also briefly discussed.

### Introduction

Many tropical gardens and parks have a well-developed lichen flora. In certain areas botanical gardens are the best place for lichenological studies. This paper will emphasize the importance of such gardens for lichens. Cultivation of these cryptogams is perhaps also possible and is discussed briefly, but the author is primarily aiming at the indigenous lichen flora that the particular habitats of gardens have attracted. Most of the thoughts presented here are also applicable to other parts of the tropics, as well as to temperate areas.

## Favourable factors

Several factors favour a diverse lichen vegetation in botanical gardens. Some of the most important are as follows:

**1.** Gardens often exhibit remnants of forests or groves and in many places they form a contrast to their deforested agricultural or industrial surroundings. Two examples can be given here: the botanical garden at Pamplemousses in Mauritius is completely surrounded by extensive fields of sugar-cane and in Java, arable land with rice-fields and tea plantations encircles the garden at Bogor. Both these places have an interesting epiphytic lichen flora being one of the few places in the areas where corticolous lichens can establish.

**2.** Well-established tropical gardens often have a long ecological continuity. For instance, the garden at Pamplemousses is over 200 years old, as is the Royal Botanic Gardens at Peradeniya in Sri Lanka. Hortus Botanicus Cibodasensis (Java) dates from 1862. This continuity is of special significance for the survival of 'old-forest' lichens (Rose 1976) and gardens which include original forest habitats are, of course, of particular value in this respect. For instance, rare woodland lichens such as *Erioderma tomentosum* and various species of *Menegazzia* and *Pseudocyphellaria* can still be found in remnants of rainforest at Cibodas. However, even some introduced phorophytes can become suitable substrates for scarce lichens as shown in the botanical garden at Furnas in São Miguel, the Azores. Here, *Leptogium corticola* was discovered on a shaded trunk of *Liriodendron tulipifera* (Arvidsson 1990). This lichen is regarded as a Tertiary relic and the garden at Furnas appears to be the only current location in Europe (Jørgensen and James 1983).

**3.** An important fact is that gardens present a variety of habitats, such as groves, solitary trees, avenues, ravines, rocks, walls, naked soil, ponds and water-falls. In addition, the great number of phorophyte species offers many different types of bark. At Cibodas (Fig. 13.1), lichens grew (1979) on all kind of substrates, from leaves of evergreen bushes to trunks of giant rainforest trees, as well as on fence-posts, concrete, and roofing-tiles. The presence of foreign trees is interesting in another way. If, for instance, the native trees in the area have acid bark, the introduction of alien species with a basic cortex creates new possibilities for lichens. The rather open conditions in gardens with lawns and thinly planted trees also make it easier for lichen diaspores to become dispersed into the surrounding area, or into the garden.

**4.** In areas influenced by air pollution, the outer vegetation of a garden functions as a filter. This results in better conditions for the growth of lichens in the central parts. However, if the pollution is too intense, this effect will not be observed. At Bogor in western Java, the

Fig. 13.1. The Botanical Garden at Cibodas, western Java, with a variety of microhabitats and phorophytes. Photo: Lars Arvidsson, 22 March 1979.

trees facing the heavy traffic just outside the garden showed a very poor lichen vegetation (1979). However, 100–200 m into the garden, the lichen flora was well-developed on palms in the avenues, with, among other things, *Coccocarpia palmicola* and species of *Heterodermia*. A similar situation was also observed in the temperate garden at Göteborg, SW Sweden. Here, small specimens of *Bryoria* and *Usnea* were observed in central parts of the garden, but not elsewhere on similar phorophytes in avenues and smaller parks in the city.

Introduced trees with a high buffer-capacity bark can also help the indigenous lichens to survive in areas affected by acid rain. To take one example from a temperate area: in the botanical garden at Göteborg, many foliose and fruticose lichens are now absent from trees with a low buffer-capacity, such as species of *Pinus*, *Picea*, *Larix*, *Betula*, and *Tilia*. This is probably due to an increase of acid rain during recent years. However, the lichens formerly present on the phorophytes mentioned are today found on many new substrates, such as species of *Amelanchier*, *Carya*, *Ligustrina*, *Magnolia*, *Ostrya*, *Pterocarya*, and *Zelkova*, none of which are native to Scandinavia (Arvidsson and Lindström 1980). *Pseudevernia furfuracea* and *Xanthoria polycarpa* are now present only on introduced phorophytes, although they were formerly known from several indigenous trees.

It is possible to improve the pH-status of the bark by low-dose liming. This was tried by Hallingbäck and Ingelög (1989), using a collar with limestone placed several metres up around the tree trunk.

5. Many endangered plants are often grown in botanical gardens. The bottle palm *Hyophorbe lagenicaulis* from Mauritius (Fig. 13.2), and several other endemic trees have practically vanished from the indigenous forests of the island. However, in the botanical garden at Pamplemousses, there are some magnificent avenues of mascarene palms which were covered in lichens in 1979 when the author visited the garden. It was not possible to collect from these rare phorophytes and

**Fig. 13.2.** The Botanical Garden at Pamplemousses, Mauritius. Avenue with the extremely rare, endemic bottle palm *Hyophorbe lagenicaulis*, covered in lichens. Photo: Lars Arvidsson, 16 April 1979.

the nature of this vegetation is still unknown, even though some of the macrolichens seemed to be widespread species of *Coccocarpia*, *Dirinaria*, *Heterodermia*, and *Pyxine*. However, one important question must be posed in connection with this: Are there any lichens confined to these palms—or more generally—to bark of other endangered trees of the Asian tropics? We lose species every day and once extinct, their epiphytes are also lost. Arboricultures like that at Pamplemousses can be a last foothold. The search for specialized lichens should be intensified before it is too late.

## Problems and unfavourable factors

Are botanical gardens safe places for lichens? Can the existence of a particular garden be guaranteed for the future? These questions are difficult to answer even in European countries. Economic circumstances and population problems are, or might become, serious threats even to well-established botanical gardens. International co-operation and control is perhaps a possibility for some of the most important places.

Another problem is that in many gardens epiphytes are regarded as unaesthetic or even believed to damage the host tree. Thus, lichens and mosses are constantly brushed off. Frequently the trunks are also painted white. Publicity on the non-harmful effect of lichen epiphytes will eventually resolve this problem.

Another question to be solved is the use of biocides to control various diseases and parasites affecting ornamental plants. Precise data on the effects of most herbicidal and pesticidal sprays on lichens are not available, but field observations indicate that they can be harmful—in some instances eliminating all lichens (Hawksworth and Rose 1976). The impoverished lichen flora of some gardens might be explained by an intense use of chemicals. Moderate doses of fertilizers such as superphosphate enrich the bark and promote growth of species of the *Xanthorion*. Excessive application, in contrast, leads to hypertrophication resulting in loss of lichens and extensive growth of *Pleurococcus*. The use of agricultural chemicals in the regions surrounding gardens can also be negative. A 'clean' zone around gardens is advantageous to epiphytes.

## Cultivation of lichens in gardens

Apart from the 'passive' lichen vegetation discussed above, cultivation of lichens in gardens might be considered as well. We know from experiments with lichen transplants in temperate areas that at least some species can be moved to new locations and continue to grow there. For instance, it was shown by Hallingbäck and Ingelög (1989), that a rare and threatened species in Sweden (*Lobaria amplissima*), could be transplanted. In this case, thin pieces of bark with the lichen on were removed and glued to the trunk of a new phorophyte over 100 kilometers away. After one year, the vitality of the *Lobaria* species was still good and some thalli had even grown out over the new phorophyte. However, for another taxon (*L. scrobiculata*), this technique proved less successful (Hallingbäck and Ingelög 1989). Another method was tried by Hallingbäck when he used soredia and thallus fragments of *Lobaria pulmonaria* (Hallingbäck 1990). Thalli of the adult lichen were rubbed on the surface of rough bark into small fragments, about 1 mm diam. In less than one year hundreds of primordia were seen on the new phorophyte. After one year the largest thalli measured 6–8 mm. As stated above, a variety of habitats, such as suitable phorophytes, are present in most gardens. If the environmental conditions of the garden conform with the ecological demands of the lichen in question it will probably grow there. If the natural habitat of a threatened species is likely to be destroyed, transplantation of the lichen must be a possibility. The transplant should, of course, be moved primarily to a similar natural habitat, but, if none exists, a botanical garden is an alternative to be considered. In contrast to the situation in nature, growth of a lichen transplant can be closely followed by staff members in the garden. Many botanic gardens already function as genetic banks for vascular plants.

It should be stated here that botanical gardens cannot possibly save all kinds of lichens. For instance, to rescue a rainforest species one probably needs a rainforest. The natural habitats are of course the most suitable and should be preserved as far as possible. However, in areas radically re-moulded by man, such as many places in the Asian tropics, botanic gardens represent important refuges and they might function as genetic banks. Botanical gardens will probably become even more important in the future as the exploitation of nature accelerates.

### References

Arvidsson, L. (1990). Additions to the lichen flora of the Azores. *Bibliotheca Lichenologica*, **38**, 13–27.
Arvidsson, L. and Lindström, M. (1980). Förändringar i lavfloran i Botaniska Trädgården i Göteborg. *Svensk Botanisk Tidskrift*, **74**, 133–43.
Hallingbäck, T. (1990). Transplanting *Lobaria pulmonaria* to new localities and a review on the transplanting of lichens. *Windahlia*, **18**. (Seen in manuscript.)

Hallingbäck, T. and Ingelög, T. (1989). *Åtgärder för att bevara luftföroreningskänslig lav- och mossflora*. Statens Naturvårdsverk, Rapport 3679. Stockholm.
Hawksworth, D.L. and Rose, F. (1976). *Lichens as pollution monitors*. Edward Arnold, London.
Jørgensen, P.M. and James, P.W. (1983). Studies on some *Leptogium* species of western Europe. *Lichenologist*, 15, 109–25.
Rose, F. (1976). Lichenological indicators of age and environmental continuity in woodlands. In *Lichenology: progress and problems* (ed. D.H. Brown, D.L. Hawksworth, and R.H. Bailey), pp. 279–307. Academic Press, London.

# 14. Some foliicolous lichens in Xishuangbanna, China

J.C. WEI and Y.M. JIANG
*Systematic Mycology and Lichenology Laboratory, Institute of Microbiology, Academia Sinica, Beijing, China*

### Abstract

Twenty species of foliicolous lichens from 12 genera in 10 families, collected in October 1980 from Xishuangbanna are reported in the present paper. Amongst them seven foliicolous genera, *Asterothyrium*, *Byssolecania*, *Dimerella*, *Fellhanera*, *Gyalectidium*, *Opegrapha*, and *Trichothelium*, and 15 species including one variety are new to China, and *Byssolecania deplanata* is new to Asia.

### Introduction

Xishuangbanna, an autonomous prefecture of the Dai nationality in southern Yunnan of China, 21°10′–22°40′ N, 99°55′–101°50′ E, includes Jinghong, Mengla, and Menghai counties (Fig. 14.1, 14.2). The area of the whole prefecture is about 19 220 square kilometres with an elevation of 420–2400 m. Most of the land comprises terraces, hills, and mountains at an elevation of 540–1200 m. The area of forest is about 640 000 hectares which covers *c*. 34 per cent of the whole area.

The flora of Xishuangbanna belongs to Yunnan, Burma, and Thailand and is within the Malaysian subkingdom of the Palaeotropic Kingdom according to Wu's regionalization of the Chinese flora (Wu 1979). The flora of this area is identical to the tropical flora of Burmese Shan State, of northern Thailand, and of northern Laos. It is very rich in palaeotropical floral elements. About 5000 species of phanerogams and ferns in Xishuangbanna are equivalent to one-sixth of the plant species in the whole of China, although the area covers only one-five

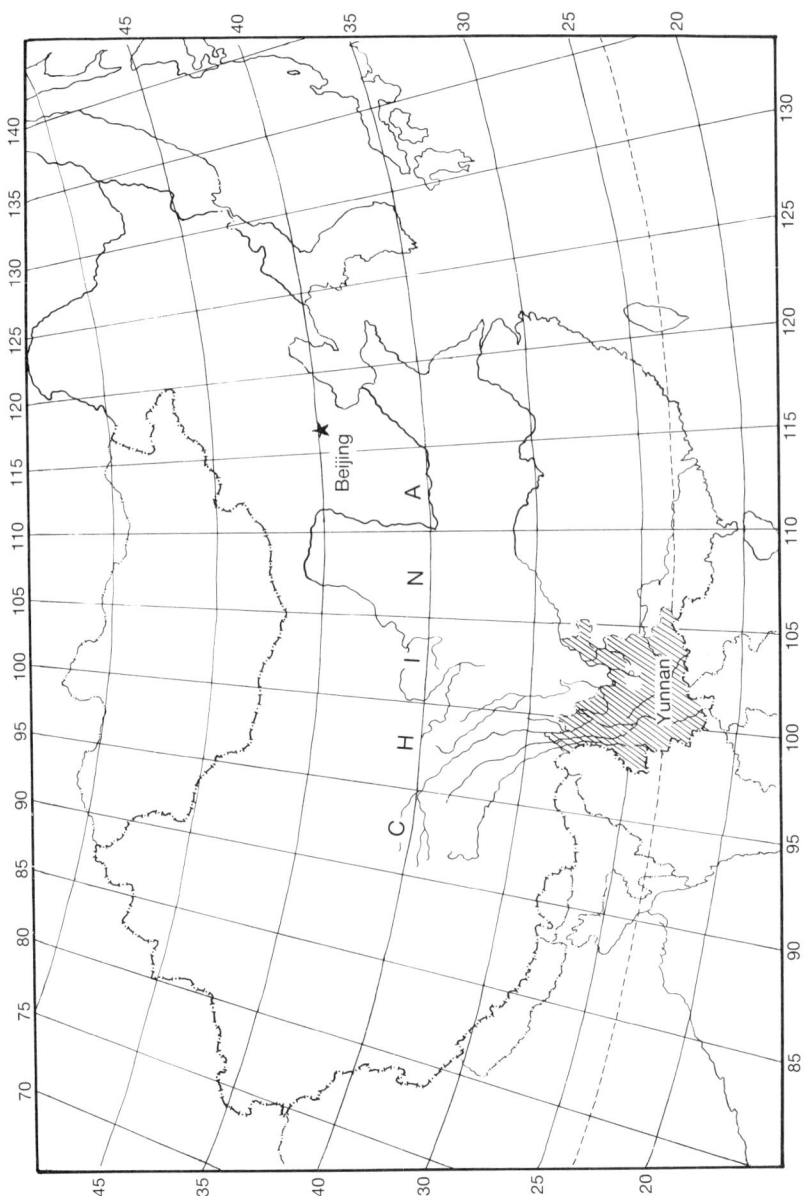

**Fig. 14.1.** Map showing the position of Yunnan province in China.

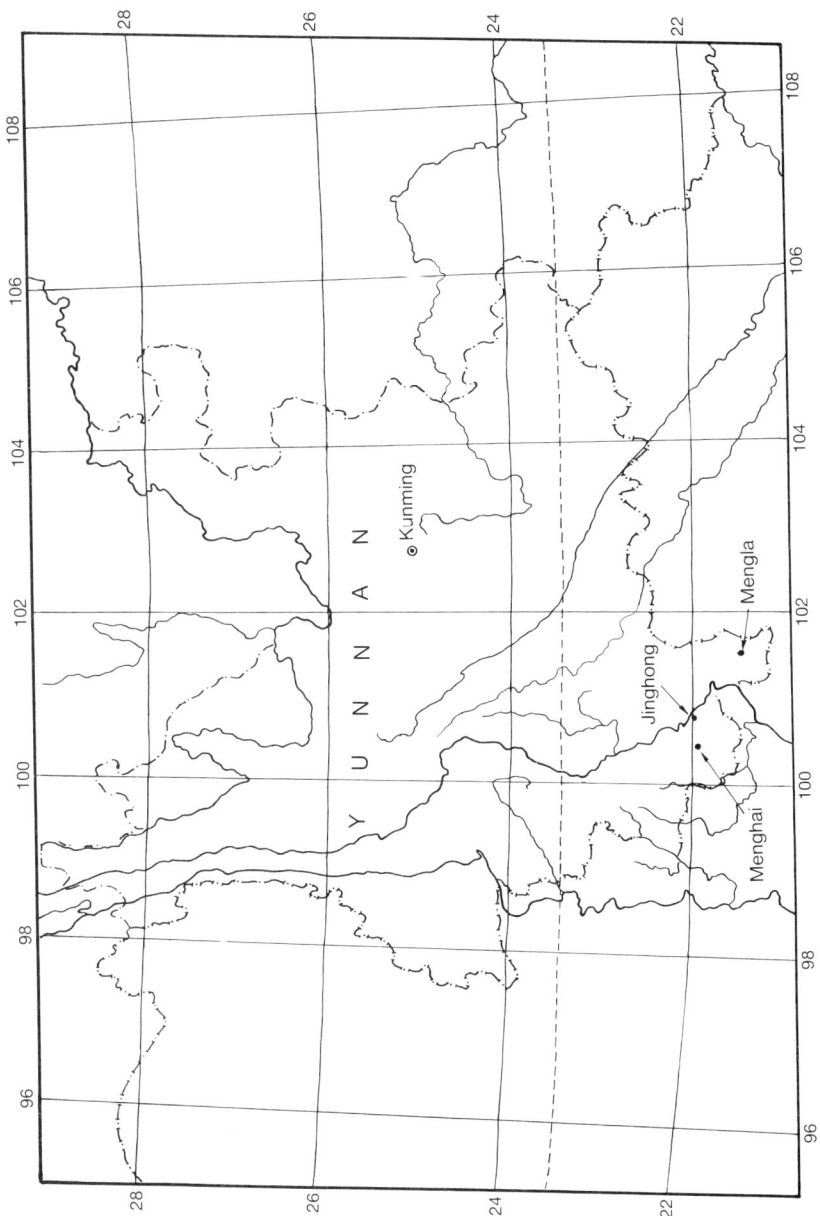

**Fig. 14.2.** Map showing the position of Xishuangbanna prefecture in Yunnan.

hundredth of the whole territory of China. Therefore, the foliicolous lichen flora must also be rich in this area.

Studies on the foliicolous lichens in this area are fragmentary. The following foliicolous genera have previously been reported from China: *Bacidia, Byssloma, Calopadia, Echinoplaca, Mazosia, Porina, Sporopodium*, and *Strigula*. In recent years, Hertel [1980], Eriksson and Wei [1980], Santesson et al. [1987] visited some of the reservations in Jinghong and collected a large number of lichens there. A series of papers concerning foliicolous lichens from China will soon be published and will bring together the current knowledge of foliicolous lichen flora in China (Wei 1990).

## Materials and methods

This paper is based in part on materials collected by Wei, mainly from the reservation near Menglun in Jinghong in October 1980. The materials of the genus *Mazosia* examined were not only from Xishuangbanna, but also from tropical Guangdong, the Mt. Dinghu shan. Observations on anatomical features of these lichens were made by light microscopy (LM) using OPTON II, and those of morphological features through a dissecting microscope (DM) using OPTON, and sometimes with scanning electron microscopy (SEM) using an HITACHI instrument. The chemical data for the testing of lichen products were obtained by standardized thin-layer chromatographic methods (TLC) (White and James 1985).

## Results and discussion

Results show that, of the 20 taxa studied in this paper, seven genera and 15 species including one variety are new to China and *Byssolecania deplanata* is new to Asia.

It is obvious that our knowledge of the Chinese foliicolous lichens is very poor, which makes it imperative that studies should proceed with some urgency, particularly when faced with the serious denudation of tropical forest today.

Only two species of the genus *Mazosia* were previously reported from China; *M. rotula* by Zahlbruckner (1930) and *M. melanophthalma* by Thrower (1988). However, the former identification seems less certain, as pointed out by Santesson (1952). After a careful re-examination of specimens collected by Handel-Mazzetti from Tonkin (no. 5 in HMAS-L and BM) in 1914, and from Yunnan (no. 1035 in BM) in 1915, and which were identified by Zahlbruckner as *M. rotula*, the following results were obtained; the radiate ridges typically found in

*M. rotula* are not seen on the whitened and felted surface of the thallus (Handel-Mazzetti, no. 5 in HMAS-L). The description given by Zahlbruckner (1930, p. 64) also indicates that the thallus is smooth ('thallus . . . laevigatus'). Another specimen from Tonkin (Handel-Mazzetti no. 5 in BM) bears not only *M. phyllosema*, but also *M. melanophthalma*, and the specimen from Yunnan (Handel-Mazzetti no. 1035 in BM) bears *M. melanophthalma* alone.

Numerous verrucose specimens of a species of *Mazosia* were collected by us from both Xishuangbanna in Yunnan and Dinghu shan in Guangdong. This lichen may be identical with *M. melanophthalma*. However, the verrucae are concolorous with the thallus. A few thalli with white or whitish verrucae were found in some collections during this study, but neither asci nor spores were seen. They seem to be dead thalli with whitened verrucae. In addition, the spores of all Chinese specimens of the species in *Mazosia* are longer then those collected from other countries (Santesson 1952). Chemical data for a few species were obtained but the compounds were not identified.

## THE SPECIES

*Asterothyriaceae*

1. *Asterothyrium pittieri* Müll.Arg.  YUNNAN, Xishuangbanna: near Menglun on the way to the reservation, on leaves of *Thunbergia grandiflora* (Roxb. ex Rottl.) Roxb. (Acanthaceae), 18 October, 1980, *Wei* 3588 (HMAS-L).

*Thallus* white, smooth, nitid, round, *c.* 1–4 mm diam., bearing usually one or sometimes several apothecia. *Apothecia* up to 1 mm diam., immersed in thallus at first, then erumpent, and later somewhat prominent, round; *disc* badious, nitid; margins formed mainly by the ragged, irregular lobes of the ruptured tissue originally covering the apothecia like the top of a pomegranate. *Paraphyses* simple, non-septate. *Asci* oval to ovoid, 2-spored, 90–99 × 27–36 μm. *Spores* hyaline, 1-septate, ellipsoid with round ends, 50–61 × 22.5–24 μm.

New to China. Pantropical.

*Bacidiaceae*

2. *Bacidia apiahica (Müll.Arg.) A.Zahlbr.*  YUNNAN, Xishuangbanna: on the way to the reservation near Menglun, on leaves of *Thunbergia grandiflora* (Roxb. ex Rottl.) Roxb., 18 October, 1980, *Wei* 3588-(1).

Most similar to *Dimerella epiphylla* (Müll.Arg.) Malme but with 3–4-septate, filiform spores.

New to China. Pantropical.

*3. Bacidia sp.* YUNNAN, Xishuangbanna: in the reservation near Menglun, on leaves of *Capparis caudata* B. S. Sun (Capparidaceae), 18 October, 1980, *Wei* 3658.

*Thallus* crustose, sordid white, partially slightly rust-coloured. *Apothecia* up to 542 μm diam., *disc* red-brown to dark brown. *Asci* 8-spored 38–50 × 8–11 μm. *Spores* needle-shaped, hyaline, 5–7-septate, 18–30.5 × 2 μm.

The species may be related to or identical with *Bacidia ziamensis* Vězda, but the spores are shorter.

New to China.

*4. Bacidia palmularis (Müll.Arg.) A.Zahlbr.* YUNNAN, Xishuangbanna: in the reservation near Menglun, on leaves of *Knema furfuraceum* (Hook. f. et Thoms.) Warb. (Myristicaceae), 18 October, 1980, *Wei* 3691-(1).

*Thallus* sordid whitish-grey, felty. *Apothecia c.* 325 μm diam., *c.* 300 μm high, only slightly constricted at base; margins almost invisible, often with aerial root-like, white mycelium threads at base, *disc* orange yellowish to yellowish brown. *Asci* 8-spored, 81–99 × 16–18 μm. *Spores* hyaline, 15–25-septate, 61–74 × 2.5–3 μm.

It may possibly be a young stage of its development.

New to China. Pantropical.

*Ectolechiaceae*

*5. Sporopodium leprieurii Mont. var. leprieurii* YUNNAN, Xishuangbanna in the reservation nearly Menglun, on leaves of *Knema furfuracea*, 18 October, 1980, *Wei* 3697.

*Thallus* thin, felty, verrucose, sordid whitish, irregular, *c.* 10–20 × 6–10 mm; verrucae concolorous with thallus. *Hypothallus* faintly visible. *Apothecia* sublecanorine, without algae in the thalline margin, *c.* 1 mm diam. *Asci* single-spored, 75.5–90 × 20–21.5 μm. *Spores* muriform, 47–83 × 12.5–18 μm. Algae in epithecium *c.* 3.5–4.5 × 3–3.5 μm.

New to China. Pantropical.

*Gomphillaceae*

*6. Gyalectidium filicinum Müll.Arg.* YUNNAN, Xishuangbanna: in the forest of the Limestone Hill near the Tropical Botanical Garden of the Kunming Botanical Institute of Academia Sinica, on leaves of *Pseuduvaria indochinensis* Merr., 19 October, 1980, *Wei* 3715-(3).

*Thallus* crustose, wax-white, strongly nitid, densely verrucose, suborbicular, *c.* 4–5 × 3 mm. *Apothecia* round, dish-shaped, *c.* 380 μm diam.; margins concolorous with thallus, crenulate; *disc* wax-yellowish to wax-brownish, nitid. *Asci* single-spored. *Spores* hyaline, muriform, curved within asci, straight after discharge, 41–52 × 16–23 μm.

*Chemistry*: One unidentified substance $R_F$ class 6 (bottom) in C, giving grey-purple colour after acid and heat treatment.

New to China. Pantropical.

7. *Tricharia vainioi* R.Sant.   YUNNAN: Xishuangbanna, on leaves of *Polyalthia viridis* Craib (Anonaceae) in the Limestone Hill near the Tropical Botanical Garden of the Kunming Botanical Institute of Academia Sinica, 19 October, 1980, *Wei* 3713 and on leaves of a broadleaved tree *Wei* 3702, 3707-(1).

*Thallus* crustose, thin, membraneous, nitid, subcircular to irregular, *c.* 5 mm diam., or 12 × 4 mm, sordid whitish, furnished with *c.* 1 mm long, black and tapering hairs. *Apothecia c.* 280 μm diam., almost not constricted at base, wax-gold with brownish colour to dark brown, biatorine type; disc plane to faintly concave. *Asci* single-spored. *Spores* hyaline, muriform, *c.* 34–47 × 16–22 μm.

New to China.

*Gyalectaceae*

8. *Dimerella epiphylla (Müll.Arg.) Malme*   YUNNAN, Xishuangbanna: in forest of Limestone Hill near the Tropical Botanical Garden of the Kunming Botanical Institute of Academia Sinica, on leaves of *Pseuduvaria indochinensis* Merr. (Anonaceae), 18 October, 1980, *Wei* 3715.

*Thallus* membraneous, grey green, *c.* 1.5 cm diam. *Apothecia* round, strongly constricted at base, *c.* 0.5–1.2 mm diam.; *disc* orange yellow to pale buff brown, margins yellowish whitish, usually growing along the veins and margin of leaves. *Asci* single-spored. *Spores* fusiform, 1-septate, 5–11 × 2 μm.

New to China. Pantropical.

Literature records: This lichen was previously reported by Zahlbruckner (1930, p. 70) from Yunnan. However, Santesson (1952) pointed out that this record is still unverified. So, it seems that *Dimerella* as a genus is new to China.

*Opegraphaceae*

9. *Opegrapha filicina* Mont.   YUNNAN, Xishuangbanna: in the reservation near Menglun, on leaves of *Knema furfuracea* (Hook. f. et Thoms.) Warb. (Myristicaceae), 18 October, 1980, *Wei*.

*Asci* 8-spored, 34–40 × 14–16 μm. *Spores* 4- (rarely 5) septate, 12.5–30.5 × 4.5–6 μm.

Spores of this species are shorter than those of the specimens collected from south-east Asia, Africa, and South America, usually 4-septate, and rarely 5-septate; one end is larger than the other.

New to China. Pantropical.

*Phragmopelthecaceae*

10. *Mazosia melanophthalma* (*Müll.Arg.*) *R.Sant.* (Fig. 14.3, 14.4, 14.5) YUNNAN: Xishuangbanna, in the reservation near Menglun, on leaves of *Knema furfuracea*, 18 October, 1980, *Wei* 3565, 3572, 3593, 3686, 3705, and on leaves of *Celtis giganticarpa* Hu, in the forest of Limestone Hill near the Tropical Botanical Garden of the Kunming Botanical Institute of Academia Sinica, 19 October, 1980, *Wei* 3714. GUANGDONG: Dinghu shan, on leaves of *Tectaria sp.* in the forest near Qingyunsi, 9 November, 1980, *Wei* 3784.

*Thallus* membraneous, thin, yellow-green or grey-green to olive-green, slightly nitid, up to 5 mm diam., verrucose; verrucae concolorous with thallus. *Ascomata* scutate, *disc* plane, dark brown to the naked eye or through a DM, but hyaline and colourless through a LM. *Asci*

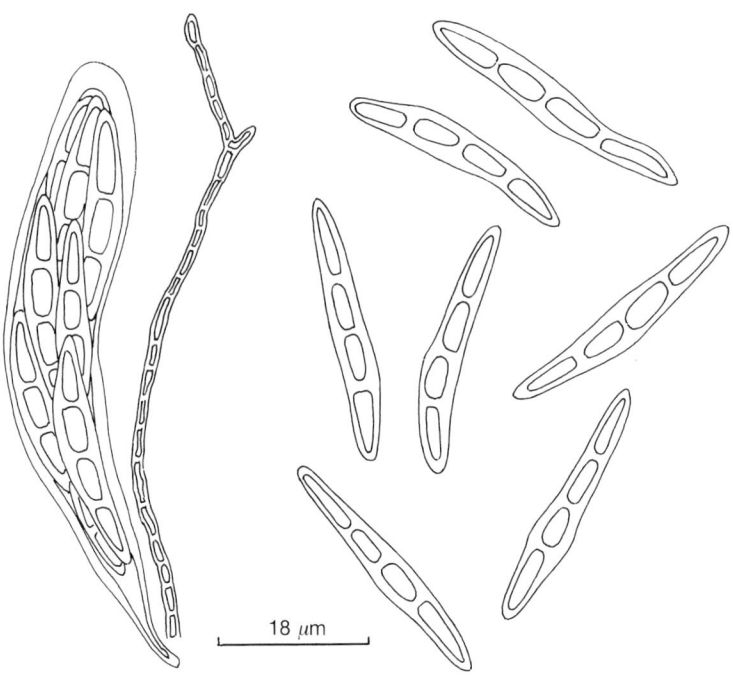

Fig. 14.3. *Mazosia melanophthalma* (*Wei* 3572), ascus, ascospores and a paraphysis.

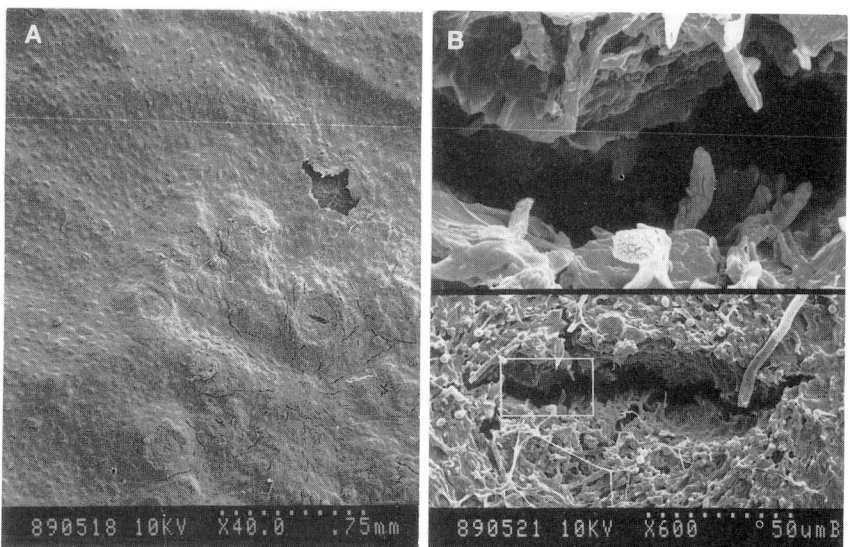

**Fig. 14.4.** (A) *Mazosia melanophthalma* (*Wei* 9074), verrucose thallus with ascomata. (B) *Mazosia melanophthalma* (*Wei* 9074), a fissure in one of the ascomata.

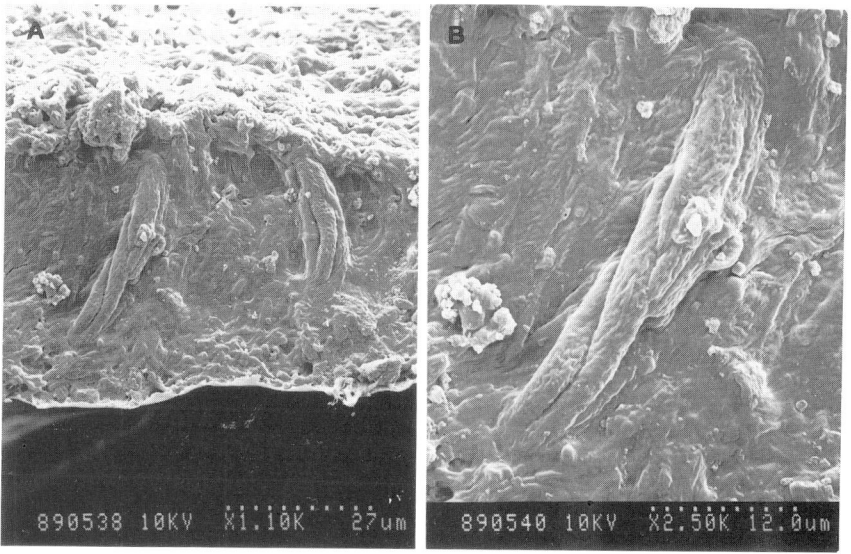

**Fig. 14.5.** (A,B) *Mazosia melanophthalma* (*Wei* 9074), asci in the hymenium.

clavate, 8-spored, $c.$ 54.5 × 10.5 µm. *Spores* fusiform, hyaline and colourless, 3-septate; the second cell from the obtuse end often larger than others, $c.$ 21–36 × 3.5–4.5 µm.

Specimens examined: YUNNAN: near Manhao, $c.$ 650 m, 6 March, 1915, *Handel-Mazzetti* no. 1035 (as *Mazosia rotula*, determined by Zahlbruckner (BM!). TONKIN: 2 November, 1914, *Handel-Mazzetti* no. 5 (as *M. rotula*, determined by Zahlbruckner, in BM!).

*Chemistry*: Three unidentified substances $R_F$ class 2–3 in C, giving yellow colour, $R_F$ class 5 (bottom) in C, giving yellowish-green, and $R_F$ class 6 (bottom) in C, giving yellowish-green colour after acid and heat treatment.

Pantropical.

Literature record: Xianggang (Hong Kong: Thrower, 1988, p. 119).

*11. Mazosia paupercula (Müll.Arg.) R.Sant. (Fig. 14.6)* YUNNAN, Xishuangbanna: in the reservation near Menglun, on leaves of *Knema furfuracea*, 18 October, 1980, *Wei* 3641.

*Thallus* crustose, thin, smooth, grey-green, usually $c.$ 2.5–3 mm, sometimes up to 20 mm diam. *Apothecia* up to 1 mm diam., prominent. *Asci* 8-spored, $c.$ 63–88 × 18–21.5 µm. *Spores* hyaline, 5–7-septate, 25–44 × 3.5–7 (–9) µm.

New to China. Pantropical.

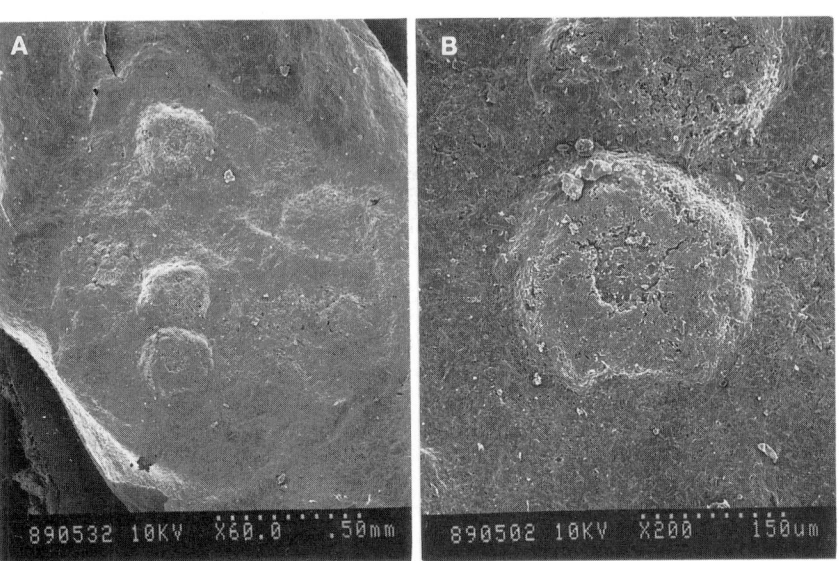

Fig. 14.6. (A) *Mazosia paupercula* (*Wei* 3641), smooth thallus with ascomata. (B) *Mazosia paupercula* (*Wei* 3641), two ascomata.

12. *Mazosia phyllosema (Nyl.) A.Zahlbr. (Fig. 14.7)* YUNNAN, Xishuangbanna: in the reservation near Menglun, on leaves of *Knema furfuracea*, 18 October, 1980, *Wei* 3568, 3571, 3697.

*Thallus* thin, membraneous, grey-green, smooth. *Spores* 3-septate, 16–27 × 3.5–5 μm.

Specimens examined: *Tonkin*: 2 November, 1914, *Handel-Mazzetti* no. 5 (as *Mazosia rotula*, determined by Zahlbruckner, in HMAS-L and BM, respectively). The radiate ridges are invisible on the whitened and felted surface of the thallus of the specimen, which is preserved in HMAS-L. The surface of the specimen thallus preserved in BM, is also completely smooth, in addition to having *Mazosia melanophthalma* on the same collection. It seems that both specimens of the same number labelled as *Mazosia rotula* are probably *Mazosia phyllosema*, as considered by Santesson (1952, p. 115).

New to China. Pantropical.

13. *Mazosia rotula (Mont.) Massal. (Fig. 14.8)* YUNNAN, Xishuangbanna: in the reservation near Menglun, on leaves of *Knema furfuracea*, 18 October, 1980, *Wei* 3570, 3572-(1), 3686; on leaves of *Prosartema stellaria*, *Wei* 3598-(1).

*Thallus* yellowish-green with olive green tint, *c*. 1–3.5 mm diam., sometimes several different small individuals confluent, with numerous

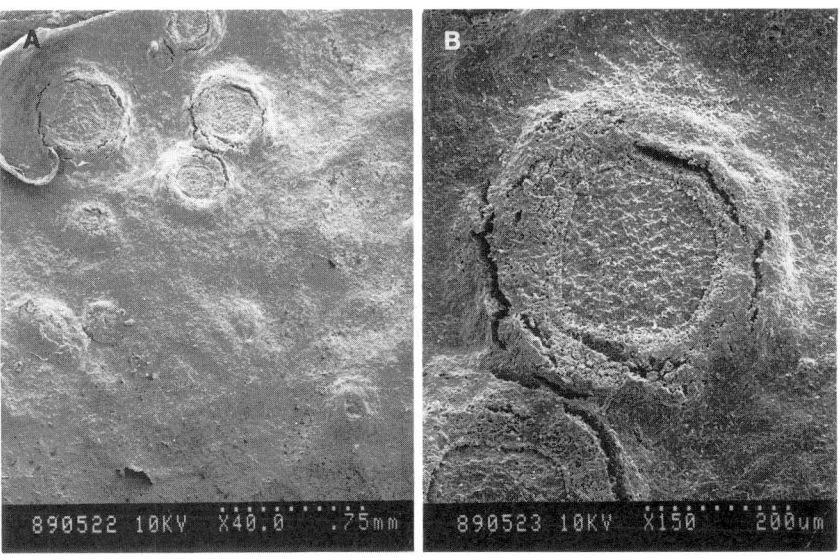

Fig. 14.7. (A) *Mazosia phyllosema* (*Wei* 3571), smooth thallus with ascomata. (B) *Mazosia phyllosema* (*Wei* 3571), two ascomata.

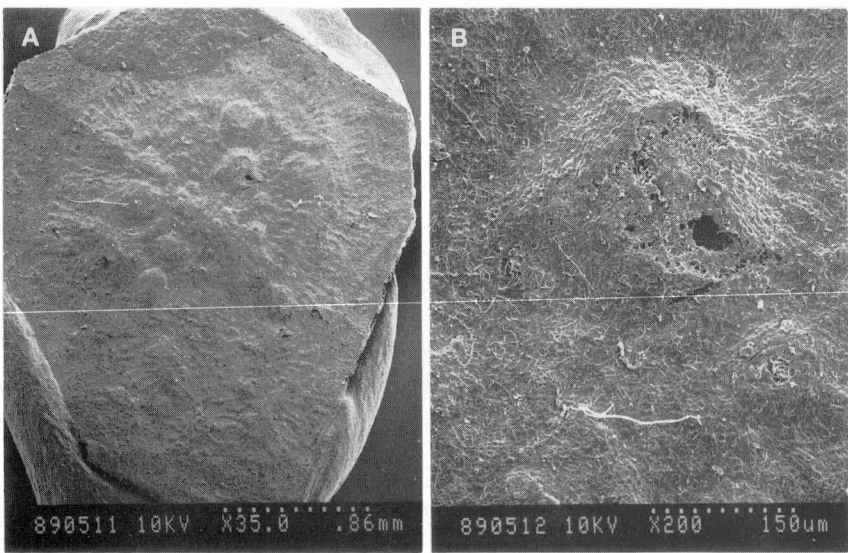

Fig. 14.8. (A) *Mazosia rotula* (without a number on leaves of *Knema furfuracea*), thallus with radiate ridges and ascomata. (B) *Mazosia rotula* (as in Fig. 14.8(A)), two ascomata.

gentle and thin radiate ridges, simple or rarely slightly branched, lighter in colour than thallus. *Ascomata* scutiform; *disc* plane, non-pruinose, leaden black to the naked eye or through a DM, but hyaline and colourless through an LM, $c.\,300\,\mu$m in diam. *Asci* clavate, 8-spored, $c.\,38$–$54 \times 10.5$–$14.5\,\mu$m. *Spores* usually 3-septate, fusiform, ends usually rather obtuse, one of the median cells often larger than the others, $c.\,21$–$26 \times 4\,\mu$m, the larger cell among them about $5\,\mu$m wide. *Paraphyses* thin and branched.

*Chemistry*: One unidentified substance $R_\mathrm{F}$ class 5 (bottom) in C, giving yellowish-green colour after acid and heat treatment.

New to China.

*Pilocarpaceae*

14. *Byssolecania deplanata (Müll.Arg.) R.Sant.* YUNNAN, Xishuangbanna: in the reservation near Menglun on leaves of *Cleidion brevipetiolatum* Pax. et Hoffm., 18 October, 1980, *Wei* 3645.

*Thallus* thin, membraneous, dimly seen, grey-green, 10–20 mm diam. *Apothecia* oblate, totally adnate, $c.\,448$–$471\,\mu$m diam., very regularly circular, white, byssoid hyphae of the thallus around the apothecia; *disc* dark brown, margins not seen. *Asci* 8-spored, *c*. 45–65 × 14–20 μm. *Spores*, 5–7-septate, 23–40 × 3.5–4.5 μm.

The spores are longer than those of this species collected from tropical Africa and America (Santesson 1952).

New to Asia. Pantropical.

*15. Fellhanera semecarpi (Vainio) Vězda*   YUNNAN, Xishuangbanna: *Jiang* 960-(3).

*Thallus* sordid whitish. *Apothecia* biatorine, numerous, small, *c*. 240 μm diam., *disc* brownish to yellowish-brown. *Asci* 8-spored 23–32 × 7–11 μm. *Spores* hyaline, 1-septate, 9 × 3.5 μm.

New to China. Pantropical.

*Strigulaceae*

*16. Strigula elegans (Fée) Müll.Arg.*   YUNNAN: Xishuangbanna, in the reservation near Menglun on leaves of *Phoebe lanceolata*, 18 October, 1980, *Wei* 3587, and on leaves of *Celtis giganticarpa* (Ulmaceae) in the Limestone Hill near the Tropical Botanical Garden of the Kunming Botanical Institute of Academia Sinica, 19 October, 1980, *Wei* 3714.

Pantropical.

Literature records: Yunnan (Zahlbruckner 1930, p. 31; Santesson 1952, p. 160), Guizhou, Hubei, Hunan (Santesson 1952, p. 160), Fujian (Zahlbruckner 1930, p. 31; Santesson 1952, p. 160), Taiwan (Zahlbruckner 1933, p. 205; Wang and Lai 1973, p. 97; Santesson 1952, p. 160).

*17. Strigula macrocarpa Vainio*   YUNNAN: Xishuangbanna, in the reservation near Menglun, 18 October, 1980: on leaves of *Amaplelopris sp.* (*Wei* 3592), on leaves of *Prosartema stellaria* (*Wei* 3598), and on leaves of *Tetrastigma sp.* (Vitaceae) (*Wei* 3626).

The species is similar to *Strigula elegans* but differs by its large ascomata.

New to China. Tropical species of the old world.

*18. Strigula melanobapha (Krempelh.) R.Sant.*   YUNNAN: Xishuangbanna, in the reservation near Menglun on leaves of *Phoebe lanceolata* (*Wei* 3584) and of *Drypetes indica* (Müll.Arg.) Pax. et Hoffm. (Euphorbiaceae) (*Wei* 3615), 18 October, 1980 and on leaves of a broadleaved tree in Limestone Hill near the Tropical Botanical Garden of the Kunming Botanical Institute of Academia Sinica, 19 October, 1980, *Wei* 3712-(1).

*Pycnidia* black, *c*. 144 μm diam., containing numerous bacillar, hyaline conidia, 4.5 × 1.5 μm.

Pantropical.

Literature records: Fujian (Zahlbruckner 1930, p. 31, as *Strigula fibrilosa*, Chung 488—lectotype, Chung 319, 487b—syntypes; Santesson 1952, p. 188).

*19. Strigula subelegans Vainio*  YUNNAN: Xishuangbanna, on leaves of a broad-leaved tree in Limestone Hill near the Tropical Botanical Garden of the Kunming Botanical Institute of Academia Sinica, 19 October, 1980, *Wei* 3706.

New to China. Tropical and subtropical species in Asia and America.

*Trichotheliaceae*
20. *Trichothelium annulatum* (Karsten) *R.Sant.* (Fig. 14.9, 14.10)

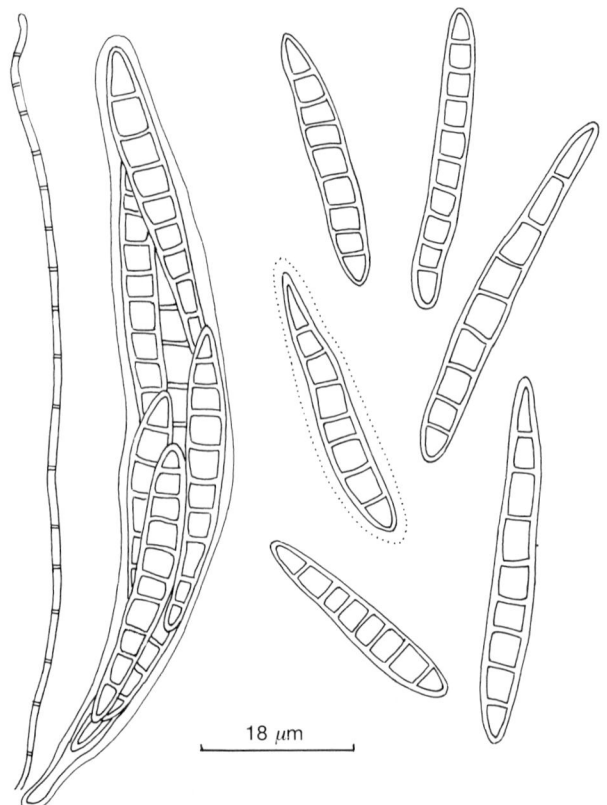

**Fig. 14.9.** *Trichothelium annulatum* (*Wei* 3714), ascus, ascospores, and a paraphysis.

YUNNAN: Xishuangbanna, on leaves of *Celtis giganticarpa* (Ulmaceae) in Limestone Hill near the Tropical Botanical Garden of the Kunming Botanical Institute of Academia Sinica, 19 October, 1980, *Wei* 3714.

*Thallus* crustose, thin and smooth, faintly nitid, olive green to badious, subcircular to irregular, *c.* 5 mm or sometimes up to 10–12 mm diam., very similar to *Trichothelium epiphyllum* Müll.Arg. but differing in the olive-greenish to badious thallus. *Spores* with 7 to 9 transverse septa and the apical setae of perithecia are no more than 12 'rays'. *Ascomata c.* 283 µm diam.; *Asci* 8-spored, *c.* 81–99 × 11.5–18 µm. *Spores* hyaline, *c.* 21.5–43 × 3.5–5 µm.

This species usually has 8–15-septate spores, but the specimens studied from Yunnan seem to be young.

*Chemistry*: Two unidentified substances $R_F$ class 5 (bottom) and 6 (bottom) in C, both of which give a yellowish-green colour after acid and heat treatment.

New to China. Pantropical.

## Acknowledgements

We are deeply grateful to Professor R. Santesson (Uppsala) for checking the identification of some collections and for his very helpful

Fig. 14.10. (A) *Trichothelium annulatum* (*Wei* 3714), apical setae of an ascoma. (B) *Trichothelium annulatum* (*Wei* 3714), top of one seta.

revisions, and to Dr Zhuang W.Y. for revising the English language of the manuscript. Our thanks are also due to Mr Dong G.J. for technical assistance in making scanning electron microscopic investigations; to Ms Yuan L.C. for printing the photographs, and to Ms Han Z.F. for inking the line drawings.

## References

*Hawksworth, D.L. (1972). A new species of *Tricharia* Fée em. R.Sant. from Hong Kong. *The Lichenologist*, **5**, 321.
Santesson, R. (1952). Foliicolous lichens. I. A revision of the taxonomy of the obligately foliicolous, lichenized fungi. *Symbolae Botanicae Upsalienses*, **12**, 1–590.
Thrower, S.L. (1988). *Hong Kong lichens*. An urban council publication.
*Vězda, A. (1977). Beitrag Zur Kenntnis foliikoler Flechten. *Vietnams Casopsis slezskeho Musea Serie A*, **26**, 21.
*Vězda, A. (1983). Foliicole Flechten aus der Kolchis (West-Transkaukasien, UdSSR). *Folia Geobotanica et Phytotaxonomica, Praha*, **18**, 45.
*Vězda, A. (1984). Foliicole Flechten der Insel Kuba. *Folia Geobotanica et Phytotaxonomica, Praha*, **19**, 177.
Wang-Jang, J.-R. and Lai, M.-J. (1973). A checklist of the lichens of Taiwan. *Taiwania*, **18**, 83–104.
Wei, J.C. (1990). *An enumeration of the lichens in China* (to be published).
Wu, C.Y. (1979). The regionalization of Chinese Flora. *Acta Botanica Yunnanica*, **1**(1), 1–22.
White, F.J. and James, P.W. (1985). A new guide to microchemical techniques for the identification of lichen substances. *Bulletin of the British Lichen Society*, **57** (supplement), 1–41.
Zahlbruckner, A. (1930). Lichens. In *Symbolae Sinicae* III. Handel-Mazzetti. pp. 254.
*List of Plants in Xishuangbanna* (ed. Yunnan Institute of Tropical Botany). Academia Sinica.

* Used in specimen identification.

# 15. Observations on the composition and distribution of the '*Lobarion*' in forests of South East Asia

P.A. WOLSELEY

*Nettlecombe Studios, Williton, Taunton, Somerset TA4 4RS, UK*

### Abstract

Members of the Lobariaceae contribute to a distinctive and well-developed epiphytic community of lichens and bryophytes associated with forests in temperate and tropical regions. Forest macrolichens in Indonesia, Malaysia, and Thailand are compared with those found within the European alliance *Lobarion pulmonariae*. Factors affecting the location and distribution of the '*Lobarion*' in a range of tropical forests are considered. Aspects of the distribution of some critical genera are discussed. Present practices of forest management and widespread degradation in South East Asia have profound affects on the distribution of this characteristic lichen community.

### Introduction

The alliance *Lobarion pulmonariae* Ochsn. defined by Barkman (1958), and elucidated for British sites by James *et al.* (1977), is a western European community of foliose lichens and bryophytes associated with ancient forests dominated by fagaceous phorophytes of the genera; *Quercus*, *Fagus*, and *Castanea*. In Europe, the *Lobarion* is found in oceanic and montane environments (Rose 1988) and is as rich in constant species in the Pyrenees and Appenines as it is in oceanic sites in the West of Britain and Norway. This alliance forms a species-rich epiphytic community with an unusually high number of faithful species, especially in the families Lobariaceae, Pannariaceae, Collemataceae,

*Tropical Lichens: Their Systematics, Conservation, and Ecology* (ed. D.J. Galloway), Systematics Association Special Volume No. 43, pp. 217–43. Clarendon Press, Oxford, 1991. © The Systematics Association, 1991.

and Peltigeraceae (James *et al.* 1977). In Europe, it is found in areas of high rainfall (over 900 mm per annum) occurring throughout the year, a seasonal climate, and primary forests with a long history of ecological continuity. The *Lobarion* is a light-demanding community and is tolerant of a wood–pasture management scheme that leaves scattered ancient phorophytes, but is intolerant of forest management that increases shading, such as coppicing and even-aged stands. Nor can it tolerate atmospheric acidification or eutrophication. Formerly widespread, this alliance is now becoming rare because of: (1) loss and fragmentation of primary forest sites; (2) changing land management in forestry and agriculture; (3) increasing atmospheric acidification. Rose (1976) suggested that species of the *Lobarion* and of the ancient dry bark association *Lecanactidetum premnae* that are restricted to primary forest sites be used to construct an Index of Ecological Continuity. The Revised Index of Ecological Continuity (RIEC) has been used to identify and evaluate British woodlands of high conservation status and to elucidate older patterns of forest management (Rose 1976; Wolseley and O'Dare 1989).

Although the lowland dipterocarp forests of South East Asia have few lichen taxa in common with the Fagaceous forests of western Europe, the montane forests of the tropic support genera and occasionally species that are found in the European *Lobarion*. On Mt. Kinabalu in Borneo, in the forests over 1000 m altitutde, the phorophytes are unfamiliar, yet *Pseudocyphellaria crocata* and *P. intricata* are abundant together with other species of *Pseudocyphellaria* and *Sticta* in a familiar '*Lobarion*'[1] community. Many of the phorophytes are tropical species of the Fagaceae in the genera *Quercus*, *Lithocarpus*, and *Castanopsis*, these genera which dominate montane evergreen forests throughout South East Asia (Whitmore 1975; Santisuk 1988). In forests visited between latitudes 18°N and 8°S characteristic genera of the '*Lobarion*' were found. Preliminary collections made usually on single visits to the sites shown in Fig. 15.1 and 15.2, form the basis of this paper and on this slender evidence the composition, location, and biogeography of the South East Asian and European *Lobarion* communities are compared.

Members of the Lobariaceae are a dominant feature of all '*Lobarion*' communities in constancy, cover value, and numbers of associated species. Component macrolichen genera of the European *Lobarion pulmonariae* from Rose (1988) and of sites in South East Asia are shown in Table 15.1. Additional macrolichen genera occurring throughout include *Parmeliella*, *Pannaria*, *Collema*, *Leptogium*, *Menegazzia*, and *Hetero-*

---

[1] '*Lobarion*' in this paper refers to undescribed communities of South East Asia with a high proportion of dominant genera characteristic of the *Lobarion pulmonariae* Ochsn.

*dermia*. Genera not found in the western Europe alliance include *Psoroma*, *Coccocarpia*, and *Lieoderma*. *Nephroma*, used as an indicator for the European associations *Nephrometum lusitanicae* and *N. laevigatae* (Barkman 1958), is a comparatively rare component of the South East Asian community. *Phyllopsora*, a component of the European alliance and of tropical rainforests elsewhere (Sipman and Harris 1989), was not found in the South East Asian sites visited.

A feature of the palaeotropical *'Lobarion'* is the high proportion of species with cyanobacterial photobionts, enabling them to photosynthesize at low light intensities and also to fix atmospheric nitrogen. This latter process is an important source of nutrients in tropical rainforests on impoverished soils, as demonstrated by Forman (1975) in Columbia.

## Location and features of sites

Sites visited lie between 18°N and 8°S, and between 98°E and 118°E (Fig. 15.1). The majority of these sites lie within the equatorial region of South East Asia dominated by a climax vegetation of endemic dipterocarp rainforests. These rainforests are physiographically distinct from other regions of equatorial forest in Africa and S. America in two major features: (1) the existence of isolated montane regions with a distinct and characteristic montane flora (van Steenis 1934–6, 1957, 1964, 1972) and (2) the fragmented nature of this tropical rainforest belt due to post-glacial rises in sea level, so that this region has the remnants of an ancient link between the Northern Hemisphere Laurasia and the Southern Hemisphere Gondwanaland, the tectonic boundary of which runs along the spine of Java and the Sunda shelf. Montane elements of the Northern Hemisphere continue into the North of Thailand as foothills to the Himalayas, but are separated from the equatorial mountain regions by the lowlands of S. Thailand and Peninsular Malaysia. Closer montane connections are found between Kinabalu and mainland China via the Philippines and Formosa. In the field, the impression of montane floras being islands in a sea of lowland rainforest is confirmed by phanerogamic and cryptogammic floras.

## Climate in South East Asia

Lowland climates are of two types: (1) aseasonal, with temperatures $c.\ 26\ °C$ throughout the year, and rainfalls between 2000–7000 mm; (2) Monsoon climates with a pronounced dry spell and rainfalls of less than 2000 mm per annum confined to the rainy season. Distribution of these climates is shown in Fig. 15.2. However within these climatic

**Table 15.1.** Macrolichen components of *Lobarion* in Europe and of the '*Lobarion*' in South East Asia

| | In Europe (see Rose 1988) | No. of taxa found in S E Asia () no. identified | Sites in South East Asia | | | | | | | No. of species with cyanobacteria | No. of species with green algae |
|---|---|---|---|---|---|---|---|---|---|---|---|
| | | | Doi Inthanon Doi Suthep | Kinabalu | Danum | Mulu | Genting Highland | Cibodas | G. Arjuno G. Semeru | | |
| **Lobariaceae** | | | | | | | | | | | |
| *Lobaria* | 4 | 11(6) | 4(1) | 2(2) | 1 | 2(1) | — | 4(3) | 2 | 5 | 6 |
| *Sticta* | 4 | 22(6) | 7(2) | 10(4) | 1 | 1 | — | 11(8) | 3 | 17 | 5 |
| *Pseudocyphellaria* | 4 | 14(9) | 1 | 6(3) | 1 | 1 | 1 | 6(4) | 5(2) | 11 | 3 |
| **Pannariaceae** | | | | | | | | | | | |
| *Parmeliella* | 4 | 10(1) | 4 | — | 1 | — | 1 | 1 | — | — | — |
| *Pannaria* | 5 | (5) | 3 | 2 | 1 | — | 1 | 2 | — | 10 | — |
| *Psoroma* | — | 1 | — | 1 | — | 1 | — | 1 | — | — | 1 |
| *Coccocarpia* | — | 3(3) | 1 | 1 | 3 | 1 | 3 | 2 | 1 | 3 | — |
| *Erioderma* | — | 2(2) | — | 1 | — | — | — | 1 | — | 2 | — |
| *Leioderma* | — | 2(2) | — | 1 | — | 1 | 2(1) | 1(1) | — | 4 | — |
| **Collemataceae** | | | | | | | | | | | |
| *Physma* | — | 1(1) | — | 1(1) | — | — | — | — | — | 1 | — |
| *Collema* | 3 | 4 | 1(1) | — | — | — | — | 2(2) | — | 4 | — |
| *Leptogium* | 4 | 14(7) | 7(3) | 4 | 1 | 1 | 3 | 8 | 5 | 14 | — |

| | | | | | | | | | | |
|---|---|---|---|---|---|---|---|---|---|---|
| Parmeliaceae | | | | | | | | | | |
| *Menegazzia* | 1 | 3(1) | — | 1 | — | — | 1 | — | — | 3 |
| *Nemphromopsis* | — | 1 | — | 1 | — | — | 1 | — | — | 1 |
| *Cetraria* | — | 1(1) | — | 1 | — | — | — | — | — | 1 |
| *Cetrelia* | 1 | 2(2) | — | — | — | — | — | — | 2 | 2 |
| Physciaceae | | | | | | | | | | |
| *Anaptychia* | 1 | — | — | — | — | — | — | — | — | — |
| *Heterodermia* | 1 | 16(10) | 8 | 5(1) | — | — | 3 | 9(1) | 5 | 16 |
| Peltigeraceae | | | | | | | | | | |
| *Nephroma* | 4 | 1(1) | 1 | 1 | — | — | — | — | 1 | — |
| *Peltigera* | 2 | 5 | 2(1) | 1 | — | — | — | 1 | 2(1) | 6 |
| Phyllopsoraceae | | | | | | | | | | |
| *Phyllopsora* | 1 | — | — | — | — | — | — | — | — | 1 |
| Species diversity | 38 | — | 39 | 39 | 9 | 10 | 15 | 49 | 26 | 78 |
| Days spent in the field | — | — | 4 | 5 | 3 | 3 | 3 | 3 | 2 | 39 |

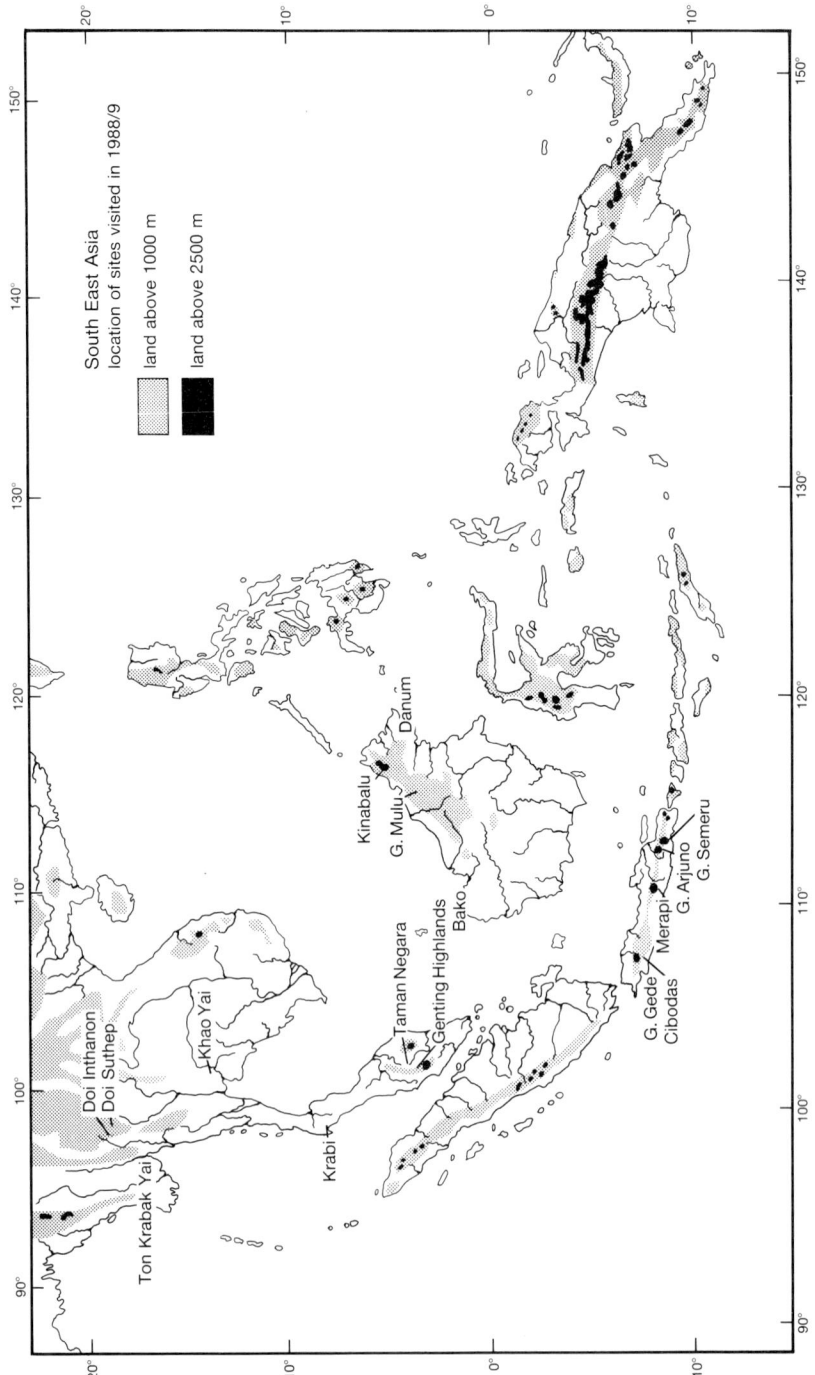

**Fig. 15.1.** Map of the South East Asian regions visited showing the location of sites and land above 1000 m altitude (redrawn from van Steenis 1934-6).

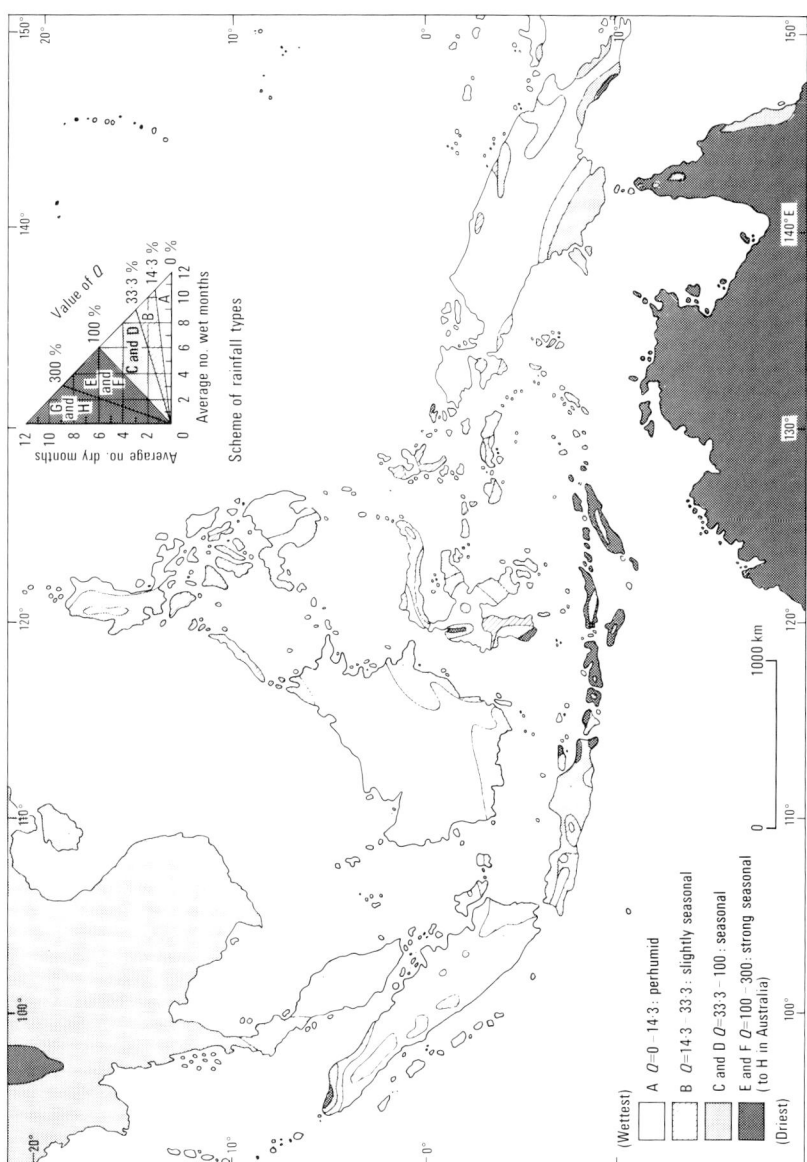

**Fig. 15.2.** Climates of the tropical Far East based on dry/wet period ratios (from Whitmore 1984).

zones there is considerable variation affecting the phanerogamic and cryptogamic floras. At Mulu the annual rainfall is between 4800–6800 mm, but there are times, usually in August and September when evaporation may exceed precipitation and drought conditions ensue (Walsh 1982). Foliose lichens are sensitive to extremes of temperature, so that the total rainfall is not as critical as the number of days without rain or when drought conditions occur. In the seasonally dry dipterocarp forests of Thailand there is a characteristic drought tolerant epiphytic community without any '*Lobarion*' species.

Montane climates tend to be cooler and wetter, the temperature falling $c$. 0.6 °C for each 100 m in altitude. This decrease in temperature is associated with the formation of a cloud zone at $c$. 1200 m where moisture-laden air rising from the lowland forests cools to form mist or rain that constantly bathes the vegetation. Above this zone there is increasing diurnal variation in temperature, freezing night conditions alternating with clear days when the temperature may be as high as 33 °C. In tropical conditions there is a strong climatic gradient between the lowland rainforests and montane forests in all countries visited.

Table 15.2. Distribution of '*Lobarion*' in forest types of South East Asia visited between September 1988 and February 1989

| Site | Latitude/ Longitude | Highest point (m) | Swamp forest | Tropical rain forest Lowland dipterocarp (often secondary) | Hill dipterocarp and variations |
|---|---|---|---|---|---|
| Doi Inthanon | 18°35′N,98°30′E | 2590 | | | |
| Doi Suthep | 18°50′N,98°54′E | 1685 | | | |
| Ton Krabak yai | 17°N,98°80′E | c.800 | | | |
| Kinabalu | 6°N,116°40′E | 4101 | | * ○ | * ○ |
| Danum | 5°5′N,118°E | c.200 | | * ○ | |
| Mulu | 4°5′N,115°5′E | 2240 | * ○ | * | * ○ |
| Taman Negara Gunung Ulu Kali | 4°30′N,102°30′E | 2189 | | * ○ | * ○ |
| Genting Highlands | 3°20′N,103°55′E | 1769 | | * ○ | ○ |
| Cibodas G.Gede | 6°55′S,107°E | 2958 | | | * ○ |
| Merapi | 7°50′S,110°40′E | 2911 | | | |
| G. Arjuno | 6°40′S,112°40′E | 3339 | | | |
| G. Semeru | 8°S,113°E | 3676 | | | |

(*) Forest type present and visited; ( + ) forest type present not visited; ( ● )species-rich '*Lobarion*' present; ( ○ ) species-poor '*Lobarion*' present. (T) Thailand; (B) Borneo; (M) Peninsular Malaysia; (I) Indonesia.

## Vegetation zones and associated epiphytic lichens

Vegetation zones associated with altitude and climate are described by Whitmore (1984) and van Steenis (1934–6, 1957, 1964, 1972) for the tropical equatorial forests and by Santisuk (1988) for the forests of northern Thailand. Terminology for forest types varies over the countries visited, as do the altitudinal limits of the vegetation types. The sequence used in Table 15.2 is taken from the above sources.

*The distribution of the* Lobarion *in temperate and tropical forests*

Factors influencing the distribution of epiphytic lichen communities are outlined in James *et al.* (1977). In Europe, the *Lobarion* is regarded as the climax community of western fagaceous woods. It is found on well-lit, south-facing trunks and lower branches of ancient trees with a bark of *c.* 5–6 pH, in sites where the humidity is high throughout the year. In areas of exceptionally high humidity it will colonize younger trunks and even rocks. In Europe, most of the *Lobarion* taxa have photobionts that require sunlight for photosynthesis, so are luxuriant in sites with long day-lengths in the summer season. However, photosynthesis and nitrogen fixation are moisture-dependent and do not

|   | Montane forests | | | Monsoon forests | | | |
| --- | --- | --- | --- | --- | --- | --- | --- |
| Hill evergreen *Lithocarpus Castanopsis* forest | Moss forest | *Leptospermum Dacrydium* | Alpine scrub | Tropical evergreen | Moist deciduous forest | Dry dipterocarp | *Pinus* zone T *Casuarina* zone I |
| * ● | * |   |   |   | * ○ | * | * |
| * ● |   |   |   |   | * | * | * |
|   |   |   |   | * ○ |   |   |   |
| * ● | * ○ | * ● | * ● |   |   |   |   |
| * ● + | * ○ | * ● | * ● |   |   |   |   |
| * | * ○ | * ● |   |   |   |   |   |
| * ● | + | + | + |   |   |   |   |
| * ● |   |   |   |   |   |   |   |
| * ● | + | + | + |   |   |   |   |
| * ● | * | + | + |   |   |   |   |

occur when the thallus is dry (Forman 1975; Jones 1989). In Europe and South East Asia taxa of the *Lobarion* cannot tolerate an extended dry season and so are absent from the monsoon climates of Thailand and Java, as they are from continental and mediterranean climates in Europe. In both regions they reappear in the moister montane climates; in Europe above 500 m (Rose 1988) and in the tropics above 1000 m. Despite the preponderance of taxa with cyanobacterial photobionts in the tropical '*Lobarion*', these are a rare component of the trunks of lowland dipterocarp forest where very little light reaches the trunks at any time. In these forests species of *Coccocarpia*, *Leptogium*, *Dictyonema*, and crustose species occur. Where light intensities are relatively high, as in the canopy or the spreading branches of emergent dipterocarps, the '*Lobarion*' is luxuriant in the rainforest, even on small twigs. In a climate that is equable throughout the year, there is no danger of exposure to the climatic extremes of temperature and wind that affect the canopies of European forests. However, if the canopy is disturbed, as it is by logging, extreme temperatures of over 30 °C occur in areas no longer protected by the forest canopy. It has been shown that temperatures over 35 °C are lethal to wet thalli, although dry thalli may survive temperatures up to 70 °C (Lange 1953). Low temperatures may affect growth rates, but in Europe these are seasonal and in the tropics they occur at night and under these circumstances do not appear to be a limiting factor. In European forests there is an association with specific phorophytes, usually *Quercus*, *Fraxinus*, and *Corylus* in Britain, extending on to *Fagus* and *Castanea* on the continent and also on to *Abies alba* in the mountains of Central Europe (Rose 1988). Characteristics of the bark affecting this association include pH, moisture retention, and roughness, as well as a tendency to flake. No information was collected in the tropics on pH, or bark types, however, the '*Lobarion*' was found on a wide variety of phorophytes within the hill evergreen and alpine scrub zones and there appeared to be a positive association with open crowns and rough bark and a negative association with dense stands and smooth bark. It is possible that in regions of very high rainfall that bark pH may not be a controlling factor. Leached sites on acid moorland in Europe have an associated *facies* of the *Lobarion* with species of *Heterodermia*, *Hypotrachyna*, and *Menegazzia* and similar communities are found in the high montane scrub in South East Asia.

## Forest zones and their associated lichens

### 1. *Freshwater swamp forest*

A dominant forest type in Borneo extending up river valleys. Only visited at Mulu, and inadequately collected. Massive and decaying trees of *Eusideroxylon conferta* and many palms such as *Eleiodoxa* spp. *Leptogium* and *Coccocarpia* present.

## 2. Lowland dipterocarp

Formerly a dominant forest type, now much reduced by logging. High canopy of mainly evergreen trees with emergent dipterocarps up to 40 m high. Rattans, palms, and young trees abundant on forest floor. Aroids, lianas, climbers, and epiphytes frequent. Forest interior cool and humid with very low light intensity. *Leptogium*, *Coccocarpia*, *Dictyonema*, and *Collema* infrequent on trunks with crustose species of *Crocynia*, *Thelotrema*, and unidentified taxa. Foliose lichens of the Lobariaceae are abundant in the canopy, and were collected from twigs fallen from the canopy or from the crowns of fallen or felled trees. *Pseudocyphellaria argyracea* and *P. junghuhniana* occur on twigs with *Sticta* spp. and *Coccocarpia* at Danum valley. It is possible that the low species number in Table 15.1 may be due to the difficulty of collecting in this habitat.

## 3. Hill dipterocarp

Found on mountain slopes between 300–700 m and characterized by fewer palms and rattans and an increasing number of fagaceous trees and *Tristania*. The canopy is not as high and more light reaches the trunks. Open-crowned dipterocarps still frequent. There is a marked increase in lichen cover on the trunks, particularly of pyrenocarpous lichens, as well as others found in the lowland dipterocarp forests. In an illegally logged area of the Genting Highlands the large branches of felled dipterocarps (*Shorea* spp.) supported large individuals of *Pannaria* and *Parmeliella*. Foliose species in the canopy are similar to those of the lowland dipterocarp forest.

## 4. Hill evergreen

Occurs over 1000 m and is dominated by members of the Fagaceae and Lauraceae. The forest structure may vary from an open savannah with scattered ancient trees, to a dense forest of evergreen trees of all sizes and ages, much of this variation being associated with management. In all the countries visited, from N. Thailand to Java, this forest was rich in epiphytes and in lichens of the '*Lobarion*', especially in species of *Sticta*, and *Lobaria*, *Pannaria*, *Parmeliella*, *Psoroma*, *Cetrelia*, and *Nephroma* when it occurred.

## 5. Moss or cloud forest

Occurs in a zone of constant mist or rain. It may occur between 1200–2500 m, depending on latitude, aspect, and local climate, throughout the regions visited. It is a dark, dripping forest with abundant epiphytes dominated by bryophytes with a poor lichen flora. Factors affecting this may be, (1) low light intensity, occasional trees emerging above the canopy (e.g. *Leptospermum flavescens*, *Schima* spp.) were festooned with species of *Usnea* and *Bryoria*, (2) increasing competition with bryophytes, (3) increasing acidity due to accumulating organic matter suggested by a lichen community of species of *Cladonia*, *Stereocaulon*, and *Leprocaulon* on ancient trunks and rocks at Paka cave, Kinabalu.

*6. Alpine scrub*

A conspicuous zone above the cloud forest of dwarfed and open forest with twisted trunks of *Leptospermum* and open crowns of species of *Dacrydium* and *Podocarpus* at $c.$ 2500 m altitude. Trunks, branches, and twigs have a well-developed '*Lobarion*' community containing genera that are associated with acid uplands of Europe, *Menegazzia*, *Cetrelia*, and *Cetraria*, as well as numerous species of *Sticta* and *Pseudocyphellaria*. With the addition of species of *Anzia*, *Hypotrachyna*, *Heterodermia* (*H. diademata*, *H. leucomelos*), this becomes a characteristic twig community of the alpine scrub.

*7. Seasonal rainforest*

Has fewer dipterocarps and emergents, but still has three tree layers, abundant climbers, and some palms present. A brief visit to Ton Krabakyai in Thailand showed the '*Lobarion*' community to be in the canopy as in tropical rainforest, and along the riverine forest. Figures not included in Table 15.1.

*8. Tropical mixed deciduous forest*

Has an extended dry period with an increasing number of deciduous trees such as *Tectona* and leguminous trees. Very little of this forest type remaining in Thailand, apparently few '*Lobarion*' species.

*9. Deciduous dipterocarp forest*

Covers large areas of Thailand where fire is a dominant form of management, Epiphytic lichen communities are restricted to canopy and upper branches. '*Lobarion*' absent.

*10. Pinus*

May become a dominant forest type in upland regions of Thailand following destruction and burning of montane forests. In Indonesia *Casuarina* formerly a colonizer of volcanic ash, may become dominant in similar situations. These forests are associated with a rich community of fruticose species of *Usnea*, *Ramalina* and, in Indonesia, *Teloschistes*. Where isolated trees of the montane oak forests remain, these are often rich in '*Lobarion*' species.

## Phytosociology

As material was collected in a rapid and subjective manner on this preliminary journey, the data are not available for community analysis. Kantvilas (1985, 1988, Kantvilas *et al.* 1985) studied the epiphytic flora of a *Nothofagus* and *Eucryphia* forest in Tasmania distinguishing four associations dominated by species of *Pseudocyphellaria*. The macrolichen composition of these associations is shown in Table 15.3. Constant associates of all four communities are from *Sticta*, *Psoroma*, and *Sphaerophorus*, while indicator species are related to the position of the community on the phorophyte; *Peltigera* at the base of the trunk, *Leptogium* on the trunk, *Nephroma* on the branches, and *Menegazzia* on the twigs. These associations occurred in a Southern Hemisphere forest at 750 m altitude. In the tropical equatorial forests visited, similar *Pseudocyphellaria*-dominated associations occurred over a much wider altitudinal range, from 1000–3000 m, on a wide variety of phorophytes. Although some phorophyte preference exists, this does not appear to be of a specific nature, but rather an association with well-lit sites, rough bark, open crowns, and an avoidance of straight or shaded trunks. It also appears that climatic conditions are more important in controlling the location of lichen communities, on Kinabalu and the Genting Highlands a similar lichen community was found on tree bases and exposed rocks.

## Biogeography of the *Lobarion*

Although the '*Lobarion*' is a distinctive association of lichen taxa, the distribution of these component taxa is rather varied including elements from the Southern Hemisphere, Northern Hemisphere, tropical, and temperate regions. Galloway discusses some implications of these distributions (Galloway 1988*a*). With the aid of this broad-based hypothesis, the distribution of four genera occurring in South East Asian '*Lobarion*' communities can be assessed.

*1. Pseudocyphellaria* (Table 15.4)

That the Southern Hemisphere is the centre of diversity of this genus was confirmed in South East Asia. The genus is a conspicuous component of the '*Lobarion*' in montane Java and Kinabalu, a dominant component in Tasmania (Kantvilas 1985, 1988), and New Zealand

**Table 15.3.** Composition of macrolichen genera in 'Lobarion' communities in a Tasmanian rainforest (from Kantvilas 1985, 1988)

| | Pseudocyphellaria | Sticta | Parmeliella | Psoroma | Leioderma | Leptogium | Parmelia | Menegazzia | Nephroma | Peltigera | Sphaerophorus |
|---|---|---|---|---|---|---|---|---|---|---|---|
| Pseudocyphellaria dissimilis—Peltigera dolichorhiza | 5 | 1 | 1 | 1 | — | 2 | — | — | — | 1 | 1 |
| P. multifida—Psoroma microphyllizans | 5 | 1 | — | 4 | — | — | 1 | — | 1 | — | 5 |
| P. billardierei—Psoroma microphyllizans | 5 | 1 | 1 | 4 | 1 | — | 1 | — | 1 | — | 2 |
| P. rubella—P. faveolata | 6 | 1 | — | 5 | 1 | — | 3 | 3 | 1 | — | 2 |

**Table 15.4.** Distribution of species of *Pseudocyphellaria* in sites visited

| | Doi Suthep and Doi Inthanon | Kinabalu | Danum | Mulu | Genting Highlands | Cibodas | G. Arjuno and G. Semeru | Photobiont |
|---|---|---|---|---|---|---|---|---|
| *P. argyracea* | | | * | | * | * | * | cyanobacterial |
| *P. aurata* | | * | | | | * | | green |
| *P. crocata* | | * | | * | | * | * | cyanobacterial |
| *P. intricata* | | * | | | | | | cyanobacterial |
| *P. junghuhniana* | * | * | | | | | | cyanobacterial |
| *P. knightii* | | | | | | | * | cyanobacterial |
| *P.* cf. *multifida* | | * | | | | | | green |
| *P. sulphurea* | | * | | | | * | * | green |
| *P. desfontainii* | | * | | | | * | | cyanobacterial |
| *P.* a (unidentified) | | | | | | * | | cyanobacterial |
| *P.* b (unidentified) | | | | | | | | cyanobacterial |
| *P.* c (unidentified) | | | | | | | * | cyanobacterial |
| Total | 1 | 7 | 1 | 1 | 1 | 6 | 5 | |

(Galloway 1988b), while in Peninsular Malaysia and Thailand it is rare, as it is in Europe.

### 2. *Sticta* (Table 15.5)

A tropical genus with eight species in South America, six species in Australia, 13 species in New Zealand (Galloway 1988a), and 16 species in Papua New Guinea (Streimann 1986). The Malesian collection of 19 taxa shows an apparently high degree of endemism; on Kinabalu four out of 10 taxa were not found elsewhere, and at Cibodas eight out of 11 species were not found elsewhere. Of the 22 taxa distinguished in the collection, only five had a green photobiont. Of the seven stalked species four were readily identified. However, among the nine taxa with a dichotomously-branching thallus and a cyanobiont, there is considerable variation in form and location of isidia, tomentum, cyphellae structure, and in the colour of the medulla. James and Henssen (1976) show that *Sticta* species with different photobionts may produce a range of morphotypes that in some cases have been given separate taxonomic identities. James (*personal communication*) also suggests that this may be an adaptation to low light intensities, allowing an increased ecological amplitude in a variety of situations.

### 3. *Lobaria* (Table 15.6)

A tropical and Northern Hemisphere genus, absent from the southern latitudes of S. America, and of limited occurrence in the Southern Hemisphere (Galloway 1988a, b). This genus was well represented in Thailand in species number and cover value, whereas in the tropical equatorial forests it is replaced by species of *Sticta* and *Pseudocyphellaria*. Chemical analysis has facilitated the distinction of races in morphologically similar taxa, such as the *L. retigera* group distinguished by Yoshimura (1971). The distribution is shown in Fig. 15.3 extending to E. Africa and the Pacific coast of America, and to New Zealand.

### 4. *Leptogium* (Table 15.7)

A pantropical genus of great adaptability extending into Northern and Southern Hemispheres, and into a wide range of habitats (Jørgensen and James 1983). It is a conspicuous component of all forest types in South East Asia from lowland and rainforest to alpine scrub. It occurs in a species-poor community on trunks in the lowland rainforest, and in a species-rich one in the canopy and the alpine scrub.

## Fagaceous forests and the '*Lobarion*'

The family Fagaceae occupies a peculiar geographical position in the

**Table 15.5.** Distribution of species of *Sticta* in sites visited

| | Doi Inthanon and Doi Suthep | Kinabalu | Danum | Mulu | Genting Highlands | Cibodas | C. Arjuno and G. Semeru | Photobiont |
|---|---|---|---|---|---|---|---|---|
| *Sticta boschiana* | * | | | | | | | cyanobacterial |
| *S. cyphellulata* | | * | | | | * | | cyanobacterial |
| *S. fuliginosa* | * | | | | | | * | cyanobacterial |
| *S. sayerii* | | * | | | | | | green |
| *S. weigelii* | | | | | | * | * | cyanobacterial |
| *S. a* (unidentified) | * | * | | * | | * | * | cyanobacterial |
| *S. b* (unidentified) | * | * | | | | * | | cyanobacterial |
| *S. nylanderiana* | * | | | | | | | green |
| *S. d* (unidentified) | | * | | | | | | cyanobacterial |
| *S. f* " | * | | | | | | | cyanobacterial |
| *S. g* " | | * | | | | | | cyanobacterial |
| *S. h* " | * | * | | | | | | cyanobacterial |
| *S. k* " | | | | | * | | | green |
| *S. m* " | | | | | | * | | cyanobacterial |
| *S. n* " | | * | | | | * | | cyanobacterial |
| *S. p* " | | * | | | | | | cyanobacterial |
| *S. r* " | | | | | | * | | cyanobacterial |
| *S. s* " | | | | | | * | | cyanobacterial |
| *S. w* " | | | | | | * | | green |
| *S. x* " | | | | | | * | | cyanobacterial |
| *S. y* " | | | | | | * | | green |
| *S. z* " | | * | | | | * | | cyanobacterial |
| Total | 6(3) | 10(4) | 0 | 1 | 1 | 11(8) | 3 | |

**Table 15.6.** Distribution of species of *Lobaria* in sites visited

| | Doi Inthanon and Doi Suthep | Kinabalu | Danum | Mulu | Genting Highlands | Cibodas | G. Arjuno and G. Semeru | Photobiont | Height above sea level (m) |
|---|---|---|---|---|---|---|---|---|---|
| *L. discolor* | * | | | | | * | | green | c.1500 |
| *L. isidiophora* | | * | | | | * | | green | 1800 |
| *L. isidiosa* | | * | | | | * | | cyanobacterial | 1430 |
| *L. lobulata* | | | | | | * | | green | 1400–2500 |
| *L. retigera* | * | | | * | | | * | cyanobacterial | 2000–2565 |
| *L. a* (unidentified) | * | | | | | | * | cyanobacterial | 1000–1500 |
| *L. b* (unidentified) | * | | | | | | | green | |
| *L. c* (unidentified) | | | * | | | * | * | green | 200 and 1430 |
| *L. d* (unidentified) | | | | * | | | | cyanobacterial | 3000 |
| Total | 4 | 2 | 1 | 2 | — | 2 | 3 | | |

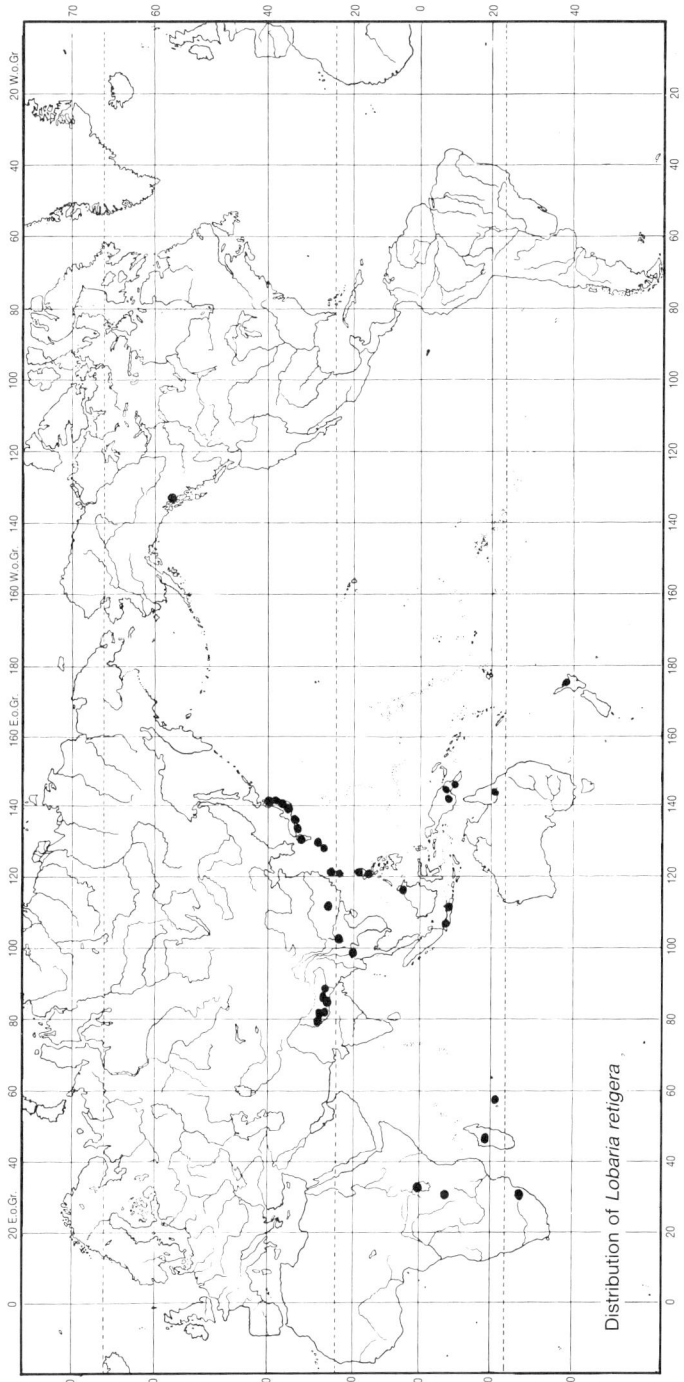

**Fig. 15.3.** Distribution of *Lobaria retigera* (Bory) Trev. (after Yoshimura 1971 with additional records).

**Table 15.7.** Distribution of species of *Leptogium* in sites visited

| | Doi Inthanon and Doi Suthep | Kinabalu | Danum | Mulu | Genting Highlands | Cibodas | G. Arjuno and G. Semeru | Ton Krabakyai |
|---|---|---|---|---|---|---|---|---|
| *Leptogium azureum* | | * | * | * | * | * | * | |
| *L. brebissonii* | | * | | | | | * | |
| *L. caespitosum* | | | | | * | | * | |
| *L. corticola* | | | | | | * | | * |
| *L. cyanescens* | * | * | | | | * | | |
| *L. marginellum* | | | | | | | | * |
| *Leptogium* sect. *Mallotium* | * | * | | | * | * | * | |
| *L. javanicum* | | | | | | * | * | |
| Total | 2 | 4 | 1 | 1 | 3 | 5 | 5 | 2 |

earth's forests, in that it dominates large areas of cool temperate forests in the Northern Hemisphere, scattered land masses in the Southern Hemisphere, and in the tropics it dominates isolated montane forests with a cool wet climate (Fig. 15.4). The family is conspicuously absent from monsoon climates with a dry season, but is found again in montane regions with a cool wet climate above 1000 m. In South East Asia fagaceous forests are associated with many tropical genera and, particularly, the Lauraceae. Members of this tropical family are also present in Macaronesia, Iberia, and Western Eire, indicating a relict status for these forests from the warm wet climate of the Miocene (P. James *personal communication*). The genus *Quercus*, a dominant of European primary forests, does not cross Wallace's Line, whereas *Lithocarpus* occupies an ancient and central position in South East Asia, extending to New Guinea and the West coast of America (Soepadmo 1972). Troll (1978) considered these montane forests to be relicts of the cool temperate flora of the Cretaceous and early Tertiary, prior to the breaking up of Gondwanaland. The *Lobarion* is a characteristic epiphytic community of ancient European *Quercus* and *Fagus* forests (Rose 1976, 1988). That the montane fagaceous forests of the tropics have a similar '*Lobarion*' community of macrolichens, with many common genera, is further evidence of this antiquity. Furthermore, the comparatively lower diversity of this community in the dipterocarp forests may indicate a migration from the fagaceous forests. Table 15.1 shows nine species collected from the primary lowland dipterocarp forest at Danum in E. Sabah, 15 species collected from the hill dipterocarp forest at Genting Highlands in Malaysia, and an average of 35 species from montane forests (excluding Cibodas and Mulu). This hypothesis is based on a rapid survey of just a few sites so that further work is needed in South East Asia to determine the components of the tropical '*Lobarion*' and the distribution in the forest, and forest types.

## Forest management and its effects on epiphytic lichen communities

Although collections of lichens were made in relatively undisturbed National Parks, Heritage Sites or reserves, these were often surrounded by large areas where the forest had been destroyed for a variety of reasons. A brief investigation of these areas showed that there were features in common with European forest management and its effect on lichen communities. Some of the differences and similarities are outlined below.

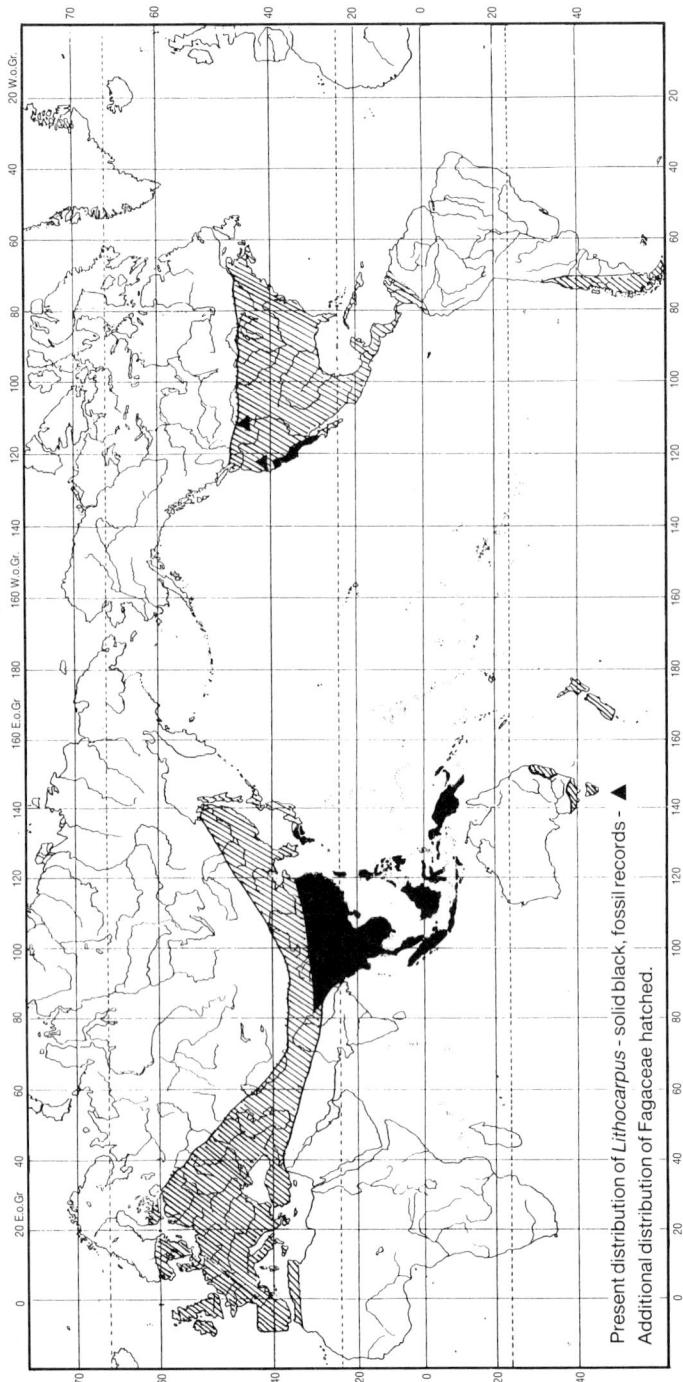

**Fig. 15.4.** Distribution of the Fagaceae, and of *Quercus* and *Lithocarpus* genera (from Soepadmo 1972).

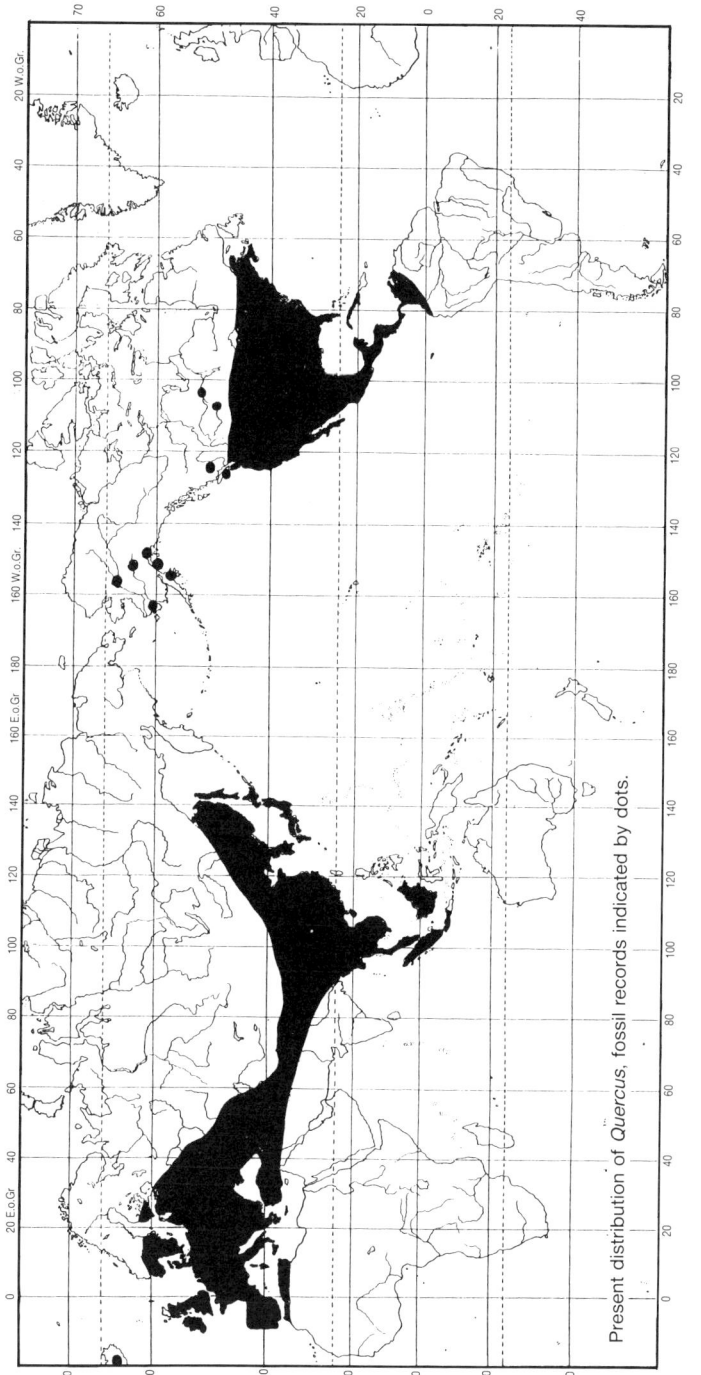

Present distribution of *Quercus*, fossil records indicated by dots.

## 1. Logging

Present methods of logging involve high line felling and bulldozed extraction roads, the first removing most of the tree cover and the second resulting in severe erosion on sandy lateritic soils. In lowland rainforest logging has a dramatic effect on the lichen communities that may remain after logging, as these are exposed to climatic extremes once the forest cover is removed. At Danum in E. Sabah crustose material collected in a recently felled area of lowland forest was already dead and had lost all spores. Secondary forest in this area is characterized by a dense growth of lianas and creepers and an invasion of tree species such as *Anthocephalus* and *Macaranga* that are unsuitable phorophytes.

In Sarawak an area adjacent to primary forest, logged for a rubber plantation, and abandoned for around 60 years had many native species established including dipterocarps. Although the twigs supported a variety of foliose lichens there were few lichens on the trunks and no '*Lobarion*' elements.

## 2. Slash and burn

Traditional slash and burn involves the clearance of small areas of forest, the cultivation of hill rice for a season, followed by a *c*. five-seven year fallow period in which the forest trees grew up. Isolated larger trees are often left standing, and care is taken not to burn adjacent forest.

Extensive areas of forest are not cut as this would cause soil erosion. The well-lit forest margins and isolated trees provide an ideal situation for epiphytic macrolichen communities, much as around small ancient settlements in European forests. Other situations in which a species-rich macrolichen community is associated with ancient trees and a more open habitat include; Cibodas mountain garden, a park-like habitat adjacent to primary forest, and the savannah edges of montane *Lithocarpus* forest in Doi Inthanon and Doi Suthep.

## 3. Fire

Annual burning is used extensively in monsoon climates of Thailand and Java and produces a savannah forest of fire-tolerant trees with a grassy herb layer. Temperatures may reach 70–90 °C up to 1 m above the ground (Stott 1986) and cause extensive damage to epiphytic communities on trunks. Although the lower trunk was devoid of lichens in the deciduous dipterocarp forests of N. Thailand, the upper trunk has a characteristic lichen community with species of *Brigantiaea*, *Dirinaria*, *Lecanora*, *Parmotrema chinense*, *Cryptothecia obtecta* and species of *Graphis* and *Graphina*. On the branches there is an exposed community dominated by species of *Usnea* and *Ramalina*. In Java, *Casuarina junghuhniana*, part of a natural fire succession on volcanic ash, has become dominant over large areas where fire is a limiting factor. It supports a number of foliaceous lichens with abundant *Usnea* and *Teloschistes* on the upper trunk.

*4. Agriculture*

Although extensive areas of lowland forests have been converted into agricultural land, or into rubber and palm oil plantations in the last two hundred years, the pressure for land is now being felt in the montane forests. Expansion of the human population into the montane regions of Java and N. Thailand has been accompanied by a change in farming methods to one of continuous cropping with the widespread use of fertilizers and insecticides. In the Doi Inthanon area of N. Thailand migrant hill tribes settled on the ridges in 1973, and started to convert montane forest into agricultural land, often on steep slopes. Five families had grown to 101 by 1988, and with large injections of foreign aid three crops of cabbages are produced a year, as well as staple crops of rice and maize. Apart from the loss of the species-rich montane forest, there are serious ecological problems caused by the loss of soil and water resources. The increasing desertification of Thailand testifies to the role of forests in soil and water conservation.

## Conclusion

Species-rich '*Lobarion*' communities are a feature of ancient fagaceous forest sites. In Europe and South East Asia, and probably elsewhere, these forests and their associated epiphytic vegetation represent a climax vegetation that has survived major changes in the earths climate since the early Tertiary. Older forms of forest management, such as wood-pasture in Europe and a sustainable slash and burn technique in Asia have allowed ecological continuity for the many component species of these communities. However, recent changes in forest management and in agricultural practices are threatening to destroy this habitat.

## Acknowledgements

This project would not have been possible without the help and support of people and institutions in all the countries visited. I would like to thank Chamblong Pengklai, Mr Rachan Pooma, and Dr Tem Smitinand of the Royal Forestry Department of Thailand, Associate Professor

Obchant Thaithong of Chulalongkorn University* Bangkok, Sarawak National Parks Department,* particularly David Labang and Wan Sepik—my guide at Mulu, Sabah National Parks Department at Kinabalu,* and the Danum Valley Field Centre. The Forestry Institute of Malaysia* and the University of Malaya, The Indonesian Institute of Sciences at Bogor Herbarium,* particularly Dr M. Riswan and Professor A.J.K. Kostermans. The collection is at the BM (Natural History) (duplicate collections at * above), and I would like to thank all those in the lichen section particularly Peter James and David Galloway for their support on my return, and the provision of facilities at The Natural History Museum to produce this paper.

## References

Barkman, J.J. (1958). *Phytosociology and ecology of cryptogamic epiphytes.* Van Gorcum, Assen.

Forman, R.T. (1975). Canopy lichens with blue-green algae: a nitrogen source in a Colombian rainforest. *Ecology*, **56**, 1176–84.

Galloway, D.J. (1988a). Plate tectonics and the distribution of cool temperate Southern Hemisphere macrolichens. *Botanical Journal of the Linnean Society*, **96**, 45–55.

Galloway, D.J. (1988b). Studies in *Pseudocyphellaria* (lichens). I. The New Zealand species. *Bulletin of the British Museum (Natural History)* Botany series, **17**, 1–267.

James, P.W., Hawksworth, D.L., and Rose, F. (1977). Lichen communities in the British Isles: a preliminary conspectus. In *Lichen ecology* (ed. M.R.D. Seaward), pp. 295–414. Academic Press, London.

James, P.W. and Henssen, A. (1976). The morphological and taxonomic significance of cephalodia. In *Lichenology: progress and problems* (ed. D.H. Brown, D.L. Hawksworth, and R.H. Bailey), 27–77. Academic Press, London.

Jones, K. (1989). Interactions between desiccation and dark nitrogen-fixation in tropical *Nostoc commune*. *New Phytologist*, **113**, 1–6.

Jørgensen, P.M. and James, P.W. (1983). Studies on some *Leptogium* species of Western Europe. *Lichenologist*, **15**, 109–26.

Kantvilas, G., James, P.W., and Jarman, S.J. (1985). Macrolichens in Tasmanian rainforests. *Lichenologist*, **17**, 67–83.

Kantvilas, G. (1985). Studies on Tasmanian rainforest lichens. Ph.D. thesis University of Tasmania. Hobart.

Kantvilas, G. (1988). Tasmanian rainforest lichen communities: a preliminary classification. *Phytocoenologia*, **16**, 391–428.

Lange, O.L. (1953). Hitze-und trockenresistenz der flechten in beziehung zu ihrer verbreitung. *Flora* (Jena), **140**, 39–97.

Rose, F. (1976). Lichenological indicators of age and environmental continuity in woodlands. In *Lichenology: progress and problems* (ed. D.H. Brown,

D.L. Hawksworth, and R.H. Bailey), pp. 279–307. Academic Press, London.

Rose, F. (1988). Phytogeographical and ecological aspects of *Lobarion* communities in Europe. *Botanical Journal of the Linnean Society*, **96**, 69–79.

Santisuk, T. (1988). An account of the vegetation of Northern Thailand. *Geoecological research*, **5**, 1–101.

Sipman, H.J.M. and Harris, R.C. (1989). Lichens. In *Tropical rainforest ecostems*. (ed. H. Lieth and M.J.A. Werger), pp. 303–9. Elsevier, Amsterdam.

Soepadmo, E. (1972). Fagaceae. *Flora Malesiana*, **7**, 265–403.

Streimann, H. (1986). Catalogue of the lichens of Papua New Guinea and Irian Jaya. *Bibliotheca Lichenologica*, **22**, 1–145.

*Swinscow, T.D.V. and Krog, H. (1988). *Macrolichens of East Africa*. British Museum (Natural History), London.

Troll, C. (1978). Der Asymmetrische Vegetations und landschaftsbau der nord- und sudhalbkugel. *Erdwissenschaftliche Forschung*, **11**, Wiesbaden; Steiner, 10–28.

Steenis, van C.G.G.J. (1934–6). On the origins of the Malaysian Mountain Flora. *Bulletin Jardin Buitenzorg*, ser. 3, **13**, 135–417; **14**, 56–72.

Steenis, van C.G.G.J. (1957). Outline of vegetation types in Indonesia and some adjacent regions. *Proceedings of the Pacific Scientific Congress*, **8**(4), 1–97.

Steenis, van C.G.G.J. (1964). Plant geography of the mountain flora of Mt. Kinabalu. *Proceedings of the Royal Society*, ser. B, **161**, 7–37.

Steenis, van C.G.G.J. (1972). *The mountain flora of Java*. Brill, Leiden.

Stott, P. (1986). The spatial pattern of dry season fires in the savannah forests of Thailand. *Journal of Biogeography*, **13**, 345–58.

Walsh, R.P.D. (1982). Climate, Gunung Mulu National Park Sarawak. *Sarawak Museums Journal*, **30** (51). Special issue no. 2, 29–67.

Whitmore, T.C. (1984). *Tropical rainforests of the Far East*. 2nd edn. Clarendon Press, Oxford.

Wolseley, P.A. and O'Dare, A.M. (1989). *Exmoor woodland lichen survey 1987–88*. Somerset Trust for Nature Conservation, Taunton.

Yoshimura, I. (1971). The genus *Lobaria* of Eastern Asia. *Journal of the Hattori Botanical Laboratory*, **34**, 231–364.

*Zahlbruckner, A. (1943). Flechtenflora von Java. I. *Beihefte Feddes Repertorium*, **127**, 1–80.

*Zahlbruckner, A. and Mattick, F. (1956). Flechtenflora von Java. II. *Willdenowia*, **1**, 433–528.

---

* Lichen floras used in the field in south-east Asia.

# 16. Spore ontogeny of *Sphaerophorus diplotypus* and *S. fragilis*

M. WEDIN
*Department of Systematic Botany, Uppsala University, PO Box 541, S-751 21 Uppsala, Sweden*

## Abstract

The spore ontogeny of *Sphaerophorus diplotypus* and *S. fragilis* has been investigated and the results are in agreement with studies of other species in *Sphaerophorus*. Concentric bodies are reported from young stages of ascospores in both species. Relationships within *Sphaerophorus sensu lato* are discussed, and the results indicate that at least two taxonomic groupings deserve generic rank.

## Introduction

All species of *Sphaerophorus sensu lato* (Caliciales) have varying degrees of rough and irregular spore ornamentation. Tibell (1981) demonstrated two different developmental processes for this spore ornamentation in *S. globosus* and *S. murrayii*. In *S. globosus*, ornamentation is formed from epiplasmatic material deposited on the spore surface inside the asci. In *S. murrayii* spores remain without ornamentation until the asci disintegrate. When the spores are released into the mazaedium, a dark amorphous substance adheres to the spore surface. The two species are classified in different subgenera by Ohlsson (1974). These processes of forming spore ornamentation were both shown to occur in other species of the two subgenera (Tibell 1984, 1985). This paper is part of a larger ultrastructural study aimed at investigating how consistent the type of spore ontogeny is within proposed subgenera of *Sphaerophorus sensu lato*.

## Material and methods

Parts of apothecia of *S. diplotypus* (Papua New Guinea, Morobe, Streimann 19696, H) and *S. fragilis* (Norway, Nordland, *Tibell* 16261, Caliciales exs. 172, UPS) were dissected and fixed in 2.5 per cent glutaraldehyde buffered with sodium cacodylate at pH 7.2 for several days. The material was then postfixed in 2 per cent osmium tetroxide at 4 °C for 2 h. Specimens were embedded in Epon and sections were double-stained with lead citrate and uranyl acetate.

## Results

*1. Sphaerophorus diplotypus Vainio*

*S. diplotypus* [subgenus *Bunodophorus* (Massal.) Ohlsson], described from Madagascar (Vainio 1898), is a subtropical and tropical–montane species widely distributed in the Indian Ocean, Australasia, and the Pacific north to Japan. It is distinguished from other species in the subgenus by a hollow, subterete to almost terete thallus with fertile branches only slightly larger than the sterile ones, and by subterminal ascocarps. It is often very pale, almost white, in colour.

When young, ascospores of *S. diplotypus* have a thick electron-lucent wall (Fig. 16.1A). Ascospores are arranged in a single row in the ascus and spore contents are already highly vacuolate. No ornamentation is deposited on the spore surface inside the asci.

The spore protoplast of young spores often contains several concentric bodies (Fig. 16.1B), organelles frequently observed in hyphae of lichenized fungi (Peveling 1974; Galun 1988) but only reported from ascospores in the genus *Microcalicium* (Tibell 1978).

When the spores are released into the mazaedium, an amorphous and electron-dense substance accumulates and is deposited in varying amounts on the spore surface to form an uneven ornamentation (Fig. 16.1C). The origin of this substance is unknown but it could well result from a process involving carbonization and disintegration of asci and paraphyses. The spores adhere to each other and to remnants of paraphyses and, because of the ornamentation, spores are often associated in aggregates of variable size.

*S. diplotypus* has thus the same type of spore ontogeny as *S. murrayii* and *S. notatus* (Tibell 1981, 1984, 1985) which are classified in the subgenus *Aghimus* Ohlsson.

*2. Sphaerophorus fragilis (L.) Pers.*

*S. fragilis* (subgenus *Sphaerophorus*) is widely distributed in the Northern

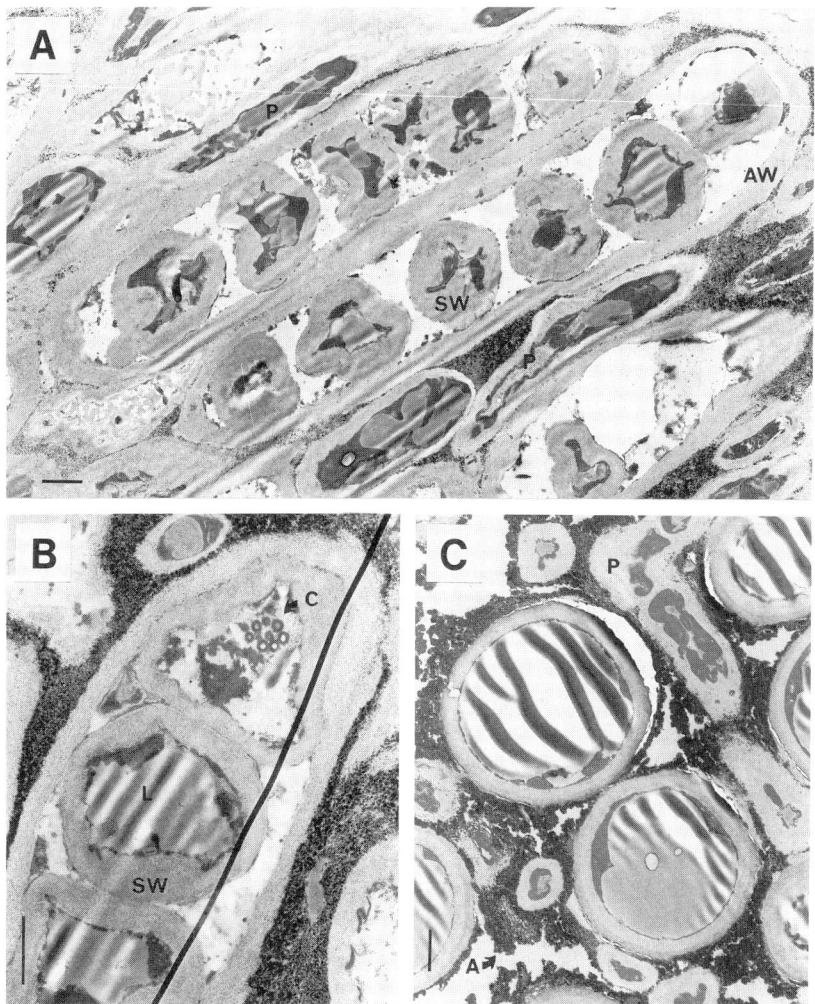

Fig. 16.1. *Sphaerophorus diplotypus*, Streimann 19696 (H). (A) Section of mazaedium showing young asci. Spores are arranged in a single row and no ornamentation is deposited on the spore surface inside the asci. (B) Detail of ascus showing a spore with several concentric bodies. (C) Mazaedium with mature or semi-mature spores embedded in an amorphous and electron-dense material. The spores adhere to each other and to adjacent paraphyses. A weak differentiation of the spore wall into two layers can be seen; A = amorphous, electron-dense material, AW = ascus wall, C = concentric bodies, L = lipid, P = paraphysis, SW = spore wall. Scale = 1 μm.

Hemisphere occurring particularly in oceanic and arctic–alpine regions. It is distinguished from other species in the subgenus by a comparatively small and terete thallus with little difference in size between primary and secondary branches, and an I- medulla.

Young ascospores of *S. fragilis* have a thick, electron-lucent wall, on top of which an irregular and electron-dense ornamentation is deposited while the spores are still within the asci (Fig. 16.2A). Concentric bodies are frequent in the protoplast of young spores in *S. fragilis*.

At a later stage, when the asci are disintegrating, the thickness of the ornamentation increases (Fig. 16.2B). The ornamentation is, at least in part, composed of tubular elements. Large elongated lacunae are formed between the spore wall and the ornamentation, as well as within the ornamentation. These lacunae could, however, be structural artefacts caused by fixing and embedding the material.

The spore ontogeny of *S. fragilis* is thus very similar to that of *S. globosus* and *S. stereocauloides* (Tibell 1981, 1984, 1985), both classified in the subgenus *Sphaerophorus*.

## Discussion

The process of development of spore ornamentation is constant within the investigated subgeneric groupings in *Sphaerophorus sensu lato*. The type of spore ontogeny in subgenus *Sphaerophorus* occurs in this group only, while the type in subgenus *Aghimus* is shown here to occur also in subgenus *Bunodophorus*. The two groups within *Sphaerophorus sensu lato*, characterized by these spore ontogenies are, in my opinion, not closely related to each other. The difference in spore ontogeny is well correlated with other characters. Subgenus *Sphaerophorus* is an homogeneous group with a morphology, thallus and ascocarp anatomy, spore shape and colour, as well as secondary product chemistry quite different from the other subgenera. *S. stereocauloides* Nyl. belongs to *Sphaerophorus* subgenus *Sphaerophorus*, although it has been placed in the monotypic genus *Thysanophoron* (Stirton 1883), mainly because of the presence of cephalodia. The cephalodia and the thick and rigid holdfast that attaches the thalli of *S. stereocauloides* to the branches on which it grows made Galloway (1985) retain the genus *Thysanophoron*. Tibell (1984, 1985, 1987) stressed, however, the similarities between *S. stereocauloides* and other species of *Sphaerophorus* subgenus *Sphaerophorus* and placed it in this subgenus, hence following Du Rietz (1925) and Ohlsson (1974) in regarding *Thysanophoron* as synonymous with *Sphaerophorus*. In my opinion this is correct, as *S. stereocauloides* shares all the important characters of the group presently classified as *Sphaerophorus* subgenus *Sphaerophorus*. It also has the same spore ontogeny as *S. globosus* and

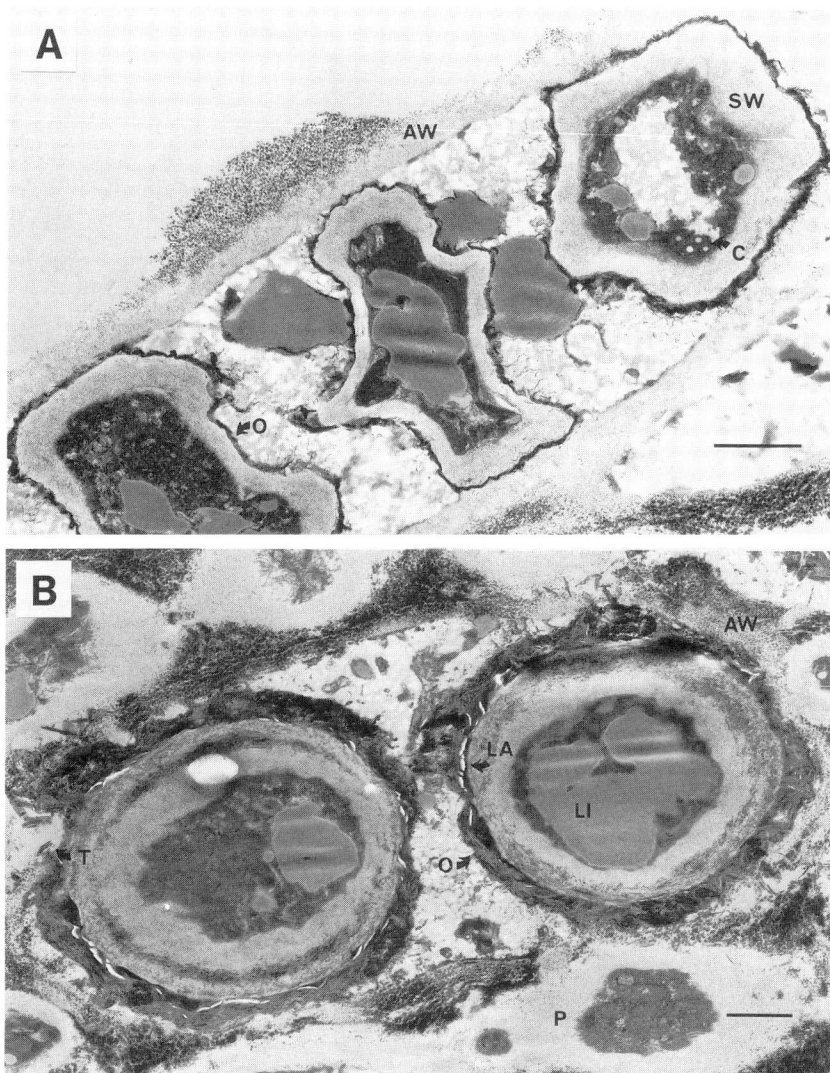

**Fig. 16.2.** *Sphaerophorus fragilis*, *Tibell* 16261 (UPS). (A) Young ascus with semi-mature spores. An irregular and electron-dense ornamentation is deposited on the spore wall. Concentric bodies are frequent in the spores. (B) Disintegrating ascus with semi-mature spores. The ornamentation has increased in thickness and consists, at least in part, of tubular elements. Elongated lacunae are seen between the spore wall and the ornamentation as well as within the ornamentation; AW = ascus wall, C = concentric bodies, LA = lacunae, LI = lipid, O = ornamentation, P = paraphysis, SW = spore wall, T = tubular elements. Scale = 1 μm.

*S. fragilis* (Tibell 1984, 1985, and verified by myself). I regard subgenus *Sphaerophorus* as a monophyletic group, with an apomorphic spore ontogeny. This group is very homogeneous and should best be treated as a genus of its own (*Sphaerophorus sensu stricto*).

Several important characters, such as ascocarp ontogeny, still have to be investigated before any definitive statements can be made about relationships among the remainder of the species of *Sphaerophorus sensu lato*, presently classified in three subgenera. The taxonomy at species level is also still very unclear in several species groups. Tropical species, in particular, are poorly understood and in great need of collection and revision. It is, however, quite clear that the genus *Sphaerophorus* as presently circumscribed is heterogeneous and includes at least two distinct groups deserving generic rank.

## Acknowledgements

I would like to thank Dr Leif Tibell for all his help and constructive criticism of the manuscript. Comments on the manuscript have also kindly been given by Dr Roland Moberg, and Miss Antona Wagstaff revised the English text. I am indebted to Dr A. v. Hofsten (Uppsala University Unit of Biological Structure Analysis) for placing TEM facilities at my disposal and to Mrs A. Axén for skilful assistance with embedding and sectioning of the TEM-preparations and for technical assistance with the illustrations. The curators of the herbaria of Uppsala University (UPS) and Helsinki University (H) are thanked for the loan of investigated material. The investigation was funded by the Swedish Nature Science Research Council as a part of the project 'Taxonomy of Caliciales'. Financial support was also given by 'Lemans stipendiefond' and 'Olof O:son Wijks minnesfond'.

## References

Du Rietz, G.E. (1925). Flechtensystematische Studien. VI. *Botaniska Notiser*, **1925**, 362–72.

Galloway, D. (1985). *Flora of New Zealand lichens*. Wellington.

Galun, M. (1988). Effects of symbiosis on the mycobiont. In *Handbook of lichenology* Vol. II (ed. M. Galun), pp. 145–51. Boca Raton, Florida.

Ohlsson, K. (1974). A revision of the lichen genus *Sphaerophorus*. Unpublished Ph. D. Thesis. Michigan State University.

Peveling, E. (1974). Fine structure. In *The lichens* (ed. V. Ahmadjian and M.E. Hale), pp. 147–82. New York and London.

Stirton, J. (1883). On lichens (1) from Newfoundland, collected by Mr A. Gray, with a list of species; (2) from New Zealand etc. *Transactions and Proceedings of the Botanical Society of Edinburgh*, **14**, 355–62.

Tibell, L. (1978). The genus *Microcalicium*. *Botaniska Notiser*, **131**, 229–46.
Tibell, L. (1981). Formation of spore ornamentation in two *Sphaerophorus* species. *Nordic Journal of Botany*, **1**, 333–40.
Tibell, L. (1984). A reappraisal of the taxonomy of Caliciales. *Nova Hedwigia Beiheft*, **79**, 597–713.
Tibell, L. (1985). Comments on Caliciales exsiccatae. III. *Lichenologist*, **17**, 189–204.
Tibell, L. (1987). Australasian Caliciales. *Symbolae Botanicae Upsalienses*, **27:1**, 1–279.
Vainio, E.A. (1898). Lichenes quos in Madagascaria centrali Dr C. Forsyth Major a. 1896 collegit. *Hedwigia*, **37** (Beiblatt 2), 33–7.

# 17. Tropical pyrenocarpous lichens. A phylogenetic approach

A. APTROOT
*Institute of Systematic Botany, Heidelberglaan 2, 3508 TC Utrecht, The Netherlands*

### Abstract

The study of the phylogenetic relationships of tropical pyrenocarpous lichens was undertaken in view of their present doubtful systematic position. The pyrenocarpous lichens as a whole are probably not monophyletic, but have evolved several times from non-lichenized fungi. Before preparing a monograph of some selected, presumably monophyletic, genera within, or closely related to, the Pyrenulaceae, a cladistic analysis was made of most groups within the pyrenocarpous lichens. For this purpose material was studied of most genera and of some species with uncertain affinities. Interpretation of the resulting cladogram leads to hypotheses about character transformations and relationships, as well as to inferred monophyletic groups. A key to genera investigated is provided in an appendix.

### Introduction

The study of tropical pyrenocarpous lichens was undertaken particularly because special problems arise in this group in regard to their phylogeny.

Pyrenocarpous lichens may be classified into five, more or less well-characterized groups: Arthopyreniaceae, Pyrenulaceae, Trichotheliaceae, Trypetheliaceae, and Verrucariaceae. All of these groups, except for the Verrucariaceae, have their greatest diversity in the tropics, where the variation and numbers of species and genera are much higher than elsewhere. Some groups are even dominant lichens in certain habitats, such as the Trichotheliaceae on leaves and tree trunks in dense rainforests and the Trypetheliaceae on branches in savannahs and in canopies. The main differences between these groups are the nature of several internal structures such as the hamathecium, the asci, and ascospores.

Besides these rather well-defined groups, a number of genera (partly undescribed) occur in the tropics that can not be satisfactorily placed in any of the families mentioned above. These include rather odd foliicolous lichens such as *Phyllobathelium* and *Aspidothelium*; genera of unknown relationships as *Thelenella*; as well as genera which show most characters of a certain group but also some of others. Examples of the latter are *Polymeridium* and the *Trypethelium uberinum*-group, both most likely related to the Trypetheliaceae, but characterized by euseptate spores and an inconspicuous thallus in the case of *Polymeridium*, and by the brown, very large spores in the case of the latter.

The first aim of the present study is to trace the systematic position of several of these genera which are not so far satisfactorily classified. From the beginning of lichen taxonomy and even taxonomy as a whole, much stress was attached to differences rather than to similarities between taxa. This has also been the case with the pyrenocarpous lichens, especially when the status of higher taxa is concerned. A second aim is, therefore, to evaluate the position of higher taxonomic categories of pyrenocarpous lichens within the whole ascomycete system, based on a phylogenetic method. Members of the Verrucariaceae are included as well as some genera of uncertain position such as *Macentina. Microtheliopsis*, *Normandina*, and *Pocsia* have many characters in common with members of this family and are, therefore, best compared with it.

It must be stressed that the present paper is only preliminary and represents a possible approach to the systematics of pyrenocarpous lichens. Therefore, no formal classification is proposed at present based on the cladistic analyses. This must await results from current searches for additional characters and the examination of some groups which are not treated here.

## History

The study of pyrenocarpous lichens begins with Acharius (1803), who compared two of the three pyrenocarpous lichen genera he recognized, *Endocarpon* and *Verrucaria*, with the fungus *Sphaeria* and even referred some species from the genus *Lichen* to it. However, as the study of pyrenocarpous lichens became more difficult, mainly because of the growing number of species involved, it became increasingly isolated from the taxonomy of related fungi. Species were accommodated in genera, which are for the greater part rather artificially delimited. Müller Argoviensis, especially, described many species and genera of pyrenocarpous lichens. Zahlbruckner (1907) gave an account of accepted genera and families of lichens, but his treatment of the pyrenocarpous lichens was largely based on the descriptions of Müller Argoviensis. His system was a standard for decades, although it became increasingly clear that it does not reflect phylogenetic relationships but is highly artificial. This was not felt to be a disadvantage at that time, since various authors thought lichens were specially created and not evolved from each other. Creation could well have been in regular schemes as proposed by Eschweiler (1824), who gives a matrix as a natural system, in which horizontal characters such as, perithecia immersed or not, are used and also vertical characters like perithecia aggregated or not. The result is a matrix with known lichen genera in most positions.

At the beginning of this century various authors, notably Watson (1929) worked with an evolutionary concept of lichen classification, but ignored the pyrenocarpous fungi. Occasionally some mycologists looked at some lichen groups, which is actually quite natural since half of the ascomycetes are in fact lichenized. But still, in 1931, Clements and Shear in their '*The Genera of Fungi*' include all pyrenocarpous lichens in one order, the Verrucariales, with exclusively lichenized genera.

The most promising project towards integration of lichen and fungal taxonomy is the *Systema Ascomycetum*, intended as a kind of forum for lichenologists and ascomycete mycologists. A recent issue, containing the assignment of most ascomycete genera to higher taxa (Eriksson and Hawksworth 1988) was the starting point of the present study. These authors, however, express a mycological point of view regarding thalline characters of lichens, which, for instance, brings them to deny acceptance of the recent subdivisions of *Parmelia*, whereas they accept most of the recent subdivisions of *Lecidea*. Their argument for this is basically that these small, parmelioid genera are not supported by generative (apothecium-) characters. They ignore the considerable evidence of thallus characters, indicative of the various parmelioid genera. A similar discussion involves one of the rather well-characterized groups of pyrenocarpous lichens, the Trypetheliaceae. Harris (1986) argues against including several fungal genera in this family, whereas Eriksson and Hawksworth (1988), not only include these groups, but add even more fungal genera to it. Their resulting family Trypetheliaceae is much more heterogenous than the lichen family in the circumscription of Harris.

The present treatment focuses on lichens, but it attempts to incorporate all non-lichenized relatives in order to deal with evolutionary rather than artificial groups. It is now commonly accepted that lichen taxonomy should be based on the mycobiont only. In pyrenocarpous lichens this is not a great inconvenience, as the photobiont is often very scarce or absent and, therefore, probably rather unimportant to thallus morphology.

Therefore, it is not surprising that several pyrenocarpous lichens were also described as fungi by mycologists. An example is *Trypethelium tropicum* which was described, three separate times, in *Zignoella!*

## Delimitation of taxa

In lichenology, as in most areas of taxonomy, the most commonly used method to detect species is to study specimen morphology and to delimit species as groups of morphologically similar members. Depending on the amount of similarity in a species, it can lead to a broader or narrower species concept, especially if chemical races or taxa with vegetative propagules are treated as species.

More sophisticated methods such as hybridization and cultivation are virtually impossible, although recently some pyrenocarpous lichens were obtained in culture at the Centraalbureau voor schimmelcultures at Baarn, The Netherlands. This allows no objective testing of a given species concept. Therefore, we have to use the morphological species as an hypothesis for the evolutionary one.

Geographical races are often given the rank of subspecies. Generally, I follow this procedure and, in my opinion, the treatment as subspecies of two near relatives with different geographical distributions gives more information than treating them as totally separate. In the former case implicit information is provided about the nearest relative of both taxa and their supposed common ancestor. In the latter case, especially if a large genus is involved, the relationship is not obvious and differences tend to be stressed more than similarities.

All taxa of ranks above species level are, in my opinion, arbitrary. As long as they accord with the presumed phylogeny, they are of equal value. *In concreto*: given a presumed evolutionary history of a group with a certain number of species, the question of whether all these belong to the same genus (or family etc.) or to several genera (or families etc.) can not be answered on scientific grounds. The degree of similarity within a family or genus is not decisive and is not agreed upon, neither in lichenology nor elsewhere in taxonomy.

At present there is a tendency in lichenology and in taxonomy in general, towards small genera, which are assumed to be monophyletic (i.e. the common ancestor and its descendants). Recent examples are the work of Hafellner (1984) and Hale (numerous publications). In this study I have not followed this tendency in all cases, but I accept some genera in a very broad, and also presumably monophyletic sense, like *Verrucaria* and the genus *Pyrenula* in the circumscription of Harris (1989).

The choice between one large monophyletic, or several small monophyletic genera is arbitrary and based on convenience. As long as the phylogeny (and the taxonomy) at higher levels is far from clear, the choice of large genera is convenient.

## Phylogenetic method

Although phylogenetic analyses have been used several times in lichenology, various methods were used. Therefore, a brief introduction to the method used here is given.

In order to approach a natural classification, it is fundamental to trace evolution in a certain group of organisms. As it is found that the course of evolution leaves a trace in the form of shared characters in the descendant organisms (Wiley 1981) one can deduce, to a certain extent, the phylogeny from the characters of presently-living organisms. The study of fossils could add important information, but is not a prerequisite in order to achieve valuable results. This is fortunate for lichenology, as virtually no fossil lichens are known.

Not all characters are equally useful for this purpose, however. In general, dissimilarities are less important than similarities. A character that is not shared with any other species will give no clue to relationships. It is important to distinguish within a group of organisms relatively (within the group) original (primitive) character states, known as plesiomorphies, and relatively derived (advanced) character states, called apomorphies. It is obvious that a phylogenetic tree should be based only on shared, derived character states (synapomorphies) because, in that case, stress is given to the characters formed by evolution from the ancestral species. Within the group such characters might well be retained by all living descendants of a common ancestor and these species are, therefore, more related to each other than with any other species. It is postulated elsewhere (Wiley 1981) that it is not necessary to find out which characters are original and which derived before starting the deduction process, if enough characters are involved.

The procedure of reconstructing a phylogenetic tree is as follows: For each genus a species is selected, not necessarily the type species, but rather one in which all characters are to be found. Species descriptions are meticulously transformed into a data matrix with characters and species (Table 17.1). This is a critical step in the analysis. In order to

**Table 17.1.** Pyrenocarpous lichens

| | |
|---|---|
| Phyllop | 00101001000000000100000000001111 |
| Belonia | 00101000010000001010000000001111 |
| Protothe | 00101000010001011000000000001110 |
| Blastode | 10000100010001000100000000011111 |
| Pyrenula | 11100120100011011000000010001111 |
| Trypethe | 01110020100001200101000110001111 |
| Porina | 00101000010000000110001000001111 |
| Verrucar | 00000102000111012010000000001111 |
| Mycoporu | 00010000001011010000000000111111 |
| Tomasell | 00000001000011010000000000110011 |
| Leptorha | 00000000010010010000000000111011 |
| Arthopyr | 00000100100000010100000100011111 |
| Plagioca | 01000010100000011000000000001111 |
| Polymeri | 00000000100001200101000000001111 |
| cfSplanc | 11100010010000200000000000002111 |
| Lithothe | 01000120100000000100000000001111 |
| Litfalkl | 11000020100001101100000000001111 |
| Ciferrio | 00000001000000200110000100011111 |
| Santesso | 00000001000010010000000000111111 |
| Ditremis | 00000001000000100100010100001111 |
| Acrocord | 00000001000000100100000100001111 |
| Pyrgillu | 11100110100000000000000010001111 |
| Aspidoth | 00101000010100000100100000001111 |
| Pseudopy | 01000020100001200101000100001111 |
| Mycopyre | 11000120100000011000000000001111 |
| Eopyrenu | 10101000010001001100000000001111 |
| Strigula | 00101000010000001100000000001111 |
| brownDit | 10000001000000100000010000001111 |
| Phylloba | 00101000010000001001000000001111 |
| Thelenel | 00101000010002000000000000001111 |
| Julella | 00000000010012001100001000011111 |
| Chromato | 00000000011001001100000000001111 |
| Clathrop | 00101000010000001000010000011111 |
| Anthraco | 10100100001001011000000000002211 |
| Massaria | 11000020100001000100000000001111 |
| Micmicro | 10000101000000000100000000000211 |
| Dipyraus | 11000011000001101100000000001111 |
| Dipyrguy | 11000111000000011000000000001111 |
| Micropsi | 10101000100110120100000000001011 |
| Micauror | 11100121000011101000000000001111 |
| Splanchn | 11000010010012001000001000002111 |
| Mycoglae | 00001000010000001000000000001011 |
| Mycomicr | 10000001000000100000010000001111 |
| Pyrenoco | 00000001000002001000000000011111 |

| | |
|---|---|
| Porinula | 00101001000000100110000000001011 |
| Trichoth | 00101000100000001010000001111 |
| Campylot | 11100000001001200101001110001111 |
| Monoblas | 10000002000000100100010100001111 |
| Laurera | 11110000001001200101001101002111 |
| Astrothe | 01110020100001200101000111001111 |
| Cryptoth | 01110000001001200101000110002111 |
| Pleurotr | 00000000100001100000000100001111 |
| Turgidos | 00001002000111010010000000001111 |
| Phyllobl | 00100000001000100010000000001011 |
| Dipyrpap | 11000111000001200100010000001111 |
| Parathel | 01100011000000200100000000002111 |
| Pocsia | 00101000010011012010000000001011 |
| Macentin | 00101000010111012010000000001011 |
| Normandi | 00100000010011012010000000001101 |
| Agonimia | 00100000001011012010000000001111 |
| Endocarp | 00101000010111012010000000001101 |
| Dermatoc | 00101002000011012010000000001101 |
| Catapyre | 00100002000011012010000000001101 |
| Thrombiu | 00001002000101001011000000001110 |

Species

*Phylloporis phyllogena*
*Belonia herculina*
*Protothelenella sphinctrinoides*
*Blastodesmia nitida*
*Pyrenula dermatodes*
*Trypethelium ochroleucum*
*Porina mastoidea*
*Verrucaria muralis*
*Mycoporum elabens*
*Tomasellia eschweileri*
*Leptorhaphis epidermidis*
*Arthropyrenia rhyponta*
*Plagiocarpa illota*
*Polymeridium albidum*
cf. *Splanchnonema uberinum*
*Lithothelium cubanum*
*Lithothelium falklandicum*
*Ciferriolichen cinchonae*
*Santessoniolichen punctiformis*
*Ditremis limitans*
*Acrocordia gemmata*
*Pyrgillus javanicus*

*Aspidothelium fugiens*
*Pseudopyrenula subgregaria*
*Mycopyrenula coryli*
*Eopyrenula leucoplaca*
*Strigula wilsonii*
brown *Ditremis* sp. ined.
*Phyllobathelium epiphyllum*
*Thelenella modesta*
*Julella fallaciosa*
*Chromatochlamys muscorum*
*Clathroporina exocha*
*Anthracothecium prasinum*
*Massaria inquinans*
*Microthelia microsperma*
*Dipyrenis australiensis* ined.
*Dipyrenis guyanensis* ined.
*Microtheliopsis uleana*
*Microthelia aurora*
*Splanchnonema argus*
*Mycoglaena myricae*
*Mycomicrothelia subfallens*
*Pyrenocollema halodytes*

*Porinula tanzanica*
*Trichothelium epiphyllum*
*Campylothelium amylosporum*
*Monoblastia rappii*
*Laurera aurata*
*Astrothelium galbineum*
*Cryptothelium sepultum*
*Pleurotrema polysemum*
*Turgidosculum complicatulum*
*Phylloblastia dolichospora*
*Dipyrenis papuana* ined.
*Parathelium megalosporum*
*Pocsia septemseptata*
*Macentina perminuta*
*Normandina pulchella*
*Agonimia tristicula*
*Endocarpon pulvinatum*
*Dermatocarpon miniatum*
*Catapyrenium cinereum*
*Thrombium epigaeum*

Characters

0   Spores colourless = 0, brown = 1
1   Spores euseptate Only = 0, distoseptate = 1
2   Thallus endosubstratic = 0, episubstratic = 1
3   Ascocarps simple or fused = 0, in pseudostromata = 1
4   Ascocarp wall carbonized = 0, not so = 1
5   Clypeus absent = 0, present = 1
6   Spore cell lumina cylindrical = 0, rounded = 1, diamond-shaped = 2; ordered character
7   Spores more than 1-septate = 0, 1-septate = 1, simple = 2; ordered character
8   Spores not consistently 3-septate = 0, 3-septate = 1
9   Spores not transversely more than 3-septate = 0, more = 1
10  Spores not muriform = 0, muriform = 1
11  Periphyses absent = 0, present = 1
12  Paraphyses or paraphysoids present = 0, absent = 1
13  No gelatine in hamathecium = 0, gelatine = 1
14  Hyphae in hamathecium simple = 0, branched above only = 1, anastomosing = 2; unordered character
15  Asci cylindrical = 0, clavate = 1
16  Hamathecial gelatine IKI negative = 0, IKI + blue = 1, IKI + red = 2; unordered character
17  Ocular chamber absent = 0, present = 1
18  Asci thick-walled (seemingly bitunicate) = 0, thin-walled = 1
19  Hamathecium not inspersed = 0, inspersed = 1
20  Ascocarps not ornamented = 0, ornamented = 1
21  Spores not ornamented = 0, ornamented = 1
22  Spores IKI negative = 0, IKI + violet = 1
23  Spores without gelatinous sheath = 0, with = 1
24  Thallus without lichexanthone = 0, with = 1
25  Thallus without anthraquinone = 0, with = 1
26  Hamathecium filamentous = 0, non-filamentous = 1
27  Hamathecium hyphal = 0, cellular = 1
28  Spores small (less than 5 μm long) = 0, medium = 1, large (more than 100 μm long) = 2; ordered character
29  Ascomata small (less than 100 μm diam.) = 0, medium = 1, large (more than 4 mm diam.) = 2; ordered character
30  Thallus squamulose to foliose = 0, crustose = 1
31  Ascus tip IKI + blue = 0, IKI negative = 1

---

avoid the necessity of judgement in advance over original or derived characters, both the presence and the absence of a certain character are scored.

The inclusion of a certain, but not closely related taxon, called an outgroup, in the matrix is necessary to direct the tree. The choice of this outgroup is the object of much discussion, but my opinion is that it does not make much difference at all, as long as the outgroup is only distantly related to the treated group. The choice of an outgroup from a very closely related group of organisms gives potentially more information, but the risk is that this related group has in fact been evolved from the ingroup (the group under study). This should be avoided in all cases. This is supported by the present study, where several different outgroups were used during the analysis, which always gave comparable results.

The deduction is carried out by computer, using one of the currently available programs (in this case Hennig 86). For information on the selection criteria and other details of the program see Farris (1988). The choice of this program implies that analyses should be made of, at most, $c.$ 40 taxa at a time.

Finally, the resulting most probable (the most parsimonious) phylogenetic trees are selected, using such arguments as, presumably, more or less readily changed characters. This is done by an automatic character weighing. The resulting trees give not only the presumed evolution, but infer also direct information about the character changes which have occurred.

## Characters

It is probable that evolution within a group like the pyrenocarpous lichens has no sense or direction and, moreover, there seems to be no such thing as a constant rate of character change. Therefore, it is highly possible that certain characters have hardly changed during a certain period of evolution, whereas other characters may have changed considerably. At another time or in another group the opposite might have happened.

This leads to the following statement: There are no characters more fundamental than others. Of course there are certain characters which are heavily influenced by the environment and they should be handled with care. An example from the present study is the degree of immersion of an ascocarp in the substrate, which is largely correlated with the substrate rather than with taxa. But there is no *a priori* scientific reason to presume that a character like ascus tip structure is more fundamental and less readily changed during evolution than spore septation or hamathecium tissue.

The use of ontogenetic characters, much stressed during recent decades, is as limited as other methods. These can even lead to wrong conclusions, caused by such complications as neotenia (the retaining of juvenile characters) and paedomorphosis (the return of ancestral characters) or simply by the fact that the ontogeny does not necessarily reflect the phylogeny.

In order to compare various taxa, it is most important to ensure that the characters are comparable. In the available literature on pyrenocarpous lichens many ambiguities exist about certain characters, for instance, simple paraphyses with a few branches at the tips are sometimes treated as branched and sometimes as simple. In other cases, relevant characters are not cited at all. Therefore, it proved to be necessary to examine material of all the treated groups. For each genus one representative is taken, to make comparisons possible with several species, partly undescribed and partly wrongly assigned to certain genera, without known relatives.

The observations have been transformed in a data matrix (Table 17.1), but characters which occur in only one taxon have been omitted because they can not give any information about similarities and some characters have been combined into classes (measurements etc.). Finally, it must be decided whether so-called multistate characters are ordered, such as spore septation or unordered, such as the Iodine reaction.

Characters of the anamorph (pycnidia, conidia) are omitted because data were not available for many taxa and, if at any time data on anamorphs should be complete, they could be used in testing the cladograms obtained on the basis of teleomorphs.

The resulting data matrix gives a representation of the pyrenocarpous lichens as presently understood. In addition, a key is presented to the recognized taxa. As the data matrix is based only on characters of one species from each genus, the key gives additional information on the generic concept.

## Phylogenetic trees

For the phylogenetic analysis, taxa are divided in three groups for the following reason: The group treated within one analysis should preferably be monophyletic. Although it is never certain whether a certain group of taxa forms a monophyletic group or not, the presence of (or a combination of) a set of characters for the whole group is a good criterion.

*Group A*: comprises the Arthopyreniaceae *sensu lato*, the Trichotheliaceae and some presumably related groups, characterized by pyrenocarpous ascocarps, the presence of paraphyses or paraphysoids, and euseptate spores. These fungi are often lichenized.

*Group B*: comprises the Verrucariales, characterized by pyrenocarpous ascocarps, the absence of paraphyses, euseptate spores, and thin-walled asci. The hymenial gel is usually IKI + red. These fungi are nearly always lichenized.

*Group C:* comprises the Pyrenulaceae and the Trypetheliaceae, characterized by pyrenocarpous ascocarps, the presence of paraphyses or paraphysoids, the usually distoseptate spores, and thick-walled asci. These fungi are nearly always lichenized.

For the analyses of each separate group one genus of one of the other groups is each time taken as an outgroup.

## Interpretation of trees

The most parsimonious cladogram from group A after automatic weighting of characters (useful because this group represents only part of the original data matrix) is the following:

Some preliminary interpretations can be made from this cladogram. A constant group in all analyses, which is also present in this cladogram, is the group containing the genera *Santessoniolichen*, *Tomasellia*, *Mycoporum*, and *Leptorhaphis*, with *Arthopyrenia sensu stricto* as sister group. This group is characterized by the non-filamentous, cellular hamathecium and could be referred to as Arthopyreniaceae *sensu stricto*.

The other group of genera which were formerly placed in *Arthopyrenia* all show up in the vicinity of this group. They could be considered to belong in at least one order. For the group with *Monoblastia*, *Mycomicrothelia*, *Ditremis*, and the unnamed genus 'brown *Ditremis*' the family name Monoblastiaceae is available (Dr R.C. Harris, *personal communication*).

*Porina*, *Strigula*, and related genera are placed together with the other members of the Trichotheliaceae and could be treated as a separate family as well.

The position of the remaining genera (*Thrombium*, the unnamed genus with *Microthelia microsperma*, *Phylloblastia* and the genera that were formerly part of *Microglaena*) is still unclear. It is remarkable that the presence of IKI + blue-reacting parts of the ascus tip in *Thrombium* and *Protothelenella*, a character that unites these genera and probably should lead to direct comparing with non-pyrenocarpous ascomycetes, is not stressed in the cladogram.

The most parsimonious cladogram for group B (without weighting) is the following:

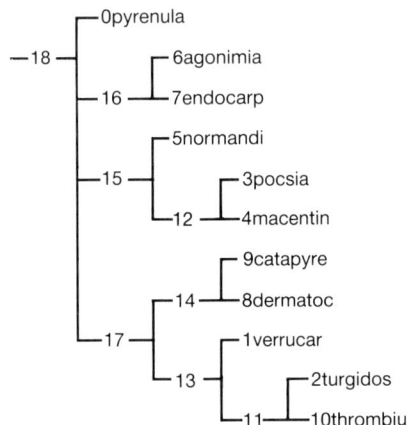

It is obvious that this group is very homogeneous and it is surprising that genera such as *Pocsia*, *Macentina*, *Turgidosculum*, and *Normandina* are not always believed to belong to the Verrucariaceae.

It is clear from the cladogram that spore septation is one of the main diversifying characters within the group. The position of *Thrombium*, which has been included in both subset A and B has not been clarified by cladogram A, but it seems a genuine member of the Verrucariaceae in this cladogram.

The consensus tree (a combination of several most parsimonious trees) after weighting from group C is the following:

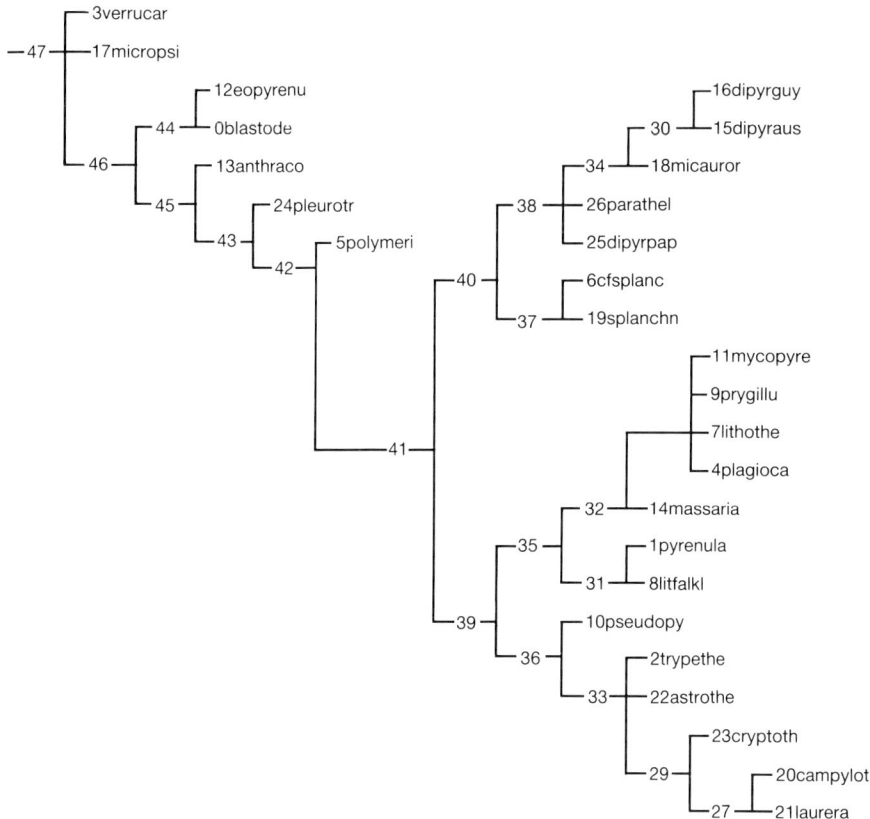

The Trypetheliaceae are supported by this cladogram only in a very restricted sense, including only *Pseudopyrenula*, *Trypethelium*, *Astrothelium*, *Cryptothelium*, *Campylothelium*, and *Laurera*. This can be interpreted as an intermediate solution to the controversy surrounding this family. The non-lichenized genera *Massaria* and *Splanchnonema* are both placed in different groups and the less characterized lichenized ('reduced') genera *Pleurotrema* and *Polymeridium* are even placed outside these groups.

The Pyrenulaceae are all placed together, except *Anthracothecium sensu stricto*.

A third family, for which no name seems to be available, except perhaps Pleomassariaceae, consists of lichenized and non-lichenized *Splanchnonema* and various *Dipyrenis* species.

*Microtheliopsis* is best accommodated in the Verrucariaceae, and was only included to avoid doubts.

As the character distribution of the presumed ancestors are provided by the cladogram, character transformations can be hypothesized. The distoseptate spore originated, according to the cladogram, in the common ancester of these three groups. The common ancestor of the Pyrenulaceae and the Trypetheliaceae had diamond-shaped lumens of the spore cells, the ancestor of the 'Pleomassariaceae' had rounded lumens. The ancestor of the Pyrenulaceae had brown spores, whereas the ancestor of the Trypetheliaceae had colourless spores.

The general conclusion is that a cladistic analysis results in a more sophisticated view of families that for the most part were already accepted. The results of the analysis of group C (Pyrenulales) were the most successful.

## Acknowledgements

I would like to thank Dr H.J.M. Sipman for general help with the project. Dr R.C. Harris for an introduction to the taxonomy of pyrenocarpous lichens and Mr R.V. Hensen for considerable help with the phylogenetic analysis. Dr R.Ch. Kruyt, Dr S.R. Gradstein and Professor Dr E. Hennipman are thanked for reading (parts of) the manuscript.

## References

Acharius, E. (1803). *Methodus qua omnes detectos lichenes* . . . F.D.D. Ulrich, Stockholm.

Clements, F.E. and Shear, C.L. (1931). *The genera of Fungi*. The Wilson Company, New York.

Eriksson, O.E. and Hawksworth, D.L. (1988). Outline of the Ascomycetes—1988. *Systema Ascomycetum*, **7**, 119–315.
Eschweiler, F.G. (1824). *Systema lichenum genera exhibens.* Norimbergae.
Farris, J.S. (1988). Hennis 86 Reference, version 1.5. unpublished.
Hafellner, J. (1984). Studien in Richtung einer natürlichen Gliederung der Sammelfamilien Lecanoraceae und Lecideaceae. *Beiheft zur Nova Hedwigia*, **79**, 241–371.
Harris, R.C. (1986). The family Trypetheliaceae (Loculascomycetes: lichenized Melanommatales) in Amazonian Brazil. *Suppl. Acta Amazonica*, **14**, 55–80.
Harris, R.C. (1989). A Sketch of the Family Pyrenulaceae (Melanommatales) in Eastern North America. *Memoirs of the New York Botanical Garden*, **49**, 74–107.
Watson, W. (1929). The classification of lichens. *The New Phytologist*, **28**, 97–116.
Wiley, E.O. (1981). *Phylogenetics.* John Wiley and Sons, New York.
Zahlbruckner, A. (1907). Ascolichenes. In *Die natürlichen Pflanzenfamilien* no. 1 (ed. A. Engler and K. Prantl), pp. 49–236.

# Appendix

*Key to treated taxa of pyrenocarpous lichens and related non-lichenized* Ascomycotina (only selected synonyms mentioned)

| | | |
|---|---|---|
| 1a | Ascomata in pseudostromata or aggregated with fused ostioles, may be completely immersed: dissect carefully | 2 |
| b | Ascomata solitary or aggregated without fused ostioles | 15 |
| 2a | Ascomata in pseudostromata, ostioles not fused | 3 |
| b | Ascomata aggregated with fused ostioles, sometimes also in pseudostromata | 8 |
| 3a | Mature spores transversely septate | 4 |
| b | Mature spores muriform | 7 |
| 4a | Spores distoseptate and euseptate | 5 |
| b | Spores euseptate only | *Tomasellia* |
| 5a | Mature spores colourless, only old spores becoming brown | *Trypethelium* |
| b | Mature spores brown | 6 |
| 6a | Paraphyses anastomosing, mature spores more than 40 μm long | cf. *Splanchnonema* (*Trypethelium uberinum* aggr.) |
| b | Paraphyses simple, mature spores less than 40 μm long | *Pyrenula sensu lato* (*Melanotheca*) |
| 7a | Thallus well-developed, lichenized | *Laurera* |
| b | Thallus only a whitish patch, non lichenized | *Mycoporum* |

| | | |
|---|---|---|
| 8a | Mature spores transversely septate | 9 |
| b | Mature spores muriform | 14 |
| 9a | Mature spores colourless, only old spores becoming brown | 10 |
| b | Mature spores brown | 13 |
| 10a | Mature spores 1-septate | cf. *Splanchnonema*(*Pleurotrema verrucosum*) |
| b | Mature spores more than 1-septate | 11 |
| 11a | Thallus well-developed, ascomata usually in pseudostromata or deeply immersed | *Astrothelium* |
| b | Thallus less developed, pleudostromata absent | 12 |
| 12a | Ascocarps free, only ostioles fused | *Plagiocarpa* |
| b | Ascocarps as well as ostioles fused | *Lithothelium* |
| 13a | Paraphyses anastomosing; mature spores more than 60 μm long | cf. *Splanchnonema* |
| b | Paraphyses simple, mature spores less than 60 μm long | *Pyrenula sensu lato* (*Parathelium* p. max. p., *Pleurothelium* and *Pyrenastrum*) |
| 14a | Mature spores colourless, only old spores becoming brown | *Cryptothelium* |
| b | Mature spores brown | *Pyrenula sensu lato* (*Parmentaria*) |
| 15a | Most ostioles lateral or skewed | 16 |
| b | Most ostioles apical | 25 |
| 16a | Mature spores transversely septate | 17 |
| b | Mature spores muriform | 23 |
| 17a | Spores distoseptate and euseptate | 18 |
| b | Spores euseptate only, septa thin | 20 |
| 18a | Mature spores colourless, only old spores may be brown | *Plagiocarpa* |
| b | Mature spores brown | 19 |
| 19a | Paraphyses simple; mature spores less than 100 μm long | *Pyrenula sensu lato* (*Pyrenastrum*) |
| b | Paraphyses anastomosing; mature spores more than 100 μm long | cf. *Splanchnonema* |
| 20a | Mature spores colourless | 21 |
| b | Mature spores brown | *Plagiocarpa* |
| 21a | Mature spores 1-septate | 22 |
| b | Mature spores 3-septate | *Pleurotrema sensu stricto* (i.e. *P. polysemum*) |
| 22a | Paraphyses anastomosing; spores more than 50 μm long | *Parathelium megalosporum* |
| b | Paraphyses only branched at the tips, spores usually less than 50 μm | *Ditremis* (incl. *Anisomeridium* and *Pleurotrema* p. max. p.) |

| | | |
|---|---|---|
| 23a | Mature spores colourless, only old spores brown | 24 |
| b | Mature spores brown | *Pyrenula sensu lato* (*Parmentaria*) |
| 24a | Thallus well developed, lichenized | *Campylothelium* |
| b | Thallus only a whitish patch, non-lichenized | *Thelenella* |
| 25a | Ascocarp wall with conspicuous spines, warts or a plate-like structure externally | 26 |
| b | Ascocarps wall smooth | 27 |
| 26a | Ascocarp wall with black to whitish spines externally | *Trichothelium* |
| b | Ascocarp wall with warts or a plate-like structure externally | *Aspidothelium* |
| 27a | Mature spores not septate, roundish to ellipsoid | 28 |
| b | Mature spores septate | 33 |
| 28a | Paraphyses present | 29 |
| b | Paraphyses absent | 30 |
| 29a | Ascus tip IKI + blue, spores not ornamented | *Thrombium* |
| b | Ascus tip IKI negative, spores ornamented | *Monoblastia* |
| 30a | Thallus squamulose to foliose | 31 |
| b | Thallus crustose | 32 |
| 31a | Thallus foliose | *Dermatocarpon* |
| b | Thallus squamulose | *Catapyrenium* (incl. *Placocarpus*, *Placopyrenium* etc.) |
| 32a | Ascocarp wall not carbonized; growing on marine algae | *Turgidosculum* (syn. *Mastodia*) |
| b | Ascocarp wall carbonized, often with clypeus or involucrellum; on other substrata | *Verrucaria sensu lato* (incl. *Amphoridium*, *Bagliettoa*, *Protobagliettoa* etc.) |
| 33a | Mature spores transversely septate only | 34 |
| b | Mature spores muriform | 73 |
| 34a | Paraphyses absent; hymenial gelatine IKI + red | 35 |
| b | Paraphyses present; hymenial gelatine IKI + blue, orange, negative or absent | 39 |
| 35a | Ascocarp wall carbonized, ascocarps medium sized (>0.3 mm) | 36 |
| b | Ascocarp wall not carbonized, ascocarps minute (<0.2 mm), thallus epiphyllous | 37 |
| 36a | Thallus well-developed, in most species even squamulose; ascocarps without clypeus or involucrellum | *Normandina* (incl. *Heterocarpon* p. p.) |
| b | Thallus endolithic; ascocarps often with clypeus or involucrellum | *Verrucaria sensu lato* (*Involucrothele* and *Thelidium*) |
| c | Thallus parasitic, spores 1-septate, with more than 10 per ascus | *Muellerella* |

| | | |
|---|---|---|
| 37a | Spores consistently 3-septate, becoming brown | *Microtheliopsis* |
| b | Spores 5- or more -septate, colourless | 38 |
| 38a | Periphyses present | *Macentina* |
| b | Periphyses absent | *Pocsia* |
| 39a | Spores distoseptate and euseptate | 40 |
| b | Spores euseptate only, septa thin | 51 |
| 40a | Spores 1-septate | *Dipyrenis* (currently under revision; several species are treated separate. *Microthelia aurora* belongs also here) |
| b | Spores more than 1-septate | 41 |
| 41a | Mature spores massing in a calyx outside the ascocarp, on top of it | *Pyrgillus* |
| b | Mature spores massing in the ascocarp or remaining in the asci | 42 |
| 42a | Mature spores colourless; paraphyses anastomosing | 43 |
| b | Mature spores brown; paraphyses anastomosing or not | 44 |
| 43a | Spore lumina rounded | *Pseudopyrenula* (syn. *Plagiotrema*) |
| b | Spore lumina angular to diamond-shaped | *Trypethelium* (*T. tropicum*) |
| 44a | Mature spores consistently 3-septate | 45 |
| b | Mature spores more than 3-septate | 49 |
| 45a | Thallus immersed in the substrate, only a whitish patch | 46 |
| b | Thallus on top of the substrate, usually greenish to brown | 47 |
| 46a | Ascocarp wall with clypeus; hymenial gel absent; on bark | *Mycopyrenula* |
| b | Ascocarp wall without clypeus; hymenial gel IKI + blue; on rock | *Lithothelium falklandicum* |
| 47a | Ascocarp wall carbonized, usually with a thickened clypeus | *Pyrenula sensu stricto* (incl. *Bottariomyces*, *Chrooicia* and *Starbaeckiella*) |
| b | Ascocarp wall carbonized or not, without clypeus | 48 |
| 48a | Mature spores with colourless end cells; thallus lichenized | *Eopyrenula* |
| b | Mature spores completely brown; thallus not lichenized | *Massaria* |
| 49a | Ascocarp wall carbonized, usually with a thickened clypeus | *Pyrenula sensu stricta* |
| b | Ascocarp wall carbonized or not, without clypeus | 50 |
| 50a | Mature spores with colourless end cells; thallus lichenized; paraphyses simple to slightly branched | *Eopyrenula* |
| b | Mature spores completely brown; thallus not lichenized; paraphyses anastomosing | *Splanchnonema* |
| 51a | Mature spores colourless, only old spores brown | 52 |
| b | Mature spores brown | 70 |
| 52a | Most mature spores 1-septate | 53 |
| b | Most mature spores more than 1-septate | 63 |

| | | |
|---|---|---|
| 53a | Spores more than 5 times longer than broad | |
| | *Leptorhaphis* (incl. *Pleurotrema* p. p.) | |
| b | Spores less than 5 times longer than broad | 54 |
| 54a | Hamathecium parenchymatous; no paraphyses formed | |
| | *Santessoniolichen* (*Arthopyrenia punctiformis* group) | |
| b | Hamathecium filamentous; paraphyses present | 55 |
| 55a | Paraphyses simple | 56 |
| b | Paraphyses branched | 58 |
| 56a | Paraphyses conspicuously septate, very broad | |
| | *Arthopyrenia sensu stricto* (*A. rhyponta* group, syn. *Naetrocymbe* and *Pyrenillium*) | |
| b | Paraphyses slender, not conspicuously septate | 57 |
| 57a | Ocular chamber absent; growing on leaf cuticula | *Phylloporis* |
| b | Ocular chamber present; if growing on leaves, than below cuticula | |
| | *Strigula* (incl. *Raciborskiella*) | |
| 58a | Paraphyses anastomosing | 59 |
| b | Paraphyses only branched at the top | 61 |
| 59a | Thallus with cyanobacteria; usually growing on seashores | |
| | *Pyrenocollema* | |
| b | Thallus with green algae | 60 |
| 60a | Mature spores more than 100 μm long; ascus wall thickened, without ocular chamber | *Parathelium megalosporum* |
| b | Mature spores less than 100 μm long; ascus wall thin, with ocular chamber | *Ciferriolichen* (*Arthopyrenia lapponina* and *cinchonae*-group) |
| 61a | Ascocarp wall not carbonized; thallus epiphyllous | *Porinula* |
| b | Ascocarp wall carbonized (rarely pale); thallus usually immersed in substrate | 62 |
| 62a | Spores more or less constricted at septa, usually one cell larger than the other | *Ditremis* (incl. *Anisomeridium* and *Pleurotrema* p. max. p.) |
| b | Spores not constricted at the septa, cells equal | *Acrocordia* |
| 63a | Ascocarp wall carbonized | 64 |
| b | Ascocarp wall not carbonized | 68 |
| 64a | Mature spores brown | *Blastodesmia* |
| b | Mature spores colourless | 65 |
| 65a | Paraphyses anastomosing, immersed in a gelatinous mass | |
| | *Polymeridium* | |
| b | Paraphyses simple or branched, not immersed in a gelatinous mass | 66 |
| 66a | Spores acicular to filiform or fusiform, almost as long as the ascus | |
| | *Leptorhaphis* | |
| b | Spores fusiform, less than half the length of the ascus | 67 |

| | | |
|---|---|---|
| 67a | Ocular chamber absent; ascus tip with two tiny dense spots | |
| | | *Porina* (incl. *Zamenhofia*) |
| b | Ocular chamber present; ascus tip hyaline | |
| | | *Strigula* (syn. *Sagediomyces*) |
| 68a | Ascocarp wall green; paraphyses branched | *Mycoglaena* |
| b | Ascocarp wall yellow to brown or red; paraphyses simple | 69 |
| 69a | Hymenial gelatine IKI −, mostly heavily inspersed | *Porina* |
| b | Hymenial gelatine IKI + blue, not inspersed | *Belonia* |
| 70a | Most mature spores 1-septate | 71 |
| b | Most mature spores more than 1-septate | *Eopyrenula* |
| 71a | Ascocarp wall carbonized, with clypeus; spores smooth | 72 |
| b | Ascocarp wall carbonized or not, without clypeus; spores often ornamented | *Mycomicrothelia* |
| 72a | Paraphyses anastomosing; most spores more than 10 μm long | |
| | | brown *Ditremis* (no generic name available) |
| b | Paraphyses simple; most spores less than 10 μm long | |
| | | *Microthelia microsperma* |
| 73a | Mature spores colourless, only old spores brown | 74 |
| b | Mature spores brown | 85 |
| 74a | Ascocarp wall carbonized | 75 |
| b | Ascocarp wall not carbonized | 82 |
| 75a | Paraphyses absent; hymenial gelatine IKI + red | 76 |
| b | Paraphyses present; hymenial gel IKI + orange, blue, − or absent | 78 |
| 76a | Thallus squamulose | *Endocarpon* |
| b | Thallus crustose to microphylline | 77 |
| 77a | Thallus very conspicuously granular to microphylline, greenish | |
| | | *Agonimia* |
| b | Thallus smooth, usually endolithic | *Verrucaria sensu lato* (*Amphoroblastia*, |
| | | *Polyblastia* and *Staurothele*) |
| 78a | Paraphyses branched to anastomosing | 79 |
| b | Paraphyses simple, thick | *Phyllobathelium* |
| 79a | Ascocarps minute (<0.3 mm); epiphyllous | *Phylloblastia* |
| b | Ascocarps medium sized (>0.3 mm); on other substrates | 80 |
| 80a | Ascus wall thick, without ocular chamber | *Thelenella* |
| b | Ascus wall thin, with ocular chamber | 81 |
| 81a | Spores with hyaline gelatinous sheath; thallus a whitish patch; non lichenized | *Julella* (*Polyblastiopsis* p. max. p.) |
| b | Spores without gelatinous sheath; thallus usually well-developed and lichenized | *Chromatochlamys* |
| 82a | Thallus squamulose | *Endocarpon* |
| b | Thallus crustose | 83 |

| | | |
|---|---|---|
| 83a | Ascus tip IKI + blue | *Protothelenella* |
| b | Ascus tip IKI − | 84 |
| 84a | Ascocarp wall green | *Mycoglaena* |
| b | Ascocarp wall yellowish to brown | *Clathroporina* |
| 85a | Paraphyses branched | *Thelenella* |
| b | Paraphyses simple | 86 |
| 86a | End cells of spores colourless | *Eopyrenula* |
| b | End cells of spores brown | 87 |
| 87a | Spores euseptate and distoseptate, or only distoseptate | *Pyrenula sensu lato* (*Anthracothecium* p. p.; *Pleurotheliopsis*) |
| b | Spores euseptate only | *Anthracothecium sensu stricto* |

# 18. Epilogue

T.D.V. SWINSCOW
24 Monmouth Street, Topsham, Exeter EX3 0AJ, UK

In drawing the Conference to a conclusion I must begin by expressing the thanks of all of us to its organizer, Dr David Galloway. It has attracted many of the world's leading lichenologists with a special interest in the tropics and some also to whom that region is unfamiliar. As well as organizing it with exemplary precision, he has picked just the right time for it.

The study of the lichens of the tropics by scientific methods is now well under way, as many contributors have shown, so that a firm basis of experience exists from which we can get a picture of today's problems in the tropics and perhaps a glimpse of future lines of research. We are all acutely aware of how tragically uncertain is the future of natural habitats in the tropics, whether from the impact of man's tools locally, or from the influence of climatic changes that he may be bringing about. As several speakers illustrated, the rainforest is but one of its characteristic habitats that man, the most competitive species on earth, is killing off. This is the first conference on Tropical Lichens: how many more will there be?

When we look at the earth's tropical zone, we see that it contains relatively little land: most is sea. Much the largest single land mass is tropical Africa. And what makes the tropics of such great interest to lichenologists, as to biologists generally, is the enormous variation in the terrain, from sea coasts to snow-capped mountains, from rainforest to desert, from sphagnum swamps to wind-swept heaths. But in the next 50 years the most pervasive influence on the growth and distribution of lichens will not be natural forces but the pressures of human populations whose expansion is not under rational control.

Though man is believed to have evolved originally in the tropics, it is the temperate regions of the world that have subsequently been most congenial to his species. To a medical man such as myself, this is a striking feature of human existence. The reason for it is the ease with which life of all kinds flourishes in the tropics—and especially those parasites that are harmful or even lethal to man. The human species has largely failed to develop an effective immune system to overcome infections from protozoa and helminths so that, until the era of chemotherapy, profound ill health was the norm for whole populations in tropical countries. The result was, and still is, great human poverty. However, the improvement in health has burdened the tropics with another problem, namely population growth, so to deep poverty must be added high fertility and the one interacts with the other.

Many of the poorest people in the world live in the tropics. Their disease load is still heavy, their nutrition inadequate, their lives short. It is to alleviate the misery or at least uncertainty of their existence that they are cutting down their forests. Driven by dire economic necessity, they win a few years' fertility from the soil at the cost of its total waste thereafter. Obviously there is more than a problem for bankers here. It is a problem surrounded by social and political hopes, philosophical arguments, moral judgements, and religious beliefs.

In passing, perhaps I may mention that I have noticed that, of about 60 people attending this conference, only two live in the tropics—and one of those only just qualifies in that his house is actually on the Tropic of Capricorn!

I felt the complexity of these forces one evening in Uganda, when I was having dinner at Makerere University with members of the staff—all Ugandans. One young man asked me what I was doing in Uganda. I told him that I was collecting lichens to make a study of them. He replied, 'I don't see what good your studies will do my country.'

This is a remark that I have often thought about. He posed a question that I believe every research worker from a developed country visiting a poor country should consider. In essence, what that Ugandan meant was, 'How will your studies make life easier for those of us who live here?' And it would be natural for him to think of agriculture, or medicine (which had primarily taken me to Uganda), or manufacture.

But should all research be directed to obviously useful, specifically-defined ends in a poverty-stricken country? Some, no doubt, should be —as it should everywhere. Furthermore, in replying to such a question we may emphasize the 'spin-off,' the accidentally useful discoveries, that often follow from basic research. Certainly these consequences of research should get due emphasis in developing countries—perhaps more than in those that are technologically advanced. But I also believe that basic or pure research, research without a defined goal of usefulness, is a valuable human activity in itself, like dancing, or composing songs, or playing drums in religious festivals—all of which are popular in Africa for instance. And I think it wrong that poor countries should be excluded from such scientific endeavour. It is patronizing the poor to think they must be associated only with research directed to improving their material welfare.

But the consequence of this approach is that we, as explorers and researchers, must accept an obligation. In the countries we visit we should try to interest the local people in our studies. We must meet them, engage their attention, invite them to join us on our trips, seek their help, explain to them what we are doing, encourage them to do such work themselves. Then we should publish our findings and send our papers and books to interested workers and libraries out there. Failure to publish our findings may leave in ignorance the people who were our generous hosts when we visited the country. This may seem obvious, yet I have personally known two such cases which caused sorrow and vexation in East Africa.

Many of you are in academic posts and you teach young men and women. I do urge those of you who come from relatively well-off countries to explain this obligation to your students or young colleagues. Explain to them that, when they go to a poor country to carry out their research, they have obligations of an academic nature to the people who live there. They should try to arouse the interest of the people they meet and encourage them to take part in the study, and help them to carry on the work when the researcher returns home. Then the researcher should not forget to send them his papers and books and perhaps a set of specimens.

In that way, such *use* as scientific work may have for a country can be brought to the attention of its inhabitants. But more—and this is the important thing!—the *excitement* of scientific work will arouse the imagination of those people. It will open their minds to the idea that the natural environment is part of their deepest lives. And it will draw them into that great international body of friends and colleagues and rivals who constitute the scientific community—a part of which community has been meeting here this week with great enjoyment.

# Index

*Abies alba* 226
*Acacia* 130
  *koa* 38
Acanthaceae 205
Acarosporaceae 136
*Acer rubrum* var. *drummondii* 175
*Acrocordia* 271
  *gemmata* 259
*Actinoplaca* 100, 136
  *strigulaceae* 146
Admiralty Islands 69
advanced character, *see* apomorphy
Africa 5, 9, 32, 64, 89, 90, 96, 97, 114, 168, 208, 213, 219
  East Africa 30, 54, 58, 85, 89, 91, 92, 107, 111, 232
  North Africa 64, 107
  Southern Africa 90, 107, 109, 111, 114
  South-west Africa 110, 116
*Agonimia* 272
  *tristicula* 259
agricultural practices 33, 44, 197, 218, 241
agricultural monocultures 4
*Alectoria ochroleuca* 129, 131
Alectoriaceae 136
*Aleurites moluccana* 37, 43
algae, *see* photobiont
alien species, *see* naturalized species
alpine habitat
  desert 39, 130
  grassland 3, 39, 74, 75, 76, 80
  páramo 125, 129, 130, 131
  shrubland 39, 225, 226, 228, 232
*Amaplelopris* 213
Amazon basin 144, 148
Amboina Island 4
*Amelanchier* 195
America 57, 97, 213, 214
  North America 8, 64
  South America 54, 64, 65, 81, 90, 109, 111, 123, 168, 208, 219, 232
amphitropical taxa 7–8
*Amphoridium* 269
*Amphoroblastia* 272
*Anaptychia* 221
ancestral species 64, 109, 257, 266

ancient woodlands 217, 237, 240, 241
  indicator species 194
Andes 123–4, 125
*Anisomeridium* 137, 268, 271
  *foliicolum* 146
  *subprostans* 175, 178, 181, 183, 186
  *tuckerae* 175, 176, 178, 181, 183, 186
Anonaceae 207
Antarctica 64, 65
*Anthocephalus* 240
*Anthracothecium* 137, 266, 273
  *prasinum* 259
  *subochraceum* 37
antitropical taxa 7–8
*Anzia* 79, 125, 128, 228
  *afromontana* 92, 93
  *formosana* 6
  *japonica* 6
  *madagascarensis* 6
  *semiteres* 6
angiosperms 64
*Apatoplaca* 117
Aphyllophorales 138, 158
apomorphy 257, 260
apothecia, *see* ascomata
Appenines 217
appressoria 185
*Araucaria* 57, 58, 64, 65, 74, 82
Araucariaceae 64
Argentina 109
aroids 227
*Arthonia* 136, 149, 172, 187, 188
  *accolens* 97, 138, 146
  *aciniformis* 138, 146, 149
  *calamicola* 96, 97
  *cyanea* 97, 146
  *pellicula* 7
  *radiata* 174
  *rubella* 174, 176, 178, 181
  *trilocularis* 138, 146
  *tumidula* 173, 174, 178, 181, 183
Arthoniaceae 136, 173, 174, 187, 188
Arthoniales 136
*Arthopyrenia* 137, 264
  *lapponina* 271
  *punctiformis* 271

*Arthopyrenia (cont.)*
  *rhyponta* 259, 271
Arthopyreniaceae 137, 253, 262, 264
*Arthothelium* 136
*Arthrorhaphis* 81
  *alpina* 80, 131
  *citrinella* 8, 131
arctic region 172
arctic-alpine region 248
artificial groups 255, 256
asci 19–21, 98, 100, 116, 205, 206, 207, 208, 209, 210, 212, 213, 214, 215, 245, 247, 248, 249, 254, 260, 261, 262, 263, 264
ascomata 19–21, 31, 98, 99, 100, 116, 205, 206, 207, 208, 209, 210, 212, 213, 215, 245, 246, 247, 260, 261, 262, 263
  ontogeny 255
  ornamentation 19, 26, 27, 28, 32, 110, 111, 112, 113
Ascomycotina 105, 107, 116, 136, 153
ascospores 19–21, 26, 98, 99, 205, 206, 207, 210, 211, 212, 213, 214, 215, 254
  discharge 116
  ontogeny 245–50
    synapomorphic 250
  ornamentation 245, 246, 247, 248, 249, 260
  septation 100, 116, 254, 260, 261, 263, 265, 266
Asia 2, 8, 30, 81, 197, 204, 213, 214
  Eastern Asia 81, 109
  Central Asia 30
  Southeast Asia 2, 5, 8, 9, 18, 23, 24, 30, 32, 57, 58, 64, 65, 81, 82, 90, 208, 217–41
*Aspicilia* 137, 186
  *alpina* 8
  *subsorediza* 8
Aspidotheliaceae 138
*Aspidothelium* 138, 254, 269
  *fugiens* 146, 259
Asterothyriaceae 136, 205
*Asterothyrium* 136
  *pittieri* 205
*Astrothelium* 127, 137, 266, 268
  *galbineum* 259
atmospheric pollution 45, 103, 152, 194–5
  acid rain 195, 218
*Aulaxina* 100, 136
  *epiphylla* 97
  *minuta* 126, 146
  *quadrangula* 146
  *submuralis* 97, 139, 147
austral taxa 9–10
Australasia 7, 17–34, 81, 107, 109, 246
Australia 2, 4, 5, 9, 26, 28, 30, 33, 47, 48, 51, 54, 57, 58, 64, 65, 73, 75, 80, 97, 168, 232
  Queensland 6, 18, 57, 63, 64, 65, 82
Australian–Indian plate 2
*Avicennia* 75, 77
  *marina* 51
Azores 194

*Bacidia* 81, 137, 204
  *aphiatica* 100, 139, 147, 205
  *brasiliensis* 147
  *dimerelloides* 97
  *palmularis* 97, 139, 147, 149, 206
  *scutellifera* 100, 101
  *stanhopeae* 139, 147
  *subsimilis* 96
  *ziamensis* 206
Bacidiaceae 137, 138, 205–6
*Badimia* 100, 102, 137
  *dimidiata* 147, 149
*Baeomyces* 80, 125, 126
  *fungoides* 127
  *imbricatus* 127
  *rufus* 131
*Bagliettoa* 269
bamboo, *see Nastus*
bark pH 195–6, 225, 226
*Barubia* 100, 102
Basidiomycotina 138, 158
*Bazzania* 38
Belgium 63
*Belonia* 136, 272
  *herculina* 259
*Betula* 195
*Biatora* 137
*Biatorella* 136
  *conspersa* 142
biocides
  herbicides 197
  pesticides 197
    insecticides 241
biogeography 4–10, 30, 48, 63–5, 80–2, 168, 229–32
bioindicators 103
biomass 38, 39, 40, 163, 165
biota 7
bipolar element 7, 81
Birds of Paradise 83
Bismark archipelago 69
*Blastodesmia* 271
  *nitida* 259
boggy areas 38, 152
*Bombax* 73
boreal taxa 8
Borneo 4, 48, 63, 218, 224, 226

botanic gardens 193–8, 206, 207, 208, 213, 214, 215
*Bottariomyces* 270
Bougainville Island 69, 71
*Bougainvillea* 77
Brazil 5
  São Paulo State 151–69
*Brigantiaea* 137, 240
  *leucoxantha* 130, 139
Brigantiaeaceae 137
*Bruguiera* 74
bryophytes 38, 73, 74, 92, 100, 127, 217, 228
*Bryoria* 79, 195, 228
  *indonesica* 6
  *smithii* 38
*Buellia* 37
  *tabacina* 142
*Bulbothrix* 30, 32, 33, 137, 154, 160, 163, 165, 166, 168
  *goebelii* 5, 128, 155
  *isidiza* 5, 155
  *meizospora* 92
  *tabacina* 5
*Bullatina* 100
bunchgrass, *see* tussock grass
Burma 65, 201
*Byssocaulon* 136
*Byssolecania* 127, 137, 149
  *deplanata* 147, 149, 204, 212–13
  *fumoso-nigricans* 97, 139, 147
*Byssoloma* 137, 148, 149, 204
  *aeruginascens* 139, 144, 147, 149
  *laevigatula* 156
  *leucoblepharum* 97, 127, 139, 147, 149
  *olivierai* 156
  *subdiscordans* 97, 147, 149
  *tabacina* 156
  *tricholomum* 96, 147
  *usambarense* 96, 97, 98, 99

cacti 131
*Calathaspis* 80, 81
  *devexa* 6
*Calenia* 100, 127, 136, 149
  *conspersa* 147
  *submaculans* 147, 149
*Caleniopsis* 100
Caliciaceae 136
Caliciales 136, 245
*Calicium* 136
  *abietinum* 7
  *glaucellum* 7
  *hyperelloides* 5, 139
  *salicinum* 7
*Calopadia* 100, 102, 137, 145, 204

  *puiggarii* 97
*Caloplaca* 107, 113, 114–16, 117, 131, 137
  *aurea* 117
  *bonae-spei* 114, 115
  *calicioides* 117
  *carpinea* 117
  *cerina* 117
  *cinnabarina* 78
  *congrediens* 117
  *eudoxa* 114, 115, 116, 117
  *flavescens* 117
  *inconstans* 36, 37, 40
  *mauritanica* 114
  *nivalis* 117
  *ochracea* 117
  *perminuta* 41
  *poliotera* 36
  *saxicola* 117
  *variabilis* 117
  campylidia 100
*Campylothelium* 137, 266, 269
  *amylosporum* 259
Canary Islands 116
*Candelaria*
  *concolor* 7, 37, 40
*Canoparmelia* 137
canopy 18, 39, 58, 64, 77, 78, 79, 91, 149, 226, 227, 228, 232, 254
Capparidaceae 206
*Capparis*
  *caudata* 206
*Carbonea*
  *vorticosa* 8
Caribbean area 63
*Carpinus*
  *caroliniana* 174
*Carya* 195
*Castanea* 217, 226
*Castanopsis* 33, 81, 218, 225
  *acuminatissima* 74
*Casuarina* 78, 225, 228–9
  *junghuhniana* 241
*Catapyrenium* 269
  *cinereum* 259
  *lachneum* 7
*Catillaria*
  *endochroma* 128
  *ochraceonigra* 37
*Catinaria* 137
  *versicolor* 130, 139, 142
*Celtis*
  *giganticarpa* 208, 213, 215
  *laevigata* 174
*Cephaleuros* 175, 176, 178, 187
  *virescens* 176, 187
cephalodia 248
*Cephalophysis* 117

*Ceriops* 51
  *tagal* 51, 77
*Cetraria* 221, 228
  *delisei* 131
  *islandica* ssp. *antarctica* 80
*Cetrariastrum*
  *africanum* 92
  *andense* 128
  *dubitans* 128
  *ecuadoriense* 128
  *vexans* 92, 93
*Cetrelia* 79, 81, 221, 227, 228
  *braunsiana* 9
*Chaenotheca* 136, 188
  *brunneola* 7, 139
*Chaenothecopsis pusilla* 7
chemical lichen substances, *see* secondary metabolites
chemosyndromic variation 24, 25, 26, 29
chemotaxonomy 22–3
Chile 5
China 219
  Fujian 213, 214
  Guangdong 204, 205, 208
  Hunan 213
  Yunnan 201–16
*Chiodecton* 137
  *perplexum* 40
  *sanguineum* 128, 172, 185
Chlorococcaceae 174
*Chlorococcus* 188
chloroplasts 176
  grana 176
  haematochrome droplets 176, 178
*Chorisia insignis* 131
*Chromatoclamys* 272
  *muscorum* 259
*Chroodiscus mirificus* 97
*Chrooicia* 270
Chrysotrichaceae 136
*Chrysothrix* 136
  *candelaris* 7, 40, 139
*Chusquea* 129
*Ciferrolichen* 271
  *cinchonae* 259, 271
cilia 87, 89
  bulbate 30–1
*Cladia* 81, 89, 130, 137, 143
  *aggregata* 5, 79, 90, 127, 131, 139, 143
  *fuliginosa* 131
  *retipora* 7
*Cladina* 125, 130, 137, 143
  *arcuata* 131
  *boliviana* 131
  *confusa* 9, 131, 142
  *dendroides* 144
  *densissima* 144
  *halei* 9
  *polia* 127
  *skottsbergii* 38
  *spinea* 144
  *sprucei* 142
cladistic analysis 254
cladograms 117–18, 261, 262–6
*Cladonia* 74, 79, 91, 125, 129, 137, 143, 165, 228
  *aleuropoda* 131
  *andesita* 131
  *calycantha* 127
  *ceratophylla* 130, 143, 144
  *chlorophaea* 131
  *didyma* 131
  *diplotypa* 92, 93
  *furcata* 131
  *lopezii* 127
  *macrophyllodes* 131
  *mexicana* 131
  *ochrochlora* 130
  *pleurota* 7
  *pocillum* 131
  *scrabiuscula* 38, 39
  *secundana* 144
  *signata* 142, 143
  *subsquamosa* 126, 131
  *verruculosa* 131
Cladoniaceae 81, 136, 138, 143, 144, 165
*Clathroporina* 137, 273
  *exocha* 259
*Cleidion brevipetiolatum* 212
climate 2, 71–2, 81–2, 85–6, 219, 223–4
  city-desert 37, 44
  dry 2, 16, 37, 42, 142
  equatorial 71, 152
  mesic 37, 41, 42
  oceanic 71
  semiarid 2
  temperate 18, 27, 30
    continental 226
    cool temperate 2, 9–10, 64, 65
    mediterranean 226
    warm temperate 65
  tropical
    cool wet 9, 41, 237
    monsoon 2, 72, 219, 226, 228, 237, 240
    warm wet 73, 78, 237
climbers 227, 228
cloud forest 2, 10, 38, 39–40
cloud zone 38, 72, 224
Clusiaceae 126
coastal cliffs 36, 163, 166
  hills 165–6
*Coccocarpia* 78, 80, 81, 125, 127, 137, 154, 161, 163, 164, 165, 197, 219, 220, 226, 227

*adnata* 6
*aeruginosa* 9
*asterella* 155
*dissecta* 6
*domingensis* 9
*endoferruginea* 7
*epiphylla* 147
*erythrocardia* 9, 128, 139
*erythroxyli* 5, 37, 42, 77, 78, 128, 155
*flavicans* 130
*fulva* 6
*glaucina* 9, 78
*palmicola* 5, 77, 130, 155, 195
*pellita* 5, 78, 128, 155
*pruinosa* 6, 9
*rottleri* 6, 9
*smaragdina* 6, 9
*stellata* 127
*tenuissima* 147
Coccocarpiaceae 137, 155, 159
*Coccotrema* 9
*Coelocaulon muricatum* 131
*Coenogonium* 78, 136, 154
    *implexum* 172
    *leprieurii* 128, 153
    *moniliforme* 153
*Collema* 125, 131, 137, 218, 220, 227
    *actinoptychum* 9
    *coilocarpum* 6
    *conglomeratum* 132
    *glaucophthalmum* 130
    *japonicum* 9
    *kauaiense* 5
    *leptaleum* 5, 128
    *neglectum* 132
    *rugosum* 9, 37, 77
    *texanum* 132
Collemataceae 137, 155, 159, 217, 220
Colombia 123, 145, 148, 219
Combretaceae 153
Comore Islands 90
compacted sand dunes 41
*Compsocladium* 80, 81
    *archiboldianum* 6
concentric bodies 185, 246, 247, 249
conidia 21, 32, 87, 88, 89, 98, 100, 102, 213, 262
conidiomata 213
Coniocybaceae 136
conservation 33–4, 82–3, 119–20
conservation areas 83
Cook Islands 48
*Cora* 154, 161
    *pavonia* 158
cork oak, *see Quercus suber*
cortex 21, 32, 109, 110, 111, 112, 113, 114, 115, 116

    epicortex 21, 32
    lower cortex 110
    upper cortex 21, 27, 107
*Corylus* 226
cosmopolitan taxa 7, 37
creepers 240
cretaceous Southern Hemisphere range 109
*Crocynia* 137, 227
    *pyxinoides* 139
Crocyniaceae 137
*Croodiscus* 149
    *coccineus* 127, 147, 149
*Croton* 131
*Cryptothecia*
    *candida* 5
    *obtecta* 240
*Cryptothelium* 137, 266, 268
    *sepultum* 259
cultivation 193, 197–8
cushion plants 9
*Cyathea* 79
*Cyclographina* 136
cyphellae 232


*Dacrycarpus* 64, 74
*Dacrydium* 225, 228
*Dactylis glomerata* 38
data matrix 255, 257–9, 262, 263
deforestation 93, 152, 169, 194, 204
*Dendriscocaulon* 137
D'Entrecasteaux Archipelago 69
derived taxa 63
    character, *see* apomorphy
*Dermatocarpon* 172, 186, 269
    *miniatum* 259
*Deschampsia australis* 39
*Dictyonema* 79, 138, 154, 226, 227
    *glabratum* 126
    *sericeum* 128, 158
Dictyonemataceae 138, 158, 159
differentiation
    chemical 106, 107
    structural 106, 107
*Dimerella* 98, 136, 149
    *epiphylla* 5, 97, 99, 205, 206
    *fallaciosa* 97
    *flavicans* 97, 98, 99
    *hypophylla* 97, 140, 147
    *lutea* 128
    *pocsii* 97, 98, 99
    *tanzanica* 97, 98, 99
    *usambarensis* 97, 98, 99
    *zonata* 37, 42
*Diploschistes* 125, 130
    *cinereocaesius* 131
    *muscorum* 131

## 284  Index

*Diploschites (cont.)*
   *scruposus* 7
*Diplotomma alboatrum* 41
Dipterocarpaceae 63
dipterocarps 57, 226, 227, 228, 240
   forests 58, 64, 237
*Dipyrenis* 266, 270
   *australiensis* 259
   *guyanensis* 259
   *papuana* 259
*Dirina catalinariae* 9, 36
   f. *sorediata* 9
*Dirinaria* 77, 78, 91, 154, 160, 161, 163, 169, 197
   *aegelita* 36, 37, 40, 78, 142, 157, 167
   *applanata* 5, 36, 37, 40, 157
   *caesiopicta* 6
   *confluens* 5
   *flava* 157
   *picta* 5, 37, 142, 157, 161, 165
   *purpurascens* 142
   *subconfluens* 6
Discomycetes 136–7, 173
disjunct patterns 8, 109, 114, 119
dispersal 63
distribution patterns 4, 8, 9, 30, 32–3, 48, 51–65, 81, 82, 109, 111, 141–4, 233, 234, 235, 236, 238, 239
disturbance 41–2, 73, 80
   recreational disturbance 44
*Ditremis* 137, 259, 264, 271, 272
   *limitans* 259
diversity 37, 38, 40, 41, 43, 44, 80, 81, 163, 164, 221, 237, 253
   centre of diversity 229
Dothideales 137
dryland forest 43
*Drypetes indica* 213
dust 44
dwarfed shrubs 9

*Echinoplaca* 100, 136, 145, 149, 204
   *affinis* 140, 147, 149
   *diffluens* 140, 147
   *heterella* 140
   *intercedens* 96
   *pellicula* 97, 127, 140, 147
ecological continuity 194, 218, 241
   Revised Index of Ecological Continuity (RIEC) 218
Ectolechiaceae 100, 102, 137, 206
Ecuador 123–33
Eire 237
*Eleiodoxa* 227
endangered habitats 93

endangered species 33–4, 36–7, 40, 41, 93, 196
endemism 6–7, 23, 26, 30, 36, 38, 39, 40, 41, 86, 97, 110, 111, 144, 232
*Endocarpon* 138, 188, 254, 272
   *pulvinatum* 259
   *pusillum* 41
endoperidermal species 172, 173
*Enterographa* 137
environmental management 82
*Eopyrenula* 270, 272, 273
   *leucoplaca* 259
Epacridaceae 64
equatorial taxa 8
   zone 54, 85
Ericaceae 64, 74, 81
*Erioderma* 79, 81, 125, 127, 130, 137, 142, 154, 161, 164, 165, 167
   *glaucescens*, see *leylandii*
   *leylandii* 82, 92
   *meiocarpum* 93
   *pulchella* 38
   *sorediatum* 5, 9, 92, 140
   *tomentosum* 194
   *unguigerum* 155, 164
   *verruculosum* 128, 140
   *wrightii* 128, 155
*Eschatagonia* 137
*Eucalyptus* 73, 77
   *terticornis* 78
*Eucryphia* 229
*Eugenia cuminii* 37, 42
*Eumitria* 54
Euphorbiaceae 213
Europe 57, 64, 188, 194, 217, 218, 220, 226, 228, 232, 237, 241
*Eusideroxylon conferta* 227
*Evernia* 185
   *prunastri* 183
*Everniastrum* 125, 127, 128
   *arsenei* 128
   *arvidssonii* 130
   *catawbiense* 130
   *cirrhatum* 130
   *colombiense* 130
   *fragile* 130
   *latilobum* 128
   *nigrociliatum* 130
   *sorocheilum* 130
   *vexans* 130
*Everniopsis trulla* 131
evolutionary processes 101, 105–20, 261
   divergence 24, 26–9, 33, 106
   macroevolution 106, 116–18
   parallel evolution 109
   *see also* speciation
evolutionary rates 107, 118, 119

evolutionary trends
  chemical 24
  morphological 22
*Excoecaria agallocha* 51
expeditions 102-3, 124, 204
explorations 3-4, 74, 96, 124-5, 135-6

Fabaceae 38, 42
fagaceous forests 232, 237
  trees, *see* Fagaceae
Fagaceae 217, 218, 225, 227, 232, 238
*Fagus* 217, 226, 237
*Fellhanera* 137
  *bouteilli* 97
  *cateila* 97
  *semecarpi* 213
feral animals 38, 39, 40, 42, 43, 44
ferns 73, 74, 79, 96, 127, 201
fertilizers 197, 241
Fiji 3, 4, 6, 33, 54
fire 38, 39, 40, 43, 44, 74, 77, 79, 82, 141, 228, 240-1
*Flavoparmelia ecuadoriensis* 131
*Flavopunctelia flaventior* 130
floristics 101
foliicolous lichens 38, 95-103, 125, 126, 143, 144, 145-9, 172, 173, 201-16, 254
forest management practices 33, 218, 227, 237, 240-1
Formosa, *see* Taiwan
fossil records 63, 64, 116, 119, 239
*Fraxinus* 226
*Freycinetia* 74
frog, *see Leptopeltis uluguruensis*
*Fulgensia* 117
  *australis* 117
  *fulgens* 117
fungal-algal interactions 185-9

Galápagos 2, 3, 4, 9, 124
gene banks 198
genotype differentiation 106
geology 1-2, 71, 153
  geological periods 63-5, 71, 119, 239
*Gingko biloba* 119
global warming 4
*Glyphis* 136
  *cicatricosa* 5, 37
Gomphillaceae 100, 136, 206-7
Gondwanaland 2, 8, 64, 65, 81, 119, 219, 237
Graphidaceae 37, 127, 136, 138, 153, 165, 173, 174, 187, 188
Graphidales 136, 138
*Graphina* 136, 241

  *insulana* 7
*Graphis* 240
  *anfractuosa* 40
  *scripta* 174, 178, 181, 187, 188
grassland 40, 43, 74, 79
Great Britain 64, 69, 217, 226
growth rate 149, 172, 226
Guianas 135-49
  Guyana 135, 136, 138, 139, 140, 141, 142, 143, 144, 148
  French Guiana 136, 139, 140, 141, 142, 143, 144, 149
  Surinam 136, 139, 140, 141, 142, 143, 144
Guinea 97
Gyalectaceae 98, 136, 153, 207
Gyalectales 136
*Gyalectidium* 100
  *caucasicum* 97
  *filicinum* 97, 147, 206-7
*Gyalideopsis* 100, 136
*Gymnoderma melacarpum*, *see Neophyllis melacarpa*
gymnosperms 64, 81
  conifers 42, 74, 79, 81
*Gyrostomum* 136

habitat
  degradation 42
  disturbance 41-2
  preservation 103
*Haematomma* 137
  *leprarioides* 140
  *punicea* 40
Haematommataceae 137
hamathecium 254, 260, 261
  paraphysis 100, 205, 208, 212, 214, 246, 247, 249, 260, 262, 263
  paraphysoids 260, 262, 263
  periphyses 260
haustoria 172, 176, 181, 183, 185-7
  intracellular 185, 186
  intraparietal 185, 186, 187
Hawaiian Islands 2
  Hawaii 3, 4, 6, 35-45, 54
  Tahiti 3, 4, 48, 54
*Helminthocarpon* 137
*Heterocarpon* 269
*Heterocyphelium leucampyx* 5
*Heterodea muelleri* 5
*Heterodermia* 125, 127, 130, 143, 154, 160, 161, 165, 195, 197, 218, 221, 226, 228
  *barbifera* 80, 128, 142
  *casarettiana* 130
  *circinalis* 127, 130
  *comosa* 128, 157
  *corallophora* 128, 157
  *dendritica* var. *propagulifera* 157, 161

*Heterodermia (cont.)*
  *diademata* 80, 228
  *flabellata* 128
  *flavo-squamosa* 143
  *galactophylla* 130
  *isidiophora* 128
  *japonica* 9, 130
  *leucomelos* 79, 80, 127, 130, 142, 228
  *lutescens* 128
  *obesa* 37
  *obscurata* 130, 157
  *podocarpa* 79
  *speciosa* 128, 157
  *squamulosa* 157
  *tremulans* 157
  *verruculifera* 128
  *vulgaris* 128, 157
*Hibiscus tiliaceus* 51
highland rainforest
high montane element, *see* bipolar element
*Holcus lanatus* 38
homoplasious character states 117, 118
Hong Kong 210
hot spot traces 2
housing developments, *see* urbanization
Hymeneliaceae 137
*Hyophorbe lagenicaulis* 196
*Hyperphyscia adglutinata* 37
hyphophores 100
*Hypogymnia bitteri* 131
hypothallus, *see* prothallus
*Hypotrachyna* 79, 81, 91, 125, 127, 137, 142, 143, 154, 160, 163, 168, 226, 228
  *andensis* 128
  *bahiana* 156
  *bogotensis* 9, 130
  *brevirhiza* 128
  *caraccensis* 128
  *chlorina* 128
  *costaricensis* 80, 128
  *degelii* 128
  *densirhizinata* 130
  *dentella* 23, 128, 156
  *ducalis* 93
  *endochlora* 130
  *ensifolia* 128
  *formosana* 5, 80, 128, 156
  *gigas* 38, 128
  *imbricatula* 38, 128
  *laevigata* 128
  *leiophylla* 92
  *longiloba* 130
  *lugubris* 80
  *microblasta* 38, 130
  *physcioides* 130
  *prolongata* 130
  *pseudosinuosa* 90, 92
  *pulvinata* 130
  *reducens* 130
  *revoluta* 5
  *rockii* 130
  *singularis* 130
  *sinuosa* 5, 38, 39, 79, 128
  *subaffinis* 23
  *thysanota* 23

Iberia 237
India 5, 9, 26, 58, 64, 65
Indian Ocean 8, 54, 246
Indo-Malayan taxa 9
Indonesia 18, 26, 27, 28, 29, 30, 48, 224, 228, 229
in-group 261
inversion layer, *see* cloud zone
*Involucrothele* 269
*Ioplaca* 117
island arcs 2
island integration hypothesis 7
Israel 173


Japan 9, 26, 54, 57, 58, 64, 65, 188, 246
Java 3, 4, 51, 54, 57, 194, 219, 226, 227, 228, 229, 240, 241
*Jenmania* 137
Juan Fernandez 2, 3, 4, 9, 10
*Julella* 272
  *fallaciosa* 259


kapok, *see Bombax*
Kenya 90, 107
*Knema furfuraceum* 206, 207, 208, 210, 211


*Laguncularia racemosa* 164
*Lantana camara* 42
Laos 42
*Larix* 195
*Lasioloma* 100, 102, 137
  *arachnoideum* 147
Lauraceae 227, 237
Laurasia 64, 219
*Laurera* 137, 266, 267
  *aurata* 259
lava flows 40–1
Lecanactidaceae 174, 187
*Lecanactidetum premnae* 218
*Lecanactis* 137
*Lecania oahuensis* 41
*Lecanora* 137, 240
  *flavovirens* 40
  *leprosa* 37

*muralis* 39, 186
*pallida* 38, 39
*polytropa* 39, 40
*rupicola* 107
Lecanoraceae 137
Lecanorales 136, 138, 155
*Lecidea* 78, 137, 255
　*granulosa* 40
　*piperis* 137
Lecideaceae 136, 137
*Lecidella* 137
Lecotheciaceae 137
leeward lichen communities 39–40
legislation bodies 82
leguminous trees 228
*Leioderma* 79, 137, 142, 219, 220, 230
　*duplicatum* 5
　*erythrocarpum* 6
　*glabrum* 130
　*sorediatum* 6, 9, 128
*Lepra* 138
*Lepraria* 138
*Leprocaulon* 228
　*albicans* 131
　*arbuscula* 92
　*congestum* 131
　*gracilescens* 131
*Leprocollema* 137
*Leptogium* 37, 77, 78, 79, 80, 91, 125, 127, 128, 130, 137, 154, 161, 164, 165, 218, 220, 226, 227, 229, 230, 232, 236
　sect. *Mallotium* 236
　*adpressum* 93
　*andinum* 130
　*austroamericanum* 91, 155, 164
　*azureum* 236
　*brebissonii* 236
　*burgessii* 130
　*burnetiae* 92
　*caespitosum* 92, 236
　*corticola* 194, 236
　*cyanescens* 128, 236
　*diaphanum* 128
　*digitatum* 128
　*furfuraceum* 92
　*hibernicum* 92
　*isidiosellum* 155, 161, 164, 165, 169
　*javanicum* 236
　*laceroides* 130
　*marginellum* 92, 155, 165, 236
　*moluccanum* 155, 164, 165
　*olivaceum* 128
　*papillosum* 130
　*phyllocarpum* 92, 128, 155
　*punctulatum* 128
　*resupinans* 130

　*sessile* 92
　*stipitatum* 128
　*ulvaceum* 155
　*vesiculosum* 128
*Leptopeltis uluguruensis* 100
*Leptorhaphis* 137, 264, 271
　*epidermidis* 259
*Leptospermum* 225, 228
　*flavescens* 228
*Leproloma vouauxii* 8
*Letrouitia* 137
　*bifera* 9
　*corallina* 9
　*domingensis* 5
　*flavocrocea* 9
　*leprolyta* 9
　*muralis* 6
　*parabola* 9
　*pseudomuralis* 7
　*subvulpina* 5
　*transgressa* 5
　*vulpina* 5
Letrouitiaceae 116, 117, 137
*Leucaena leucocephala* 41
lianas 73, 227, 240
*Libocedrus* 74
*Lichen* 254
lichen communities 7, 197, 217–42
　climax 225, 241
　facies 226
lichen transplants, *see* cultivation
Lichenes imperfecti 138
*Lichina* 137
Lichinaceae 137, 155
Lichinales 137
life-cycle, short 101
*Linhartia patellarioides* 96
*Liquidambar styraciflua* 174, 175
*Liriodendron tulipifera* 194
*Lithocarpus* 74, 80, 218, 225, 237, 238
*Lithothelium* 268
　*cubanum* 259
　*falklandicum* 259, 270
*Lobaria* 78, 125, 128, 130, 137, 154, 164, 165, 188, 220, 227, 232, 234
　*amplissima* 198
　*clemensiae* 79
　*crenulata* 155
　*dendrophora* 79
　*discolor* 234
　*holstiana* 91
　*insularis* 78
　*isidia* 79
　*isidiophora* 234
　*isidiosa* 79, 80, 234
　*lobulata* 234
　*peltigera* 128, 155

*Lobaria (cont.)*
  *pseudopulmonaria* 79, 80
  *pulmonaria* 198
  *retigera* 6, 92, 232, 234, 235
  *scrobiculata* 79, 198
Lobariaceae 78, 137, 142, 143, 217, 218, 220, 227
*Lobarion* 217–41
  *pulmonariae* 217, 218
*Loflammea* 100, 102, 137
  *flammea* 147
*Logilvia* 100, 102, 137
  *gilvia* 140, 147
logging 4, 33, 40, 149, 226, 227, 240
*Lonicera*
  *sempervirens* 175
*Lopadium* 137
*Loricaria* 129
Louisiade Archipelago 69
lowland habitat 135–49
  coastal deserts 111, 114, 124
    forest 36, 51, 54, 141
    shrubs 109–10
  dry semi-evergreen forest 91
  rainforest 24, 37, 48, 73, 75, 78, 86, 91, 124, 232, 240
    dipterocarp forest 30, 33, 218, 219, 224, 226, 227, 237
    mixed evergreen 73, 225, 228
    monsoon deciduous 73, 124, 225, 228
      dry dipterocarp 224, 225, 228, 240
  savannah forest 124, 141, 143, 227, 240, 254
    eucalypt 58, 73, 77–8
  semi-deciduous forest 124
  scrub 40
  swamp forest 224, 226–7
*Lumnitzera* 75
  *racemosa* 51
lycopods 127

macadamia trees 37
*Macaranga* 240
Macaronesia 237
*Macentina* 254, 265, 270
  *borhidii* 97, 98
  *perminuta* 259
Madagascar 90, 246
Madeira 90
*Magnolia* 195
  *grandiflora* 173, 175
maidenhair, *see Gingko biloba*
Malaya 64, 65, 219, 224, 232
Malaysia 64, 232, 237
mangroves 3, 18, 37, 48–9, 51, 54, 65, 73, 75, 77, 90, 152–3, 159, 163–4, 165, 166, 167, 169
*Marattia* 99
*Massaria* 266, 270
  *inquinans* 259
*Mastodia*, see *Turgidosculum*
Mauritius 54, 90, 194, 196
mazaedium, *see* ascomata
*Mazosia* 127, 136, 137, 148, 149, 174, 178, 181, 183, 187, 204
  *melanophthalma* 5, 147, 204, 205, 208, 209–10, 211
  *paupercula* 210
  *phyllosema* 5, 147, 205, 211
  *pilosa* 147
  *praermosa* 144, 147
  *pseudobambusae* 140, 147
  *rotula* 147, 204–5, 210, 211–12
  *rubropunctata* 144, 147
  *tumidula* 144, 147
medula 87, 89, 232, 248
  crystals 114
*Medusulina* 136
*Megalospora* 125, 137
  *admixta* 130
  *albescens* 7
  *atrorubicans* 6
  *coccodes* 9
  *granulans* 6
  *halei* 7
  *hillii* 7
  *sulphurata* 5, 37, 38, 130
  *sulphureorufa* 7
  *tuberculosa* 5, 128
  *weberi* 6
Megalosporaceae 107, 137
megapodes 83
Melanesian rural society 82
*Melanotheca* 137, 267
*Melaspilea* 137
Melastomataceae 126
*Menegazzia* 78, 79, 81, 89, 125, 128, 194, 218, 221, 226, 228, 229, 230
  *propagulifera* 79
  *terebrata* 90, 92
methods 86, 173, 204, 246
*Metrosideros polymorpha* 38
*Micarea* 137
Micareaceae 137
*Microcalicium* 246
microclimate 164, 165, 167
*Microglaena* 264
*Microthelia*
  *aurora* 259, 270
  *microsperma* 259, 264, 272
*Microtheliopsis* 138, 254, 266, 270
  *uleana* 140, 147, 259

# Index

mining operations 82
mitochondria 176
*Monoblastia* 137, 264, 269
   *rappii* 259
Monoblastiaceae 137, 264
monophyletic groups 250, 256–7, 262
montane habitat 73, 217
  cloud forest 88, 89, 90, 91, 125, 127, 129, 130, 142, 225, 227–8
  conifer (*Pinus/Casuarina*) forest 74, 79, 82, 225, 228–9
  ericaceous heath 88, 89, 90, 91, 92, 93
  rainforest 3, 18, 38, 48, 57–8, 76, 78, 79, 81, 85–93, 95, 96, 125–7
    mixed evergreen forest 73, 89, 92, 218
    *Nothofagus* forest 82, 83
    'oak' (*Castanopsis/Lithocarpus*) forest 33, 74, 78, 225, 229, 237, 240
  savannah forest 74, 86, 142, 143, 240
  shrub communities 79, 142
moorland 226
morphology 18–22, 256
morphotype 232
*Muellerella* 269
*Mycobilimbia* 136, 137
Mycobilimbiaceae 137
mycobiont 178–82, 183
*Mycocalicium subtile* 7
*Mycoglaena* 272, 273
  *myricae* 259
*Mycomicrothelia* 127, 137, 264, 272
  *subfallens* 259
*Mycoporum* 264, 267
  *elabens* 259
*Mycopyrenula* 270
  *coryli* 259
*Myelorrhiza* 6
*Myrica faya* 41
*Myriotrema*
  *album* 9
  *cinereoglaucescens* 6
  *compunctum* 5
  *glaucophaenum* 9
  *microporum* 6
  *subconforme* 6
  *terebratulum* 9
Myristicaceae 206, 207
Myrtaceae 37, 42

*Nadvornikiana hawaiensis* 6
*Naetrocymbe* 271
Namib Desert 111, 114
Namibia 109, 119
*Nastus* 74
national parks 33, 36, 38, 39, 83, 237
native species 36

natural groups 116, 255
naturalized species 36, 37, 38, 42–3
*Neophyllis melacarpa* 6
neotenia 261
*Nephroma* 79, 125, 219, 221, 229, 230
  *helveticum* 79
  *tropicum* 92
*Nephrometum*
  *laevigatae* 219
  *lusitanicae* 219
*Nephromopsis* 9, 81, 221
  *stracheyi* 80
New Caledonia 2, 3, 4, 5, 6, 48, 54, 65
New Guinea 2, 3, 4, 6, 18, 48, 71, 73, 75, 80–2, 237
  Irian Jaya 69, 71
  Papua 26, 27, 28, 30, 33, 48, 51, 54, 57, 58, 64, 65, 69–83, 232
new combinations 9, 97
  species 86–9
New Britain 71, 72
New Ireland 71, 72
New Zealand 4, 5, 9, 54, 58, 64, 65, 89, 229, 232
*Normandina* 138, 254, 265, 269
  *pulchella* 130, 140, 259
Norfolk Island 3, 5, 54
North Solomons, *see* Bougainville
Northern Hemisphere 2, 90, 107, 219, 229, 232, 237, 246
Norway 217
*Nothofagus* 57, 64, 65, 81, 229
  *brassii* 65
Noumea 4
nuclear testing 4

Oceania 28, 30
oceanic regions 217, 248
*Ocellularia aurata* 5
*Ochrolechia* 137
oil reserves 82
*Olearia* 79
*Omphalina foliacea* 131
ontogeny 261
*Opegrapha* 137, 149, 172, 188
  *dibbenii* 174
  *filicina* 97, 144, 147, 174, 207–8
  *lambinonii* 97, 174
  *mougeotii* 174
  *rufescens* 174
  *santessonii* 174
  *viridis* 173, 174, 176, 178, 181, 187, 188
Opegraphaceae 136, 137, 138, 174, 187, 207–8
Opegraphales 137
*Opuntia megacantha* 42

orchids 74
original character, see plasiomorphy
Oropogon 127, 128, 137
  loxensis 129, 130, 140, 142
Ostrya 195
outgroup 260, 263

Pachyospora verrucosa 131
Pacific-American taxa 9
Pacific basin 1–2, 81
Pacific region 1–10, 47–65, 246
Pacifica 2, 8
paedomorphosis 261
palaeobiogeography, see biogeography
palaeotropical taxa 5–6, 9
palm trees 51, 73, 74, 153, 161, 195, 227, 228
  bottle palm 196
  rattans 74, 227
pan-austral taxa 5
pandan, see Freycinetia
Pandanus 74
  forest 36
Pannaria 78, 79, 80, 125, 137, 154, 161, 164, 167, 218, 220, 227
  brisbanensis 6
  conoplea 130
  fulvescens 6, 92
  hookeri 8
  isidioidea 155
  lurida 38, 40, 92, 155
  mariana 6, 37, 77, 78
  pezizoides 79
  prosenii 155
  rubiginosa 38, 130
  santessonii 92
  stylophora 155
  tavaresii 130
Pannariaceae 137, 142, 155, 159, 217, 220
Pannoparmelia 137
Panthalassic–Gondwanan origin 10
pantropical taxa 5, 89, 97, 100, 143, 144, 168, 205, 206, 207, 208, 210, 211, 213, 214, 215, 232
paraphyses 100
parasitism 172, 183, 185, 186
Parathelium 268
  megalosporum 259, 268, 271
Parmelia 17, 137, 230, 255
  crambidiocarpa 23
  erumpens 9
  sulcata 187
Parmeliaceae 19, 21, 22, 23, 30, 32, 131, 136, 142, 143, 155, 158, 159, 160, 165, 166, 168, 169, 221
Parmeliella 125, 154, 161, 218, 220, 227, 230

nigrocincta 10, 155
pannosa 128
Parmelina 125, 160, 165, 166, 168
  antillensis 156
  dissecta 128
  horrescens 128
  lindmanii 132
  spumosa 156
  subfatiscens 167
  versiformis 156
Parmeliopsis
  aleurites 93
  swinscowii 131
Parmentaria 137, 268, 269
Parmotrema 78, 91, 92, 125, 127, 137, 154, 157, 158, 160, 163, 166, 167, 168, 169
  andinum 6, 132
  argentinum 156, 165
  arnoldii 130
  austrosinensis 5, 91, 156, 166
  bangii 130
  cetratum 156
  chinense 5, 156, 169, 240
  conformatum 128
  consors 156, 161
  crinitum 5, 130, 156, 164
  cristiferum 5, 37, 43, 78, 80, 156, 165, 167
  degelianum 92
  dilatatum 5, 23, 156, 164, 165, 167, 169
  dominicanum 39
  expansum 156
  fasciculatum 156
  fragilescens 87–8, 89, 92
  fumarprotocetraricum 156
  gardneri 130, 156, 165, 167, 169
  hababianum 132
  hensseniae 92
  hicksii 92
  internexum 156, 164
  laciniatulum 88–9, 92
  latissimum 128
  lobulatum 156, 165
  macrocarpum 156
  madilyneae 156, 164, 165
  melanothrix 156
  mellissii 128
  michauxianum 156, 165
  mordenii 156, 169
  neotropicum 156
  peralbidum 128, 142
  permutatum 5, 156, 166
  planilobatum 88, 89
  praesorediosum 5, 142, 156, 161, 163, 165, 167, 169
  reticulatum 5, 37, 38, 40, 80, 91, 128, 156, 161, 163, 164, 165

*saccatilobum* 77
*sancti-angeli* 156, 166
*subarnoldii* 156, 169
*subisidiosum* 128, 156
*subsumtum* 132
*sulphuratum* 78, 156, 165
*tinctorum* 5, 37, 91, 132, 156, 161, 166
*ultralucens* 5, 156, 169
*viridiflavum* 128
*zollingeri* 5
*Passiflora*
  *mollissima* 38
pathogenic algae 187
  fungi 172
*Peltigera* 79, 125, 130, 221, 229, 230
  *austroamericana* 131
  *dolichorhiza* 6, 39, 230
  *laciniata* 126
  *nana* 9
  *pulverulenta* 127
Peltigeraceae 218, 221
Peltigerales 137
*Peltula* 137
  *decoticans* 6
  *euploca* 7
  *subglebosa* 6
Peltulaceae 137
*Pennisetum*
  *clandestinum* 40
  *setaceum* 40
*Pertusaria* 137
  *dactylina* 8
  *isiodophora* 38
Pertusariaceae 137
Peru 109, 111, 114, 123
*Phaeographina* 136
*Phaeographis* 136
*Phaeophyscia*
  *hispidula* 38
  *laciniata* 40
phanerogams 63, 201
*Philippia* 88, 89, 90, 91
Philippines 4, 5, 48, 54, 58, 64, 219
*Phlyctella* 136
Phlyctidaceae 136
*Phlyctis* 136
*Phoebe lanceolata* 213
photobiont 32
  cross-wall structure 178, 181
  cyanobacterial 77, 161, 163, 164, 165, 219, 226, 231, 233, 234
  green 114, 116, 158, 164, 172, 176–9, 206, 231, 232, 233, 234, 256
    trebouxioid 172, 185
    trentepohliaceous 172, 185, 188
  specificity 188–9
Phragmopelthecaceae 208–12

*Phycopeltis* 174, 175, 176, 178, 181, 187, 188
Phyllobatheliaceae 138
*Phyllobathelium* 138, 254, 272
  *epiphyllum* 147, 149, 259
*Phyllophiale* 137
  *alba* 127, 147
*Phylloporis* 137, 149, 271
  *phyllogena* 97, 140, 147, 259
  *platypoda* 140, 148
*Phyllopsora* 92, 137, 219, 221
  *corallina* var. *ochroxantha* 5
  *martinii* 91
  *mauritiana* 90, 91
Phyllopsoraceae 221
phylogeny 101, 116–18, 168, 185, 253–66
  Hennigian principles 117, 261
phylogenetic trends 116
  trees, *see* cladograms
*Physcia* 77, 137, 154, 160, 165
  *adscendens* 7
  *alba* var. *linearis* 157
    var. *obcessa* 157
  *albicans* 161
    var. *hypomela* 157
  *convexa* 158
  *coronifera* 128
  *erumpens* 77
  *integrata* 132
  *lopezii* 130
  *rolfii* 132
  *sorediosa* 132, 158
  *stellaris* 7
Physciaceae 77, 80, 125, 131, 157, 159, 160, 164
*Physcidia* 137
  *squamulosa* 140
  *wrightii* 92, 140
*Physma* 77, 80, 81, 91, 137, 161, 220
  *byrsaeum* 5, 78, 155
  *pseudoisidiatum* 37
phytosociology 229, 230
*Picea* 195
Pilocarpaceae 99, 137, 212–13
*Pilocarpon cateileum*, *see Fellhanera cateilea*
*Pinus* 195, 225, 228–9
pioneer lichen communities 44
pioneer species 98, 149
*Placocarpus* 269
*Placopsis* 9, 81, 126, 127
  *cribellans* 39, 40
  *parellina* 38
*Placopyrenium* 269
*Plagiocarpa* 268
  *illota* 259
plantations 149
  cabbage 241
  cash-crops 82

292  *Index*

plantations (*cont.*)
  cocoa 82
  coconut 82
  coffee 82, 187
  maize 241
  oil palm 82, 241
  pineapple 37, 42, 44
  rice-fields 194, 240, 241
  rubber 80, 82, 240, 241
  sugar-cane 37, 42, 44, 194
  tea 82, 187, 194
plasmodesmata 176, 178, 181
*Platanus occidentalis* 174, 175
plastoglobuli 176, 178, 181, 185
plate tectonics 7, 119
Pleomassariaceae 266
plesiomorphy 257, 260
*Pleurococcus* 197
*Pleurotheliopsis* 273
*Pleurothelium* 268
*Pleurotrema* 127, 137, 266, 268, 271
  *polysemum* 259, 268
  *verrucosum* 268
*Pocsia* 254, 265, 270
  *septemseptata* 259
Podocarpaceae 64
*Podocarpus* 74, 228
*Polyalthia viridis* 207
*Polyblastia* 272
*Polyblastiopsis* 272
*Polychydium* 137, 154
  *dendriscum* 128, 140, 142, 155
*Polylepis* 127
*Polymeridium* 254, 266, 271
  *albidum* 259
polyphyletic groups 116, 117
*Polystroma fernandezii* 144
population growth 33, 82, 197, 241, 276
*Porina* 127, 137, 148, 204, 264, 272
  *epiphylla* 5, 97, 127, 148, 149
  *epiphylloides* 97
  *fulvella* 148
  *imitatrix* 148
  *leptosperma* 140, 148
  *limbulata* 127
  *longispora* 96
  *mastoidea* 259
  *multipuncta* var. *schizospora* 96
  *papillifera* var. *rubrofusca* 96
  *phyllogena* 127
  *pseudofulvella* 175
  *pulla* 175, 176, 178, 181, 185, 186
  *rubentior* 140, 148
  *rufula* 140, 148, 149
  *sphaerocephala* 98
  *sphaerocephaloides* 97, 98
  *subpilosa* 97
Porinales 137
*Porinula* 271
  *tanzanica* 259
*Porpidia crustulata* 7
Porpidiaceae 136
primary forest 145, 218, 237, 240
primitive character, *see* plesiomorphy
*Pritchardia* forest 36
*Prosartema stellaria* 211
*Prosopis pallida* 41
Proteaceae 64
prothallus 98, 100, 206
*Protobagliettoa* 269
*Protothelenella* 264, 273
  *sphinctrinoides* 259
*Pseudephebe minuscula* 39
*Pseudevernia furfuracea* 195
pseudocyphellae 107, 108, 109, 114
*Pseudocyphellaria* 10, 78, 81, 124, 125, 127, 137, 154, 194, 218, 220, 228, 229–32
  *argyracea* 5, 78, 79, 92, 227, 231
  *arvidssonii* 128
  *aurata* 7, 78, 128, 140, 142, 155, 231
  *aurora* 155
  *bartlettii* 128
  *billardierei* 230
  *clathrata* 5, 80, 128
  *crocata* 7, 38, 39, 79, 80, 130, 218, 231
  *desfontainii* 231
  *diplomorpha* 9
  *dissimilis* 230
  *dozyana* 128
  *encoensis* 128
  *faveolata* 230
  *flavicans* 38
  *godeffroyi* 7
  *intricata* 7, 79, 92, 128, 218, 231
  *junghuhniana* 9, 227, 231
  *knightii* 231
  *lombokensis* 79
  *multifida* 79, 230, 231
  *physciospora* 10
  *pickeringii* 79
  *poculifera* 6, 7
  *reineckeana* 7
  *rubella* 230
  *rufovirescens* 79
  *semilanata* 7
  *stenophylla* 7
  *sulphurea* 6, 9, 231
*Pseudoparmelia* 125, 137, 154, 160, 166, 168
  *amazonica* 157
  *caroliniana* 157
  *crozalsiana* 157
  *intertexta* 9
  *leucoxantha* 132
  *salacinifera* 157

*sphaerospora* 92, 128, 157
*texana* 5, 132, 157
*Pseudopyrenula subgregaria* 137, 266, 270
  *subgregaria* 128, 259
*Pseuduvaria indochinensis* 206, 207
*Psidium guajava* 37, 42
*Psilolechia lucida* 41
*Psoroma* 10, 78, 81, 137, 219, 220, 227, 229, 230
  *hypnorum* 79, 137
  *microphyllizans* 230
  *sphinctrinum* 10
*Psorotichia* 137
*Pterocarya* 195
*Punctelia* 125
  *neutralis* 93
  *stictica* 131
  *subrudecta* 132
*Pyrenastrum* 137, 268
Pyrenees 217
*Pyrenillium* 271
pyrenocarpous lichens 125, 127, 137-8, 172, 173, 186, 187, 188, 227, 253-66
  key to genera 267-73
*Pyrenocollema* 271
  *halodytes* 259
*Pyrenotrichum* 137
*Pyrenula* 127, 137, 175, 257, 267, 268, 269, 270, 273
  *anomala* 173, 175, 176, 178, 181, 183, 185, 187
  *dermatodes* 159
  *nitida* 175
Pyrenulaceae 37, 137, 138, 175, 187, 253
Pyrenulales 137, 138
*Pyrgidium* 136
  *monticellum* 141
*Pyrgillus* 137, 270
  *americanus* 141
  *javanicus* 5, 259
Pyxinaceae 137, 138, 142, 143, 263, 266
*Pyxine* 77, 78, 81, 91, 137, 154, 160, 197
  *berteriana* 5, 132
  *cocoes* 5, 37, 142
  *meissneri* var. *comnectens* 158
  *minuta* 5
  *retirugella* 37
  *rhizophorae* 158
  *sorediata* 78
  *subcinerea* 80

*Quercus* 217, 218, 226, 237, 238, 239
  *suber* 188

races
  chemical 22-3, 51, 57, 58, 232, 256

  geographical 106, 256
*Raciborskiella* 137, 271
  *janeirensis* 141, 148
  *prasina* 97
*Ramalina* 47-65, 125, 137, 154, 165, 229, 241
  *celastri* 5, 39
  *cerinella* 40
  *exiguella* 37, 43
  *farinacea* 40, 51
  *intermediella* 54
  *inflata* 57, 63
    ssp. *perpusilla* 57
  *javanica* 57
  *leiodea* 6
  *microspora* 36
  *nervulosa* 51
    var. *dumeticola* 6, 51, 52, 63
    var. *lucie* 51, 52, 63, 77
    var. *nervulosa* 52
  *pacifica* 9
  *peruviana* 5, 54-7, 63
  *pocsii* 92
  *pumila* 57
  *roesleri* 92
  *sandwicensis* 40
  *soraligera* 40
  *subfraxinea* 50, 51
    var. *confirmata* 50, 51
    var. *leiodea* 50, 51
    var. *norstictica* 50, 51
    var. *subfraxinea* 50, 51
  *tenella* 6
  *tropica* 50, 51, 63, 77
  *umbilicata* 36
  *zollingeri* 51
Ramalinaceae 136, 137, 138, 142
*Ramalinopsis* 3, 6, 36
  *mannii* 37, 43
rare species 99, 194, 198
reforestation 42
*Relicina* 17-33, 77, 80, 90, 137, 154, 168
  *abstrusa* 18, 20, 21, 22, 28, 29, 30, 31, 78, 90, 157, 160
  *acrobotrys* 26, 31
  *agglutinata* 23, 24-6, 31
  *amphithrix* 18, 20, 22, 23, 25, 29, 31, 78
  *butleri* 78
  *circumnodata* 18, 20, 21, 23, 26, 31
  *columnaria* 23
  *conglutinata* 18, 20, 23, 26, 31
  *connivens* 9, 22, 23, 31, 78
  *demethylbarbatica* 18, 22, 23, 31
  *echinocarpa* 20
  *eximbricata* 30
  *fijiensis* 22, 23, 25, 29
  *filsonii* 22, 23, 27, 31

*Relicina (cont.)*
  *fluorescens* 7, 20, 21, 22, 23, 27–8, 31
  *gemmulosa* 19, 20, 21, 22, 23, 31
  *hirtifructa* 22, 23, 31
  *incongrua* 30
  *limbata* 18, 20, 21, 23, 26–7, 30, 31
  *luteoviridis* 19, 20, 21, 22, 23, 28
  *malesiana* 9, 22, 23, 31
  *neoabtrusa* 23
  *nuiginiensis* 22, 23, 25, 29, 31
  *planiuscula* 18, 20, 21, 22, 23, 28, 31, 89, 90, 92
  *precircumnodata* 20, 21, 26, 31
  *ramboldii* 18, 22, 23, 27, 31
  *ramosissima* 20, 23, 26, 31
  *relicinella* 21, 30
  *relicinula* 9, 22, 23, 25, 29, 31
  *retrospinosa* 19, 20, 21, 22, 23, 28, 31
  *samoensis* 6, 19, 20, 22, 23, 24, 25, 28–9, 31
  *schizospatha* 18, 19, 20, 22, 23, 31
  *subabtrusa* 20, 21, 22, 23, 28, 30, 31, 90
  *subconnivens* 31
  *sublanea* 23, 26, 31
  *subnigra* 18, 21, 22, 23, 27, 31
  *sydneyensis* 18, 20, 21, 22, 23, 26, 27, 31
  *terricrocodila* 22, 23, 25, 29, 31
*Relicinopsis* 32
relict species 237
relict theory 8
reserves 83, 96, 152, 169, 204, 205, 206, 207, 208, 210, 211, 212, 213
restinga vegetation 152–3, 159, 164–5, 166, 169
*Rhacomitrium lanuginosum* 39
rhizines 19, 21–2, 27–8
*Rhizophora* 51, 63, 75
  *mangle* 164
  *stylosa* 51
Rhizophoraceae 51, 63
*Rhododendron* 74, 81
*Rinodina* 137
  *hawaiiensis* 39
road construction 44, 152, 169
*Roccella* 77
Roccellaceae 137
*Roccellina*
  *badia* 9
  *limitata* 9
  *nigrocincta* 9
  *suffruticosa* 9
  *terrestris* 9
rocky shores, *see* coastal cliffs
Rubiaceae 126

Sabah 58, 64, 237, 240

*Sagediomyces, see Strigula*
*Salix nigra* 175
Samoa 3, 4, 6
sanctuaries, *see* reserves
*Santessoniolichen* 264, 271
  *punctiformis* 259
Sarawak 240
*Sarcographa* 136
*Sarrameana* 6
*Saurauia* 79
Scandinavia 195
*Scaveola coriacea* 41
*Schima* 228
*Schinus terebinthifolius* 43
*Schismatomma* 137, 174
  *rappii* 178, 181, 183, 187, 188
secondary forests 149, 224, 240
secondary metabolites 22–4, 26, 27, 32, 51–4, 87, 89, 111, 116, 207, 210, 212, 215, 248
*Seirophora* 117
semi-desert habitat 109, 130, 132
  dry deciduous forests 132
shared derived characters, *see* synapomorphy
*Shorea, see* dipterocarps
shrubland 3
*Siphula* 10, 38, 81, 89
  *decumbens* 90, 92
  *fastigiata* 131
  *pteruloides* 131
slash and burn 240, 241
soil erosion 33, 43, 44, 93
*Solorina*
  *crocea* 8
  *saccata* 131
  *spongiosa* 8
South America 7
Southern Hemisphere 2, 8, 81, 109, 119, 219, 229, 232, 237
Spain 144
speciation 105
  allopatric 7, 106, 107, 114
  continuous 106
  geographical, *see* allopatric
  gradual 114
species aggregates 114, 117, 118, 119
species concept 256
species isolation 106, 109, 119
species migration 48, 63, 64, 237
species pair 19, 20
*Sphaeria* 254
*Sphaerophorus* 10, 78, 80, 125, 229, 230, 245, 248, 250
  subg *Aghimus* 246, 248
  subg *Bunodophorus* 246, 248
  subg *Sphaerophorus* 246, 248
  *diplotypus* 6, 245–50

*fragilis* 245–50
*formosanus* 5
*globosus* 245
*kinabalensis* 6
*melanocarpus* 7, 92, 128, 130
*murrayii* 245
*notatus* 246
*stereocauloides* 248
*Sphinctrina tubaeformis* 5
Sphinctrinaceae 136
*Splanchnonema* 266, 267, 268, 270
  *argus* 259
  *uberinum* 259
*Sporopodium* 100, 102, 127, 137, 204
  *leprieurii* var. *leprieurii* 206
*Squamacidia* 137
Sri Lanka 6
*Starbaeckiella* 270
starch grains 176, 178, 181, 183
*Staurothele* 272
Stereocaulaceae 137
*Stereocaulon* 80, 125, 126, 129, 130, 137
  *crambidiocephalum* 131
  *fibrillosum* 90
  *glareosum* 131
  *meyeri* 131
  *novogranatense* 127
  *obesum* 131
  *octomerellum* 40
  *pityrizans* 134
  *ramulosum* 38, 40, 137
  *strictum* var. *compressum* 127
  *tomentosum* 131
  *verruculiferum* var. *surreptans* 131
  *vesuvianum* 131
  *vulcani* 40, 41
*Sticta* 78, 79, 92, 125, 127, 128, 130, 137, 154, 161, 164, 165, 220, 227, 228, 229, 230, 232, 233
  *ambavillaria* 92, 93
  *boschiana* 7, 233
  *cyphellulata* 233
  *filix* 79
  *fuliginosa* 38, 130, 141, 142, 233
  *heppiana* 7
  *humboldtii* 130
  *lenormandii* 128
  *nylanderiana* 233
  *papyracea* 92
  *pedunculata* 7
  *samoana* 6
  *sayeri* 233
  *sinuosa* 155
  *tomentosa* 38, 128
  *variabilis* 155
  *wiegelii* 37, 38, 92, 93, 130, 142, 155, 164, 165, 167, 233

Stictaceae 78, 155, 159, 188
*Stictina diplomorpha*, see *Pseudocyphellaria diplomorpha*
*Stirtonia* 136
  *macrocephala* 5
  *sprucei* 141, 144, 148
*Strigula* 138, 145, 187, 264, 271, 272
  *complanata* 173, 175, 176, 178, 181, 185, 187
  *concreta* 97, 148
  *elegans* 97, 127, 148, 149, 172, 178, 185, 186, 213
  *fibrilosa* 214
  *macrocarpa* 213
  *maculata* 141, 148
  *melanobapha* 141, 148, 213–14
  *nemathora* 148
  *nitidula* 97
  *subelegans* 214
  *subtilis* 127
  *subtilissima* 148, 149
  *wilsonii* 259
Strigulaceae 137, 175, 187, 213–14
Strigulales 137
subalpine habitat 8
submontane rainforest 90, 91, 95, 96
subtropical regions 97, 107, 116, 172, 187, 188
  savannah 32–3
subtropical species 127, 173, 214, 246
Sumatra 65
Sunda 219
Sweden 195, 198
symbiosis 186, see also fungal–algal interactions
synanthropic taxa 143
synapomorphy 257
*Synplagiotrema* 270
systematics 30–3, 101, 171, 254


Taiwan 26, 213, 219
Tanzania 85–93, 95–103
  Usambara Mts 85, 87, 89, 90, 91, 92, 95, 96, 97, 98, 99, 100
*Tapellaria* 100, 102, 137, 145
  *epiphylla* 127, 141, 148
Tasmania 6, 58, 65
*Tectaria* 208
*Tectona* 228
Teloschistaceae 105–20, 137, 229, 230
Teloschistales 116, 137, 138
*Teloschistes* 107, 111–14, 115, 117, 118, 125, 137, 154, 241
  *capensis* 111, 112, 114, 117, 119
  *chrysocarpoides* 113, 115, 116
  *chrysophthalmus* 132

*Teloschistes (cont.)*
  *exilis* 130
  *flavicans* 7, 40, 82, 130, 141, 165
  *hypoglaucus* 117, 132
  *lacunosus* 117
  *perrugosus* 111, 113
  *peruensis* 111, 112
  *puber* 111, 113, 114
  *scorigenus* 116
temperate regions 54, 63, 90, 97, 172, 185, 188, 193, 195, 197, 229
  sub-oceanic 78
temperate species 48, 57, 58, 64, 65, 127, 172
*Tephromela atra* 7
*Terminalia cattapa* 153, 161
terranes 2
Tertiary relics 194
Tethys Sea 2, 8, 64, 65
*Tetrastigma* 213
Thailand 57, 65, 201, 219, 224, 225, 226, 227, 228, 232, 240, 241
thallus 87, 89, 98, 99, 100, 108, 115, 205, 206, 207, 211, 212, 213, 215, 246, 247, 248, 254, 255, 260
  fissures 111, 112
  lobes 109, 110, 112, 113, 114, 116
  pigmentation 114
*Thamnolia* 81
  *vermicularis* 80, 129, 131
*Thelenella* 254, 269, 272, 273
  *modesta* 259
*Thelidium* 269
*Thelotrema* 136, 227
  *coccineum* 5
  *lacteum* 5
  *monosporum* 5
  *pilulifera* 6
  *platycarpoides* 9
  *platysporum* 6
  *porinoides* 9
  *weberi* 6
Thelotremataceae 136, 138, 173, 188
threatened species, *see* rare species
*Thrombium* 264, 265, 269
  *epigeum* 259
*Thunbergia grandiflora* 205
*Thysanophoron* 248
*Thysanothecium* 81
  *scutellatum* 9, 78, 81
*Tilia* 195
timber resources 82
*Tomasellia* 137, 264, 267
  *eschweileri* 254
tomentum 111, 112, 113, 114, 232
*Toninia* 137
tourism 4, 40, 42, 83

Trapeliaceae 137
*Trapeliopsis* 137
*Trebouxia* 172
*Trentepohlia* 172, 174, 175, 176, 178, 187, 188
  *annulata* 188
  *lagenifera* 188
  *umbrina* 188
*Tricharia* 100, 127, 136, 149
  *albostrigosa* 127
  *carnea* 148
  *dilatata* 97, 148
  *santessoniana* 141, 148
  *urceolata* 148
  *vainioi* 97, 207
Trichotheliaceae 98, 137, 175, 187, 214–15, 253, 254, 262, 264
*Trichothelium* 127, 137, 269
  *alboatrum* 97
  *annulatum* 141, 148, 149, 214–15
  *epiphyllum* 215, 259
Trimenaceae 64
*Tristania* 227
tropical hardwoods 82
tropical rainforest 32, 33, 229, 232
Trypetheliaceae 137, 138, 165, 175, 187, 253, 254, 255, 263, 266
*Trypethelium* 127, 137, 266, 267
  *aeneum* 128
  *eleuteriae* 172, 183, 185
  *ochroleucum* 173, 175, 178, 181, 185, 186, 259
  *tropicum* 173, 175, 176, 178, 181, 186, 256, 270
  *uberinum* 254, 267
*Turgidosculum* 9, 265, 269
  *complicatulum* 259
turtles 83
tussock grass 9, 38, 39, 40, 43, 129
*Tylophoron* 136
  *crassiusculum* 141
  *moderatum* 5
  *protrudens* 141

Uganda 107
Ulmaceae 213, 215
ultrastructure 171–89, 245–50
*Umbilicaria*
  *decussata* 39
  *hirsuta* 39
United Kingdom, *see* Great Britain
United States of America
  Southeast 32
unnatural groups, *see* polyphyletic groups
upper dry forest 39–40
upper rainforest, *see* montane rainforest

urban areas 153, 161, 166
urbanization 37, 38, 41, 43–4, 152, 153, 169
*Usnea* 47–65, 78, 79, 80, 92, 125, 127, 128, 130, 137, 154, 164, 195, 228, 229, 241
  *australis* 38
  *baileyi* 54–7, 63, 78, 141
  *bicolorata* 93
  *capillacea*, see *U. contexta*
  *contexta* 58, 59, 65
  *durietzii* 131
  *eburnea* 61
  *entoviolata* 40
  *flexilis* 58, 59, 63, 65
  *grandis* 54
  *gigas* 58, 61
  *himantodes* 58, 61, 63, 64
    f. *neoguineensis* 57, 58, 61
  *hossei* 57, 58, 62, 63, 64
  *intercalaris* 54
  *japonica* 54
  *longissima* 57, 58, 60
    var. *misamisensis*, see *U. misamisensis*
  *misamisensis* 57, 58, 60, 63
  *neocaledonia* 54
  *neoguineensis* var. *gracilor* 58
  *nexilis* 54
  *nidifica* 51, 53, 54, 78
  *osseoleuca* 40
  *propincua* 54
  *societatas* 54
  *sorediosula* 92
  *squarrosa* 58, 62
  *straminea* 54
  *trichodeoides* 57, 58, 60, 64

*Vaccinium* 74, 79
Vanuatu 48
vegetation kingdoms
  Holantarctic 3
    Fernandezian region 3
    Neozeylandic region 3
  neotropic 18, 30, 32, 135–49
  palaeotropic 3, 97, 143
    Indomalesian subregion 3
    Malaysian subkingdom 201
    Polynesian subregion 3
vegetative propagules 18–19, 32, 63, 194, 256
  isidia 18–19, 26, 27, 28, 89, 98, 100, 232
  laciniae 88, 89
  lobulae 18–19, 107, 109
  soredia 107, 109, 198
  thallus fragments 198
vehicles 39, 41, 44, 119

Venezuela 142
verrucae 205, 209
*Verrucaria* 138, 186, 188, 254, 257, 269, 272
  *muralis* 259
Verrucariaceae 98, 138, 253, 254, 265, 266
Verrucariales 138, 255, 262
vicariance 7, 58
Vietnam 97
Vitaceae 213

Wallace's line 237
water resources 241
watershed 38, 40, 42
windward lichen communities 36–8
Winteraceae 64
wood-rotting fungus 185
Woronin bodies 185

*Xanthodactylon* 110, 117
*Xanthoparmelia* 107, 130, 137, 154, 157, 160, 168
  *cotopaxiensis* 131
  *distincta* 131
  *farinosa* 132
  *kurokawae* 132
  *mendoza* 132
  *mougeotii* 131
  *subramigera* 36, 37, 40
  *subsorediata* 131
  *taractica* 38, 39
  *ulcerosa* 132
  *vagans* 131
*Xanthopeltis* 117
*Xanthoria* 81, 107, 117, 118
  *africana* 107
  *candelaria* 107
  *capensis* 107, 108
  *elegans* 8, 80, 107
  *fallax* 107, 117
  *flammeum* 110
  *ligulata* 109
  *mandschurica* 108, 109
  *marlothii* 108, 109
  *mendozae* 109, 119
  *parietina* 7, 107, 109, 117
  *polycarpa* 107, 195
  *turbinata* 110
*Xanthorion* 7, 197

*Zamenhofia* 272
*Zelkova* 195
*Zignoella* 256

# Systematics Association Publications

1. Bibliography of key works for the identification of the British fauna and flora, *3rd edition* (1967)
   *Edited by G. J. Kerrich, R. D. Meikle and N. Tebble*
2. Function and taxonomic importance (1959)
   *Edited by A. J. Cain*
3. The species concept in palaeontology (1956)
   *Edited by P. C. Sylvester-Bradley*
4. Taxonomy and geography (1962)
   *Edited by D. Nichols*
5. Speciation in the sea (1963)
   *Edited by J. P. Harding and N. Tebble*
6. Phenetic and phylogenetic classification (1964)
   *Edited by V. H. Heywood and J. McNeill*
7. Aspects of Tethyan biogeography (1967)
   *Edited by C. G. Adams and D. V. Ager*
8. The soil ecosystem (1969)
   *Edited by H. Sheals*
9. Organisms and continents through time (1973)[†]
   *Edited by N. F. Hughes*

Published by the Association (out of print)

Systematics Association Special Volumes

1. The new systematics (1940)
   *Edited by J. S. Huxley* (Reprinted 1971)
2. Chemotaxonomy and serotaxonomy (1968)*
   *Edited by J. G. Hawkes*
3. Data processing in biology and geology (1971)*
   *Edited by J. L. Cutbill*
4. Scanning electron microscopy (1971)*
   *Edited by V. H. Heywood*
   Out of print
5. Taxonomy and ecology (1973)*
   *Edited by V. H. Heywood*

6. The changing flora and fauna of Britain (1974)*
   *Edited by D. L. Hawksworth*
   Out of print
7. Biological identification with computers (1975)*
   *Edited by R. J. Pankhurst*
8. Lichenology: progress and problems (1976)*
   *Edited by D. H. Brown, D. L. Hawksworth and R. H. Bailey*
9. Key works to the fauna and flora of the British Isles and north-western Europe, *4th edition* (1978)*
   *Edited by G. J. Kerrich, D. L. Hawksworth and R. W. Sims*
10. Modern approaches to the taxonomy of red and brown algae (1978)*
    *Edited by D. E. G. Irvine and J. H. Price*
11. Biology and systematics of colonial organisms (1979)*
    *Edited by G. Larwood and B. R. Rosen*
12. The origin of major invertebrate groups (1979)*
    *Edited by M. R. House*
13. Advances in bryozoology (1979)*
    *Edited by G. P. Larwood and M. B. Abbot*
14. Bryophyte systematics (1979)*
    *Edited by G. C. S. Clarke and J. G. Duckett*
15. The terrestrial environmental and the origin of land vertebrates (1980)*
    *Edited by A. L. Panchen*
16. Chemosystematics: principles and practice (1980)*
    *Edited by F. A. Bisby, J. G. Vaughan and C. A. Wright*
17. The shore environment: methods and ecosystems (2 Volumes) (1980)*
    *Edited by J. H. Price, D. E. G. Irvine and W. F. Farnham*
18. The Ammonoidea (1981)*
    *Edited by M. R. House and J. R. Senior*
19. Biosystematics of social insects (1981)*
    *Edited by P. E. Howse and J.-L. Clément*
20. Genome evolution (1982)*
    *Edited by G. A. Dover and R. B. Flavell*
21. Problems of phylogenetic reconstruction (1982)*
    *Edited by K. A. Joysey and A. E. Friday*
22. Concepts in nematode systematics (1983)*
    *Edited by A. R. Stone, H. M. Platt and L. F. Khalil*
23. Evolution, time and space: the emergence of the biosphere (1983)*
    *Edited by R. W. Sims, J. H. Price and P. E. S. Whalley*

24. Protein polymorphism: adaptive and taxonomic significance (1983)*
    Edited by G. S. Oxford and D. Rollinson
25. Current concepts in plant taxonomy (1983)*
    Edited by V. H. Heywood and D. M. Moore
26. Databases in systematics (1984)*
    Edited by R. Allkin and F. A. Bisby
27. Systematics of the green algae (1984)*
    Edited by D. E. G. Irvine and D. M. John
28. The origins and relationships of lower invertebrates (1985)‡
    Edited by S. Conway Morris, J. D. George, R. Gibson and H. M. Platt
29. Infraspecific classification of wild and cultivated plants (1986)‡
    Edited by B. T. Styles
30. Biomineralization in lower plants and animals (1986)‡
    Edited by B. S. C. Leadbeater and R. Riding
31. Systematic and taxonomic approaches in palaeobotany (1986)‡
    Edited by R. A. Spicer and B. A. Thomas
32. Coevolution and systematics (1986)‡
    Edited by A. R. Stone and D. L. Hawksworth
33. Key works to the fauna and flora of the British Isles and north-western Europe, 5th edition (1988)‡
    Edited by R. W. Sims, P. Freeman and D. L. Hawksworth
34. Extinction and survival in the fossil record (1988)‡
    Edited by G. P. Larwood
35. The phylogeny and classification of the tetrapods (2 Volumes) (1988)‡
    Edited by M. J. Benton
36. Prospects in systematics (1988)‡
    Edited by D. L. Hawksworth
37. Biosystematics of haematophagous insects (1988)‡
    Edited by M. W. Service
38. The chromophyte algae: problems and perspectives (1989)‡
    Edited by J. C. Green, B. S. C. Leadbeater and W. Diver
39. Electrophoretic studies on agricultural pests (1989)‡
    Edited by Hugh D. Loxdale and J. den Hollander
40. Evolution, systematics, and fossil history of the Hamamelidae (2 Volumes) (1989)‡
    Edited by Peter R. Crane and Stephen Blackmore
41. Scanning electron microscopy in taxonomy and functional morphology (1990)‡
    Edited by D. Claugher

42. Major evolutionary radiations (1990)[‡]
    *Edited by P. D. Taylor and G. P. Larwood*
43. Tropical lichens: their systematics, conservation, and ecology (1991)[‡]
    *Edited by D. J. Galloway*

[*] *Published by Academic Press for the Systematics Association*
[†] *Published by the Palaeontological Association in conjunction with the Systematics Association*
[‡] *Published by the Oxford University Press for the Systematics Association*